W9-ADZ-754

The red clay layer containing about 100 ppm of iridium found between the Cretaceous (bottom layer) and Tertiary (above red clay) limestones in Northern Italy. There is a grey clay layer between the red clay and the Cretaceous limestone, and a laying of Lucite bonding material above and to the left of the Tertiary limestone. (By courtesy of L. Alvarez et al.).

Teaching school chemistry

Teaching school chemistry

Edited by D. J. Waddington

Unesco

The authors are responsible for the choice and the presentation of the facts contained in this book and for the opinions expressed therein, which are not necessarily those of Unesco and do not commit the Organization.

Published in 1984 by the United Nations
Educational, Scientific and Cultural Organization
7 place de Fontenoy, 75700 Paris
Typeset by Asco Trade Typesetting Ltd, Hong Kong
Printed by Imprimerie Floch, Mayenne

ISBN 92-3-102131-1
Arabic edition: 92-3-602131-x
French edition: 92-3-202131-5
Spanish edition: 92-3-302131-9

Preface

Teaching School Chemistry has been prepared as a part of Unesco's programme to improve the teaching of science, on both a disciplinary and an interdisciplinary basis, through the production of resource materials for the use of those working in the field of science education. This book is intended for the use of teachers of chemistry and for curriculum developers, teacher educators and other key personnel working in the field of chemical education.

The book has been produced by a selection of authors from different regions and countries of the world. Each chapter has been revised and considerably changed by many chemistry teachers in many Member States of Unesco. It is hoped that this book, which reflects the latest achievements in the field of chemical education, will be of help in the further improvement of chemistry teaching at the secondary and tertiary levels.

Grateful thanks are due to the editor, Professor D. J. Waddington, to the various contributing authors, to the members of the IUPAC Committee on the Teaching of Chemistry, and to all others who made helpful suggestions and comments at various stages in the development of the material.

Contents

Introduction 9

Acknowledgements 11

1 The changing face of chemistry *J. A. Campbell* 15

2 Curriculum innovation in school chemistry *R. B. Ingle and A. M. Ranaweera* 45

3 Some methods in teaching chemistry *A. Kornhauser* 115

4 Practical work and technology in chemical education: three aspects *M. H. Gardner, J. W. Moore and D. J. Waddington* 159

5 Assessment of students *J. C. Mathews* 223

6 Education and training of teachers *A. V. Bogatski, D. Cros and J. N. Lazonby* 275

7 Current research in chemical education *J. Fensham* 309

8 The future *M. H. Gardner* 343

Notes on contributors 389

Introduction

Society expects much from its schools, both in terms of general education and vocational training. Chemistry is one of the key subjects because it sits astride both, and every country is striving to improve the quality of the teaching of chemistry in schools.

Chemistry's value in general education and vocational training leads inevitably to questions being raised. How do curricula meet these demands? How do we teach a 'difficult' subject such as chemistry? How can we train teachers to teach such wide-ranging curricula? How do curriculum developers and teachers keep abreast of the developments in chemistry, for it is a dynamic subject? How, indeed, do they keep abreast of research in chemical education? These and many other equally significant questions spring to mind.

It is not possible to give definitive answers, for while each country has the same problems to a lesser or greater extent, the answers depend on a myriad of economic, cultural, historical, geographical and political factors. So that when we were given the opportunity to contribute towards a Unesco book on the subject, we decided to adopt an approach where we examined several different and interrelated areas from the viewpoint of the teacher, the teacher-educator and those closely associated with the actual teaching of chemistry at school.

Teachers are individuals plying their responsible but lonely trade who, day by day, are thinking about their work and developing new techniques and ideas. They have opinions of their own, developed with experience and training. Thus it is a difficult enough task to write a book on teaching school chemistry for a specific country that will meet the demands of such a wide-ranging corps of people. To write one for an international audience adds enormously to the difficulty of the task.

Thus, we chose authors and advisers from various countries, with different experiences but all with an international as well as a national reputation for their work in chemical education. Although all had worked in countries other than their own, they decided to write to many other chemical educators all over the world in order to increase the international flavour of their contributions. The correspondence underlying the chapters was voluminous and our thanks to our correspondents are unlimited.

We have chosen eight areas in which to explore various aspects of teaching school chemistry worldwide. Some of these areas are obvious enough in design; for example, the curriculum, methods of assessment and methods of teacher training. In all three, though, we have attempted to reflect how they are being considered in different countries. The danger of this approach is that one cannot give enough space to develop, for example, a real appreciation of the reasons why a particular curriculum was developed in a specific country, or why a particular method of assessment was used for the curriculum. Similarly, the way in which teachers are trained in a country is often dictated by underlying factors which again cannot be expressed adequately in so short a book.

Nevertheless, the authors of these chapters knew the limitations set upon them by the length of the book and have given detailed references so that, if readers are interested in following up a curriculum, an assessment technique, an examination procedure or a method of teacher training, they are able to do so. The authors have given us guiding torches; the readers can use these in any way they wish.

I am very grateful to a very large number of individuals who have corresponded with me and with the authors of the chapters. They have given us the benefit of their experiences and expertise. My thanks are also due to the authors who have given unstintingly of their time. Of these I would like to mention one, Professor Marjorie Gardner, who has given me so much help and encouragement. Finally, my thanks are due to Unesco for giving me, as editor of *Teaching School Chemistry*, such a fascinating and enjoyable task.

Any defects the reader may feel are inherent in the book will be due to me, for the authors and correspondents have nobly responded to their briefs.

D.J.W.
York, 1983

Acknowledgements

The authors of the chapters have corresponded with many science educators throughout the world and in particular wish to thank the following:

Mr R. T. Allsop (University of Oxford, United Kingdom).
Professor A. Ambrogi (Centro de Treinamento para Professores de Ciencias Exectas a Naturais de São Paulo, São Paulo, Brazil).
Professor T. A. Ashford (University of South Florida, Tampa, Florida, United States).
Professor J. K. P. Ariyaratne (University of Kelaniya, Sri Lanka).
Professor G. Aylward (UNDP, Jakarta, Indonesia).
Dr S. T. Bajah (University of Ibadan, Nigeria).
Dr D. Barlex (Stantonbury Campus, Milton Keynes, United Kingdom).
Professor M. Bartok (University of Szeged, Hungary).
Mr L. Beckett (University of London Institute of Education, London, United Kingdom).
Professor S. P. Beletskaya (University of Moscow, USSR).
Professor V. V. Bondar (University of Moscow, USSR).
Miss R. Bonner (International Baccalaureate Office, Geneva, Switzerland).
Professor M. L. Bouguerra (University of Tunis, Tunisia).
Mr D. G. Chisman (ICASE).
Mr E. H. Coulson (formerly County High School, Braintree, Essex, United Kingdom).
Mr B. Deutrom (Ministry of Education, Papua New Guinea).
Professor W. Dierks (Institut für die Pädagogik der Naturwissenschaften, Kiel, Federal Republic of Germany).
Professor M. J. Frazer (University of East Anglia, Norwich, United Kingdom).
Dr R. M. Garrett (University of Bristol, Bristol, United Kingdom).
Professor V. Gil (University of Coimbra, Coimbra, Portugal).
Mr W. G. Gregory (Joint Matriculation Board, Manchester, United Kingdom).

Professor A. Gorski (Technical University, Warsaw, Poland).
Professor A. Guerrero (Instituto de Investigaciones Biofisicas, Buenos Aires, Argentina).
Dr R. Hagenauer (Bundesgymnasium und Bundesrealgymnasium, Baden, Austria).
Mrs W. Harries (Northumberland Park School, London, United Kingdom).
Professor F. Halliwell (formerly University of East Anglia, Norwich, United Kingdom).
Dr E. H. Hegarty (University of New South Wales, Kensington, Australia).
Professor H. Heikkinen (University of Maryland, United States).
Miss L. M. de Hernandez (Centro Nacional para Mejoramiento de la Enseñanza de la Ciencia, Caracas, Venezuela).
Dr A. Hofstein (The Weizmann Institute of Science, Rehovot, Israel).
Dr J. B. Holbrook (University of Hong Kong, Hong Kong).
Professor D. Holdsworth (University of Papua New Guinea, Papua New Guinea).
Dr J. Honderbrink (SLO, Enschede, Netherlands).
Dr J. H. Hoogereen (SLO, Enschede, Netherlands).
Mr F. Jenkins (Harry Ainley CHS, Edmonton, Alberta, Canada).
Dr A. H. Johnstone (University of Glasgow, United Kingdom).
Dr A. V. Jones (Trent Polytechnic, Clifton, Nottingham, United Kingdom).
Herr G. Keller (Carl-Schurz-Schule, Frankfurt-am-Main, Federal Republic of Germany).
Dr P. Lai-Min (Hong Kong Examinations Authority, Hong Kong).
Professor M. Laffitte (Université de Provence, Marseilles, France).
Professor W. T. Lippincott (University of Arizona, Tucson, Arizona, United States).
Professor S. E. Loke (University of Malaya, Kuala Lumpur, Malaysia).
Dr H. B. Abd. Manan (Examinations Syndicate, Ministry of Education, Malaysia).
Professor L. N. Markovsky (University of Kiev, USSR).
Dr M. Murtagh (Universities Entrance Board, New Zealand).
Dr Nida Sapianchai (Institute for the Promotion of Teaching Science and Technology, Bangkok, Thailand).
Professor M. Oki (University of Tokyo, Tokyo, Japan).
Mr M. M. Owen (Ministry of Education and Culture, Barbados).
Mr F. M. Pottenger (III) (University of Hawaii at Manon, Honolulu, Hawaii, United States).
Mr Quck Bong Cheang (Examinations Syndicate, Ministry of Education, Malaysia).
Mr D. W. Ridgway (University of California, Berkeley, California, United States).

Dr C. T. Robertson (Inner London Education Authority Inspectorate, London, United Kingdom).
Miss J. Sadler (University of Leicester, United Kingdom).
Professor D. Samuel (The Weizmann Institute of Science, Rehovot, Israel).
Dr K. V. Sane (University of Delhi, India).
Mr H. Schmidt (Colegio Aleman de Sucre, Sucre, Bolivia).
Dr R. B. Shulka (National Council of Education Research and Training, Delhi, India).
Mr M. Sinclair (Ministry of Education, Nairobi, Kenya).
Dr F. Solarz (Northumberland Park School, London, United Kingdom).
Mr J. W. Steward (Ministry of Education, Georgetown, Guyana).
Mr F. G. Stewart (Joint Matriculation Board, Manchester, United Kingdom).
Mr B. J. Stokes (King's College School, Wimbledon, United Kingdom).
Professor B. Suprapto (Director, Direktorat Pendidikan Menengah Umum, Jakarta, Indonesia).
Miss D. R. Titheridge (British Council, London, United Kingdom).
Mr B. Thompson (University of Wisconsin-Eau Claire, Eau Claire, Wisconsin, United States).
Mr P. Towse (University of Zimbabwe, Harare, Zimbabwe).
Dr A. D. Turner (University of London Institute of Education, London, United Kingdom).
Dr G. Van Praagh (Universiti Sains Malaysia, Penang, Malaysia).
Dr J. G. H. Van Santen (ACLO-Sk, Enschede, Netherlands).
Mr K. Yaffe (Institute for Curriculum Development in Science and Mathematics, Boston University, Massachusetts, United States).
Professor R. Viovy (Ecole Normale Supérieure, Saint-Cloud, France).
Dr I. W. Williams (University College of North Wales, Bangor, United Kingdom).

We thank the following for permission to quote examination questions: the Hong Kong Examinations Authority; the Joint Matriculation Board; the University of London School Examinations Department; the Examinations Syndicate, Malaysia; the Universities Entrance Board, New Zealand; the Department of Examinations, Sri Lanka.

We thank the editor of the *Proceedings of the Sixth International Conference on Chemical Education* for permission to include papers written by Professor Kornhauser and her colleagues in Chapter 3, and the editor of the *Source Book for Chemistry Teachers*[1] for permission to

1. John W. Moore, 'Calculators and Computers in Chemistry', in W. T. Lippincott (ed.), *Source Book for Chemistry Teachers*, pp. 125–44, prepared in connection with the Sixth International Conference in Chemical Education, American Chemical Society, 1981.

adapt an article by Professor J. W. Moore for inclusion in Chapter 4. We are also grateful to Unesco for permission to quote extensively from a section on safety in the *Unesco Handbook for Science Teachers*. We reprint an article from *Education in Science*, with the permission of the Association for Science Education, and we use, again with permission, in an abridged form, material from the *Science Teacher Education Programme*, published by McGraw-Hill.

We are grateful to the British Council, London, for access to their invaluable collection of curriculum development materials from many parts of the world.

1 The changing face of chemistry

J. A. Campbell

One of the many challenges facing curriculum developers and teachers, at all levels, is that science is dynamic; our knowledge, in terms of both facts and theory, grows rapidly. Indeed, there are many who feel that there is an unacceptable gap now between the chemistry that is taught in many undergraduate courses and the chemistry that is being pursued, whether it be academic, industrial or environmental. This, in turn, will reflect on the content of our school courses.

The 1960s and 1970s saw many changes in school chemistry courses, both in terms of content and in methodology. We may see an added emphasis now on examining the social and economic factors that concern the implementation of chemistry in our societies.

'New' chemistry and its implications may be brought into the school classrooms and laboratories by formal changes in the curriculum or it may be simply through the initiative of individual teachers. So before looking any further into aspects of chemical education as curricula, educational technology, assessment, training and educational research, we are devoting the first section to some of the changes that are occurring in chemistry.

Professor Campbell has illustrated the changing face of chemistry with examples of his choice, mainly stimulated by the work of Nobel Prize-winners, but emphasizing the interrelationship of chemistry both with other sciences and with ourselves. In doing so, he draws to our attention, as teachers, three important guidelines. First, it is possible to describe in simple terms and with clear examples theories which are often introduced to students in an abstruse and difficult way. Second, he shows that chemistry can be academic as well as relevant. Third, he describes a few of the directions in which chemistry is moving. So often, we, as teachers, describe and discuss what has been done, what has been finished. So here are a few examples to show that chemistry is both an exciting and a living subject—vital to our very existence.

1.1 Introduction

One must constantly think of what will be retained in the audience's memory, not of what can be crammed into the lecture.

Lawrence Bragg, Nobel Prizewinner, 1915

Chemistry is the science of molecular behaviour. Chemists specialize in interpreting observations on large amounts of material in terms of the properties and interactions of individual molecules and atoms. We trace our origins from the early atomic theories of the Indian Vedas and the Greek philosophers, through the alchemists' efforts at transmutations, Lavoisier's discovery of the conservation of matter, Dalton's and Avogadro's brilliant leaps from rather poor data to excellent insights, Maxwell and Boltzmann's kinetic theory, Mendeleev's and Mayer's ordering of the chemical elements, to the discovery of radioactivity, the atomic nucleus, isotopes, and the still-increasing set of sub-atomic particles. Then we retrace to atoms, the building blocks on which chemists concentrate while fully aware of the complexity of atomic composition. And, most recently, chemists have moved toward measurements based on the number and/or energy of photons and/or electrons (quantized particles) in addition to the bulk properties of mass and volume used previously.

The great bulk of the historical evidence in current use is based on average properties. Maxwell and Boltzmann were the first to make clear that the averages covered a wide range of individual behaviours—molecular velocities and energies in their case. They were closely followed by the discovery that Mendeleev's atomic weights were averaged over isotopes. The way was open for quantum theory to tie together the existence of units of charge, units of mass, units of energy, and all the other quantized properties we have found so useful in describing and interpreting observations.

But during all this growth of ideas there were almost no observations of the behaviour of single atoms. The first such observations were probably the scintillations Rutherford and his students found so useful for detecting nuclear radiation. And it is still true that the individual sparkles seen in a spinthariscope or in a fluorescent screen viewed by microscope are among the most convincing and aesthetically satisfying proofs of quantization.

Students enjoy doing experiments. From time to time in this section a possible experiment will be outlined, an experiment which amplifies the material. Here is the first.

EXPERIMENT 1: SEEING BEHAVIOUR OF A SINGLE ATOM

Radiation connected with radioactive processes can cause emission of visible light from suitable materials. The radiation expels electrons from close to the atoms it passes. Light is emitted as electrons recombine with the atoms. Commercial, low-cost spinthariscopes may be available for observing the effect. A low-power microscope focused on a fluorescent material activated by radiation also shows the effect. 'Radium' clock and watch faces are a good source. They must be viewed in the dark for the eye to see the individual flashes.

Surely one of the greatest changes in chemistry, especially in the last thirty years, has been the increasing ability to use photons and electrons to detect and measure behaviours and detailed structures of individual atoms and molecules. It is this theme we shall follow in exploring the changing face of chemistry.

As a framework, consider the very units now used on an almost worldwide basis—the SI set (Système International) devised by an International Committee on Weights and Measures in 1965. The triple set of mass, length and time used for the preceding 100 years was replaced by a set of seven: mass (m), length (l), time (t), temperature (T), amount of substance (N), electric current (I) and light intensity (cd). This increase did not result from the discovery of four new quantities; all of them had been well known previously and had been expressed in terms of mass, length and time. A principal reason for redefining three of them was the increasing desirability of dealing with individual units of light as photons, electricity as unit charge, and atoms and molecules as units of amount of substance. For example, electric charge, Q, formerly had the dimensions of $m^{1/2} \times l^{3/2} \times t^{-1}$, but is now expressed as $I \times t$.

Measurements in six of the seven dimensions are now done routinely on individual molecules. Individual molecular masses are readily available to 1 part in 10^5, individual bond lengths to 1 in 10^4. Molecular vibrations may be timed in picoseconds (10^{-9} s) to better than 1 part per million (1 ppm = 1 in 10^6). Individual charged particles and individual photons can be detected and their number determined. And a rapidly increasing number of atoms and molecules can be detected individually and their concentrations measured at very low levels. Indeed, almost any atom or molecule can now be detected if present in as high a concentration as 1 ppm, with part per billion (or even 1 part per trillion, 1 in 10^{12}), being ever more common. Almost always the measurement is of a photon or an electron current, usually amplified to a metered electron current. It is interesting to note that temperature, the only one of the seven dimensions which depends only on average properties, is measurable to the least accuracy. Measurements can be readily made to $\pm$ 0.01 K near room temperature (say 300.00 $\pm$ 0.01 K), but absolute measurements to better than $\pm$ 0.001 K are almost never attempted. Note that bond lengths (also an average property as measured) are not very well known either. The fact that so many things (including mass, charge and energy) are quantized is responsible for much of the accuracy of measurement and prediction in science. And the increasing ability to isolate, detect and measure individually quantized changes accounts for some of the most rapidly developing areas in chemistry. Forty years ago, most chemical measurements were of mass and/or volume (bulk properties); now a large number are of the number and/or energy of photons and electrons (quantized properties).

1.2 Dinosaurs and asteroids

The great virtue of the chemical artist is ... appreciating that no art is worth its result which does not produce practical consequences from its theory.

So wrote Andreas Libavius in *Rerum Chymicarum Epistolica*, which is the first known book on chemistry, published in 1595.

Our first example involves no professional chemists in the narrowest sense. It might be described as geology. But it is an excellent example of the nature of chemistry as central to many current areas and of the desirability and need to understand and use a broad range of ideas and techniques. This utilization of skills from many areas is characteristic of most contemporary explorations, research and theorizing.

Some 65 million years ago the face of the earth changed remarkably and quickly. We know it occurred at that time by using radioactive dating (among other methods), and we know it was a cataclysm from the fossil record. There is not only a great deposit of fossils but a great change in fossil types. In a short period of time all the dinosaurs, all the marine reptiles, all the flying reptiles and large numbers of marine invertebrates became extinct. Simultaneously there were massive decreases in the population of many tiny marine animals and plants such as planktons. About half of all the then existing genera perished during the great extinction. The geological record is clear throughout the world and provides the definite boundary between the Cretaceous and Tertiary periods of the earth. What happened?

During the more than 100 years the geologic record has been known, none of the interpretations of the extinction has been widely accepted. They include explosion of a nearby supernova, flooding of the ocean surface from a large, freshwater polar lake, meteorological changes perhaps correlated with a solar fluctuation. But none fits enough of the record to be convincing.

Two of the best exposed records are in northern Italy and in Denmark. The limestone strata (see frontispiece) not only show a clear discontinuity in the fossil content but also exhibit a thin layer of clay about 1 cm in thickness. This clay attracted the attention of a team at the University of California, Berkeley. The members were Luis Alvarez (Emeritus Professor of Physics), his son Walter Alvarez (Associate Professor of Geology), Frank Asaro and Helen Michel (Senior and Staff

Scientist, respectively at the Lawrence Berkeley Laboratory [1].* Their results were published in June 1980. It was clear the clay layers had settled out of an ocean, shallower in Denmark than Italy, over a short period of time. So what was in the clay? There were no life-forms so a search was made for the chemical elements present.

One of the most sensitive methods of analysing for chemical elements is neutron activation (NA) analysis. The sample to be analysed is surrounded by a source of slow neutrons. This leads to nuclei in the sample absorbing a neutron, becoming radioactive and emitting radiation (especially gamma rays) whose energies are characteristic of the excited nucleus. Most of the chemical elements can be detected by NA, many at the ppb level, and with good accuracy (say ± 10 per cent, or better, of the actual value).

If the clay were terrestrial matter, it should have the elementary composition of earth sources. But if it were, at least partly, extraterrestrial (say from a supernova), the composition would be different. Two other possible extraterrestrial sources could be comets (mainly ice), or meteorites. Of special interest were chondritic meteorites known to contain a higher concentration of iridium—about 500 ppb (1 ppb = 1×10^{-9})—than earth, and presumed to be the source of the iridium found at about 0.3 ppb in many geological sediments. Neither comets nor supernovae serve as sources of high iridium levels.

EXPERIMENT 2: HOW MUCH IS 1 PPM?

One part per million (1 mg per 1,000 g) corresponds to say a grain of sand in 500 cm^3 of sand, or one drop of water in 50 dm^3 of water. The human eye can detect some items at corresponding concentrations. Try the following: (a) a 1 mm^3 crystal of potassium permanganate dissolved in 1 dm^3 of water. This gives a concentration of 1 ppm; (b) a 1 mm^3 crystal of copper(II) sulphate dissolved in 1 dm^3 of water. Then add some 6 M ammonia to the copper(II) solution.

If any of the above give visible evidence at 1 ppm, dilute that solution by tenfold steps to see the ultimate detectable concentration.

Try similar experiments with some dyes: natural, food or acid-base indicators.

Try flame tests on separate solutions which contain low known concentrations of an alkali metal, or other, ion giving a flame test. What is the minimum concentration that you can detect?

The clays from Italy and Denmark were treated with dilute nitric acid to remove calcium carbonate. This left approximately 50 per cent of the sample which was then subjected to NA. The Italian sample was analysed for twenty-eight elements, the Danish for forty-eight. Samples from the rock layers above and below the clay layer were also analysed. Then comparisons of composition between the clay layers and those

* Figures in brackets refer to the references at the end of each chapter.

above and below were made. For most of the elements, the variation in concentration was less than a factor of two.

What about iridium? The Italian layer showed an iridium enrichment of thirtyfold (to 9.1 ppb), the Danish layer an enrichment of 160. Recently, analysis of a similarly situated clay layer from New Zealand showed an enrichment factor of twenty. It seems clear the results are consistent with a meteoric source for iridium. But, now that neutron activation suggests a collision between earth and a meteor rich in iridium, what has that to do with extinction of species, and is there any other evidence for the meteoric source?

Well, how large might the meteor, if any, have been? The Italian data gave the amount of iridium per square centimetre of clay deposit as 8×10^{-9}. If fallout were evenly distributed over the earth's surface (3×10^{18} cm^2), the total mass of iridium would have been 2×10^{10} g. The average concentration of iridium in chondritic meteors is 0.5×10^{-6} g Ir g meteor^{-1} so the total mass deposited worldwide from the meteor would be

$$5 \times 10^{16}\,\mathrm{g} = 8 \times 10^{-9}\,(\mathrm{g\,cm^{-3}}) \times 3 \times 10^{18}\,(\mathrm{cm^2}) \,/ 0.5 \times 10^{-6}\ \mathrm{g\ Ir\ g\ meteor^{-1}}$$

The density of such meteorites is 2.2 g cm^{-3}, so the volume would be 2×10^{16} cm^3 = 20 km^3.

However, we have assumed that all the meteor went into the stratosphere and then settled out uniformly worldwide. This seems most unlikely. The only comparable (?) event for which any data are available is the explosion of Krakatoa volcano in Java in 1883. Then 18 km^3 of mountain blew into the air and 4 km^3 or about 20 per cent got into the stratosphere for worldwide distribution. If the iridium source behaved similarly, it would have had a volume of about 100 km^3 (= 20 km^3/0.20), and a diameter of about 6 km.

There are two more ways of estimating the size of the presumed meteor: (a) from the size of crater needed to produce 1 cm of dust worldwide, which gives a diameter of 7 km, and (b) from the size distribution of meteors and their frequency of collision with earth, which gives 10 km as the most likely diameter. Including separate calculations for the Danish and New Zealand values gives a probable size of 10 ± 4 km for the diameter of the meteor.

Now back to the extinctions. If such a meteor hit the earth, a tremendous explosion would occur and very large amounts of debris put into the atmosphere. The sky would darken and photosynthesis would diminish. The resulting interruption of the biological food chains could easily lead to mass extinction, especially among photosynthetic plankton and vegetarians such as most dinosaurs. The available evidence is that a meteor 10 km in diameter would reduce solar light to a level less than the light from the full moon. Photosynthesis would essentially cease at such levels. Mass extinction consistent with the observations would occur.

Where is the crater? Possibly in the oceans. Three meteor craters are known on earth of sufficient size, but two are much too old, one much too young (30 million years). But again chemistry may help in locating the site. It should have a high iridium content, should be surrounded by fallout remains containing a high concentration of iridium, and should be about 65 million years old. It is even possible that enough clay samples will be found that their distribution of thickness and iridium will tell whether the site was in the northern or southern hemisphere and in what location. Surely, without good chemistry none of the present or future evidence can be effectively obtained or evaluated. And without the quantized emission of radiation the present evidence would not have been gathered.

There have been other periods of massive extinctions in the history of earth, but evidence concerning their cause is still fragmentary. Nor can one even be sure that an iridium-rich meteor caused the one we have been discussing. But it is interesting to speculate on probable survival rates today if such a meteor did strike. The local devastation would, of course, be complete. Probably many distant earthquakes would be triggered (but they would have happened sooner or later anyway). And there would be a dust cloud obscuring the sun. The evidence from Krakatoa is that the cloud would have appreciable effects for five to ten years but major effects for only two or three years. Most standing trees and food crops would not survive the loss of sunlight. The seeds would survive and, as in past extinctions, there would be a rather rapid recovery, but with a different balance of species. Plant-eaters would be under great threat. And carnivores survive on plant-eaters so many of them would not survive either. Human food stores, even those made after the cataclysm, would almost certainly assure survival of the species, but in much reduced numbers. The degree of survival would be strongly influenced by how much chemical knowledge was used to assure minimum food loss to decay and vermin. But this is, of course, true now. If countries were able to use effectively existing food storage methods, hunger would almost disappear. Perhaps this is the message we should take from the fate of the dinosaurs 65 million years ago.

1.3 Faster and fastest

If people understood the methods of science they might carry them into their everyday lives, and their politics, and understand how difficult it is to find the truth—George Porter, Nobel Prizewinner, 1967.

The iridium evidence helped describe an event which occurred 65 million years ago and the changes which have been going on since. Most events of interest occur much more rapidly, sometimes so rapidly they are difficult to control. One of the accomplishments of chemists is the ability to control the rate of many changes. In fact, one of the reasons that so many people are unaware of chemical facts, as opposed to biological and physical ones, is the slow rate of change of most of the chemicals around them. In any large city in the world almost everything in the city has undergone chemical processing to make it last indefinitely. And people get irritated if road surfaces wear out, paint peels off, cement crumbles, cars rust, wood decays, or clothes cannot be cleaned readily. Each of these involves chemical treatments or manufacturing processes, and the utilitarian products seem rather dull and uninteresting chemicals just because they are so inert to change.

But there are times when people want fast reactions. They like, for example, fires (which are rapid chemical reactions) for cooking and warmth, drying laundry (rapid evaporation of water) to reuse their clothes, quick-acting aspirin for removing aches and pains, fertilizers for crops and quick-setting dyes to prettify their surroundings.

One of the most used ways of controlling the rate of a chemical change is to modify the temperature. Raising the temperature usually increases a rate; lowering temperature usually decreases it. The rate of drying of laundry is a good example. And chemists have learned to interpret the effect of temperature on rate in terms of a stepwise mechanism with an activation energy. If the activation energy is low, or zero, there is little effect on rate when temperature varies. But if the activation energy is high, a change in temperature greatly changes the rate. The marked change in temperature required to boil eggs or other food at higher altitudes is interpreted in terms of a large activation energy for the cooking reactions. The rather small change in temperature of the boiling water, say less than 10 °C, can lead to a tenfold increase in the time required to cook food.

Another common way of increasing reaction rates is to add a catalyst for the change. A catalyst is a chemical which provides a new and faster mechanism for the reaction. It does so by reacting with one or more of the reactants and then later being regenerated so it can enter the reaction once more. The recycling of the catalyst not only speeds the reaction, it allows a small amount of catalyst to speed the reaction of a large amount of the reactants. Enzymes are by far the most effective catalysts known. Some are able to cycle so rapidly they have a turnover number of 600,000 per second, whereas no chemically synthesized catalyst is better than 1,000. (The turnover number represents the number of cycles of reaction in which a single catalyst molecule participates each second.)

As the number of enzymes known grows (from less than 100 in 1930 to about 3,000 today), their roles in nature are better understood and their use in medicine and industry grows. Methods which fix (immobilize) the enzymes on to a solid surface permit repeated use of the enzymes because they can then be readily separated from the reaction products. Present uses include wine, beer, cheese and bread-making, tea, coffee and cocoa processing, liquefaction of starch by alpha amylase followed by further enzymes to produce sugar syrups, detoxification of blood in humans, waste disposal, enriching soya bean flour with essential amino acids, and even electrometric analysis for such substances as glucose and amino acids. Most enzymes are extracted from bacterial cultures, but, thanks to the pioneer work of R. B. Merrifield [2], synthesized and modified enzymes are becoming possible.

The effectiveness of catalysts, including enzymes, is often interpreted in terms of their effect on the activation energy, consistent with the great sensitivity of rates to changes in activation energy. For most catalysts the interpretation is valid. But for enzymes the effect on the activation energy is often small, and sometimes even in the wrong direction. The enzyme may well raise the activation energy but also considerably raise the rate of reaction. How can this be?

For any reaction to happen, at least three steps must occur: (a) the reactants must collide; (b) they must come together in such a spatial arrangement that bonds can readily shift from those in the reactants to those in the products; and (c) there must be enough activation energy present to provide the necessary push to convert reactants to products.

Moving a rock with your hand provides a good analogy. You must (a) place your hand on the rock, (b) move your hand to where you can get a grip on the rock and (c) exert some energy to move the rock. You might make it easier to move the rock by pouring some water around it to loosen the soil. This would lower the activation energy for the moving. And you could, if necessary, reclaim the water unchanged to use in moving another rock and so indefinitely. Many catalysts 'soften the soil' and so make it possible for change to occur with less activation energy.

Or, going back to the rock analogy, you could find places where your hand just fitted. The firmer grip would enable the rock to be moved more easily, even though a little more energy might be required. But the smaller the number of handholds, the slower the rock is moved. Most enzymes, even those which also lower the activation energy, work by improving the fit between the reacting substances. Of all the possible ways the reactants could come together (most of them with their reactive sites not near together) the enzyme forces the occurrence of that way which allows most rapid matching of the reactant sites and so a quicker conversion of reactants to products. But each enzyme usually has only one 'handhold'.

Just as minimizing the energy expenditure is attributed to lowering the activation energy, so maximizing the fit of the reactants is known as lowering the entropy of activation.

Entropy of activation measures the increase in disorder of the system as it passes through the state of being catalyzed by the enzyme. A highly

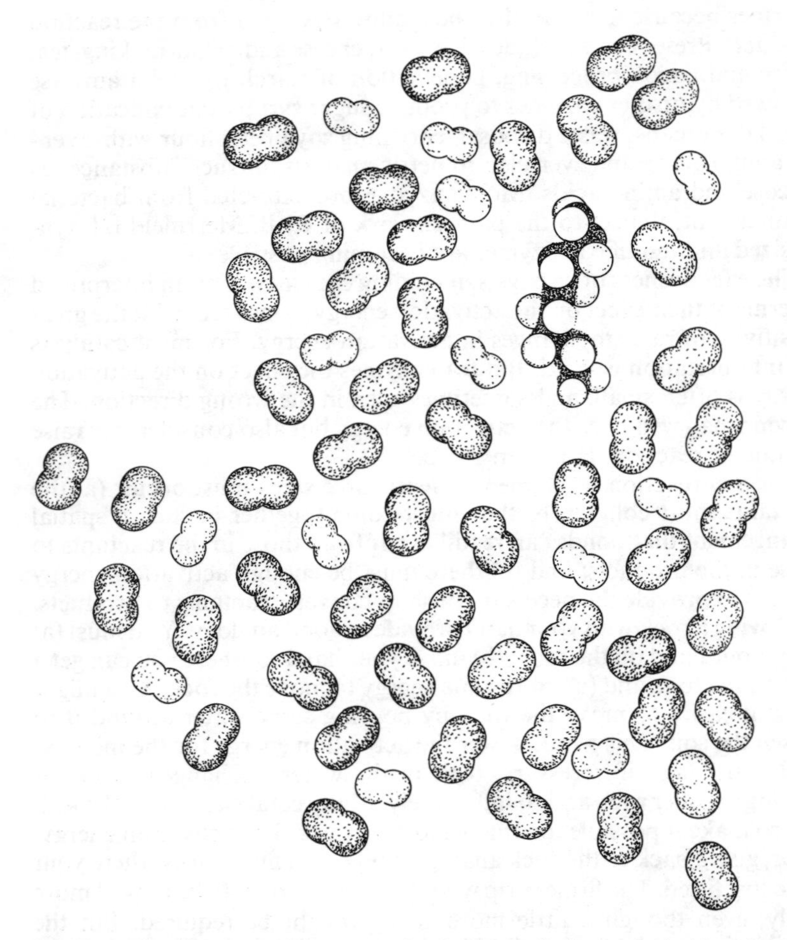

FIG. 1 (a). The reactants when octane molecules (in gasoline) are mixed with molecules of oxygen and nitrogen prior to burning to form product molecules and energy. Most reactions involve: (a) two species (molecules, ions, atoms, radicals etc.) colliding at a time, with (b) an appropriate orientation, and (c) enough energy to move from reactant bonds to product bonds.

ordered transition state leads to a low entropy of activation, a disordered one to a high entropy of activation. The latter helps increase the rate, the former keeps the rate low.

Maximizing the fit, in and of itself, actually slows the rate, because the reactants must find the one active site on the enzyme which catalyses the reaction. However, enzymes are large, complicated molecules. As they bring the reactants together, the enzymes change shape and loosen up their own structure. Simultaneously, they liberate many molecules of

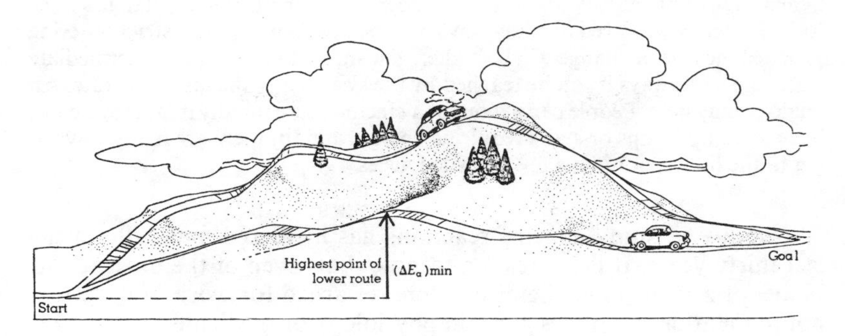

FIG. 1 (b). Car travel analogy to a chemical reaction mechanism. To proceed, the car must: (a) find the road; (b) be moving in the correct direction; and (c) have enough energy to get over the highest point in the route.

water bound to themselves. The net effect of all this is an overall increase in the entropy of activation. This is accompanied by an increased likelihood of the reaction occurring; that is, the rate of reaction increases. In terms of the rock analogy, it is almost as though placing your hand on the handhold made the rock flexible and loosened it from the surrounding soil.

For example, the activation energy for a certain pepsin-catalysed reaction is 12 kJ mol^{-1} higher than for the acid-catalysed mechanisms. But the entropy of activation (loosening-up effect) gives a value of $T\Delta S^{\ominus}$, which is *c.* 45 kJ mol^{-1} (at room temperature) more favourable in the case of the pepsin. The net effect is a more favourable (higher) rate with the pepsin corresponding to a $\Delta G^{\ominus}$ value lower by *c* (45 – 12) = 33 kJ mol^{-1}, in spite of the unfavourable shift in activation energy. In this and many other cases the entropy (loosening – up) effect is large enough to offset a simultaneous increase in the activation energy. This is contrary to past expectations that decreases in activation energy were required if a catalyst was to be effective. Often the loosening up is the larger effect. A modern treatment of rate of reaction should include this. The very best catalyst, of course, would presumably provide a mechanism which had the lowest activation energy and highest entropy of activation of any possible path. (Note that the terms energy and entropy of activation have been used here for brevity; but the ideas could be taught with much more basic language if desirable.)

For example, rates are increased by any change which lowers the energy required to convert reactants into the intermediates which form as the reactants proceed toward the products. The less energy required, the faster the reaction occurs. Rates are also increased by any process which decreases the orderliness required to achieve the intermediate state. If the intermediate state can be achieved in only one way (or a small number of ways), the rate of change will be low. If it can be achieved in many ways (that is, does not require a strict ordering process) the rate of change will be higher. The more disordered the intermediate state, the more ways it can be reached by the system and the faster the rate. An analogy may help. People can get out of a cinema more rapidly if (a) they do not have to go up ramps or stairs requiring energy, and (b) there are many ways to get to the doors, not just a few crowded aisles.

The ability to study enzyme reactions has increased markedly in the last thirty years. It is interesting to note that seven of the most recent Nobel prizewinners in chemistry were rewarded for work in this area, not to mention the prizes given in physiology or medicine.

There are, however, many reactions which are even faster than enzyme-catalysed reactions. They occur as fast as the appropriate reactants collide because both the entropies of activation and energies of activation are close to zero. Furthermore, they involve extremely short-lived intermediates. The change is almost instantaneous. But what does instantaneous mean at the level of chemical reactions? An instantaneous chemical reaction is perhaps best defined as one which occurs in the time of a single vibration. Two molecules, for example, collide, exchange atoms and/or electrons, and then recoil. They may do this in the time for a single vibration. For many molecules, bond vibrations occur in about 10^{-12} s, about 1 picosecond. Short time though this is, methods are becoming available to study such changes.

Any means of studying changes at the 10^{-12} s level must involve a rapid signal. This turns out not to be a problem. Photons travel 3×10^8 m s^{-1}. Molecules are about 10^{-10} to 10^{-11} m in diameter so a photon passes a molecule in about 10^{-18} s. For all practical purposes molecules are almost stationary as far as photons can tell. The average atoms would move through about one-millionth of a vibration, or less than 10^{-17} m, as the photon passed.

The actual problem has been to get a high enough concentration of photons so there will be one passing by when each very fast change occurs, and then to record the photon signals with a resolution time of about 10^{-12} s. Resolution is a problem in electronic circuitry which is now sufficiently advanced to permit measurements at this speed. The concentration of photons has been solved by the development of pulsed lasers. For example, a mode-locked Nd^{3+} glass laser can produce a train of 6 picosecond pulses with an interpulse separation of 7 picoseconds. Each pulse contains about 10^{14} photons over an area of 0.2 cm^2. This gives a photon concentration of about 2×10^{-6} mol dm^{-3}. This is sufficient to give quite adequate interaction with fast processes in molecules present at about the same concentration (10^{-5} to 10^{-6} mol dm^{-3}).

Our knowledge of reaction intermediates and mechanisms, even in very fast reactions, is increasing rapidly. One method is to send intermittent flashes through the reaction medium to see what concentrations are changing and at what rate. Another is to provide a flash of energy (light or shock) followed by light pulses to detect the subsequent rapid changes. In a similar fashion, molecular beam experiments which observe the scattering of products from collisions in crossed beams of gaseous molecules have produced considerable information on the details of interactions between very simple molecules such as Na and Cl_2.

The best experiments are simple and on a large scale, and their workings are obvious to the audience.

Michael Faraday's view leads us to the next experiment.

EXPERIMENT 3: LIGHT AND CHEMICAL REACTION

It is well known that plants provide the energy to cause their photosynthesis by absorbing sunlight. The process is very complicated and, though we see it occur, the changes caused by the light itself are not directly observable. They can be resolved by flash photolysis methods as described above. Here is a photochemical reaction which proceeds slowly enough to be studied in any laboratory.

Irradiate with bright sunlight, or a 500-watt lamp, a solution of 2–3 g of iron (II) sulphate in 250 cm^3 of water slightly acidified with sulphuric acid and moderately coloured with either thionine or methylene blue. Then set the container in a dark place. Repeat the process and note the relative rates of the light-induced reaction and the change in the dark. Vary concentrations of the salt and acid and light intensity individually, to see how fast the reaction can become.

Filter the light through different colours to see if the reaction is more sensitive (reacts faster) to some colours than to others. Remember, the filter does two things: it changes the colour but also cuts down the overall intensity of light entering the flask. Try to adjust for this effect and measure the extent to which colour or overall intensity is responsible for the changes.

As would be expected, the rate of most very fast reactions depends only on the concentration of one of the species. They are first order reactions. There is not enough time for molecular collisions to occur. But fast second order reactions, often occurring at almost every collision, and so controlled by the rates of molecular diffusion, are common both in gases and solution and often have half-lives of the order of nanoseconds and shorter. Work is now under way to push the time resolution to femtoseconds (10^{-15} s). No molecular processes known at present are faster than this. If some are discovered, excitement will reign. But the Heisenberg uncertainty principle will begin to limit observations in that time range ($\Delta\tau$) because the uncertainties in bonding energy (ΔE) will be as large as the bond energy itself:

$$\Delta E \times \Delta\tau \simeq \frac{h}{2\pi}$$

$$\therefore \Delta E \times 10^{-15}\ \mathrm{s} \simeq \frac{6.6 \times 10^{-34}}{2\pi}\ \mathrm{Js}$$

$$\therefore \Delta E \simeq \frac{6.6 \times 10^{-34}}{2\pi \times 10^{-15}} \simeq 1.05 \times 10^{-19}\ \mathrm{J}$$

Thus $$\Delta E \simeq 60\ \mathrm{kJ\ mol^{-1}}.$$

In other words, as we can measure the timing of an event more and more closely, it becomes even more impossible to measure the accompanying change in energy. And this is not a question of improving our measuring devices. The Heisenberg uncertainty principle describes a fundamental limitation in making measurements. Some would say time is catching up with chemistry; or is it vice versa? And again it is the quantized nature of energy and mass which allows the observations to be made, and then limits the attainable accuracy.

1.4 Interlude

> Now, in 1979, I would say that the first-year course in chemistry should give a student enough understanding of chemical structure and chemical properties to enable him to understand a good bit of molecular biology, to appreciate the significance of the double-helix structure of DNA, to follow the arguments about the molecular basis of biological specificity and in general of the nature of living organisms. I do not think that any of them would suffer from not having been exposed to molecular orbital theory during their first year in chemistry.
>
> Linus Pauling, Nobel Prizewinner, 1954 and 1962 [3].

Stay a minute, you may say (and so do I): is this material and this set of ideas, as suggested by Linus Pauling, twice Nobel Laureate, suitable for *school* chemistry? I think so; suitable, but far from sufficient. The ideas are teachable. The level must be decided by each teacher.

You will remember I write here for teachers, not primarily for students. Fundamentals are fundamentals, and must be taught and learned first. None of the above is intelligible to one with no feel for the existence and general nature of atoms and molecules and photons. But these we teach now. Yet why? And where are the students left? In the early nineteenth century with Dalton? In the late nineteenth century with Mendelefev? In the late 1920s with Bohr's first model of the atom? Surely all these ideas are important. But are they not stepping stones rather than platforms?

It is true that students become uncomfortable in the face of either new ideas or of ignorance. Most wish to have simple questions and firm answers. But the evidence is strong, even overwhelming, that stepping

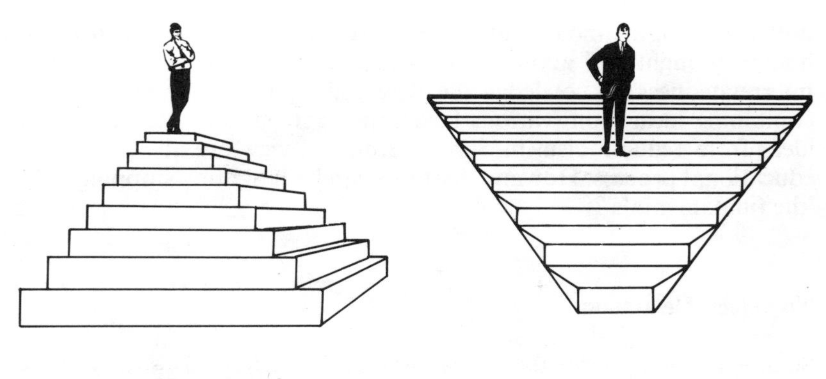

FIG. 2. Two views of the educational process.

stones are what we teach. We believe they are stepping stones from ignorance through new ideas to a rising pyramid of knowledge. What the student fears is that the steps will get ever steeper, too steep to permit the climb.

But what are the alternatives? Achieving a flat plane with no new features? A descending ramp with narrower and narrower possibilities? Or a cliff with a possibly exhilarating fall but catastrophic terminus? Much of education makes the mistake of modelling itself as a stepped pyramid which tapers upward but inward. The view gets narrower, the path more crowded, the pinnacle more isolated (pointed but pointless). Much better to consider the steps as ever expanding; the subjects more numerous, more interesting and more exciting to explore. From each step there are evermore upward possibilities. The student's ability to find and use personal capabilities increases. The pyramid is inverted. It emphasizes growth, not finality.

The Cretaceous–Tertiary discontinuity is not solved and shelved if an asteroid was the simultaneous event. We then might ask why only half the genera became extinct, and why those particular ones? Studying rates of chemical reactions at or beyond the femtosecond level may be impossible because of limitations described by the uncertainty principle. On the other hand, new observations may turn up. We can also go on to explore somewhat more complicated and more interesting reactions than the simple ones to which we are now limited.

As to the non-scientist, and these are the most abundant of people, the study of the fundamentals can come much more to life if applied to some current ideas. The course becomes more interesting. The ideas are apt to be set in a framework useful in later life. The ability to bring to bear on a single problem ideas from many fields may grow. The strength to live with ambiguities (as all must do) may increase. The comprehension of both the strengths and limitations (not really weaknesses) of science and human thought and action may mature. Further, the wholeness and interrelatedness of knowledge, the increased capabilities associated with combining ideas from various fields, the aggregation of the powerful ideas from reductionism into a more unified view are all vital to the educational process. How much of this can be taught by stopping with 'the fundamentals'?

1.5 No detectable amount

Science does not provide the mechanisms for constraining the society to use scientific discoveries wisely.

No scientist today need look on whom the mantle of responsibility for enlightening the society in science falls—it falls on each of us.

Most legislators attempt to generate laws for the public good. If nothing else this keeps them in a position to generate further laws. But, as with all of us, they are often poorly informed. A case commonly cited is the legislature that decided the value of pi (π), as in the circumference of a circle $= 2\pi r$, was too difficult to remember so some passed a law defining $\pi = 22/7 = 3.14285$, etc. Now mathematicians know that the figure has to be 3.14159, etc. for accurate work. But few legislators need to operate with an error of less than 0.06 per cent, so for legislative purposes 22/7 worked fine. There was no detectable problem!

In 1958 the Congress of the United States did a similar thing. It passed a law, now called the Delaney Amendment, which required that all substances known to contain material carcinogenic to humans or test animals be banned from human consumption. This has since been labelled the absolute zero risk concept. Any well-educated person realizes that absolute zero risk is no more attainable than absolute zero temperature or absolute zero pressure. But what now makes the 'Delaney problem' acute in many parts of the world is the tremendous increase in ability of analytical chemists to determine the presence of tiny amounts of material. The lowest detectable levels of most chemicals in 1958 were about 1 in 10^3, now they are 1 in 10^9 and going lower. This is an improvement by a factor of 1 million in chemical analysis, and still improving.

But the end is in sight. A few years ago the best detectors of chemicals seemed to be insects. Many male insects can detect molecules, called pheromones (often emanating from females as a sex attractant), at a concentration and rate of less than ten molecules per cubic centimetre per second. The males would then fly upwind, as much as 2 kilometres, to find the female and mate. This concentration is about 1 molecule of

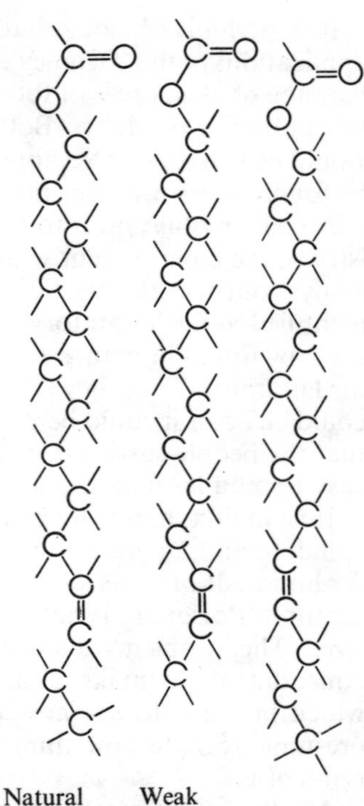

FIG. 3. A pheromone produced by the red-banded leaf roller (*Argyrotaenia velutinana*), and two similar molecules.

pheromone in 10^{18} of air. But the insects have a very limited analytical capability. Small modifications in molecular structure from that of the natural pheromone make the substitute undetectable by the insect.

Methods are now available, even to many college students, to detect and identify many elements and compounds at the level of 1 in 10^6, and 1 in 10^9 is often possible (remember the iridium and extinction problem). Chemists have developed methods which are much more sensitive than this and it now appears feasible to have methods which will detect and identify single atoms of many elements at a level exceeding the one in 10^{18} of the insects. Finding one atom or molecule per cubic centimetre seems ultimate enough. It seems clear such analyses will tell us more than we can reasonably expect to use.

(Experiment 2 can also be reviewed or amplified in connection with this discussion.)

It is perhaps obvious, but will be explicitly stated for clarity, that applications of the 'Delaney condition' in any country would lead to the banning of all sources of food, water and air. It would be impossible to be indoors or outdoors. Both mountain climbing and deep-sea diving would be forbidden. No appreciable sample of food, water or air could be found which was free of carcinogens.

A similar situation is found in many attempts to regulate the chemistry of the environment with respect to other threats. The skills of the analyst outstrip the tests of the pathologists, so that many substances identified in the laboratory as pathogens at appreciable concentrations are now found in nature at levels 10^{-3} or 10^{-6} of the test levels used in the laboratory. Are they still pathogens at these low levels? The answer could be no or it could be yes, at least for some people. But it is unlikely that the people susceptible at such low levels are identifiable. Nor is it easy to set a level of tolerable threat.

It is, in fact, highly likely that all chemicals are toxic, probably even lethal, to all humans at some dose level. This has been widely accepted for hundreds of years, ever since Paracelsus pointed out in the sixteenth century: 'Poisoning is not a matter of chemicals; it is a matter of dose size.' The problems associated with the laws such as the Delaney Amendment do make clear the distinction between legislative laws which presume to say how things *should be*, and natural laws which presumably state how things *are*. It does not pay to confuse the two types of law.

Another point is made clear by our search for improved methods of chemical analysis. New knowledge not only reveals things that please us and broaden our abilities and comprehension. New knowledge always also reveals new threats, previously unsuspected sources of harm, and may even destroy some of our comfortable ideas from the past. As with other two-edged swords, new knowledge must be used with good sense. The responses to the promises and threats are important; the new knowledge alone is useless.

Finally, it is important to note that the new techniques in science produce much more data and much more reliable data than were ever available before. A single chemist can easily do more analyses of silicate rocks each year than all the analysts working did in the hundred years prior to 1950. And most undergraduate chemists today produce and analyse more experimental data than any research student prior to 1950 did in his four or five years of graduate work. Our problem today is more and more becoming one of handling intelligently both immediately, and in store for the future, the flood of data that pour from our instruments. One of the reasons for studying science and chemistry is to learn better how to collect, understand, interpret and act on both qualitative and quantitative data.

And with all data it is important to recall the need for good data, data that are not only precise but are also accurate. One recalls the work of Rayleigh on nitrogen from different sources. He passed air over hot copper and measured the density of the resulting gas at 25 °C and 1 atm as 1.2572 g dm^{-3}. He heated ammonium nitrite and obtained a dry gas of density 1.2505 g dm^{-3} at 25 °C and 1 atm. And he burned ammonia in air to produce a dry gas of density 1.2564 g dm^{-3} under the same conditions. Many experimenters would have complained about their lack of precision and assumed they obtained pure nitrogen in every case while some unknown difficulty was the cause of the variation.

But Rayleigh and Ramsay noted that Cavendish had tried a similar experiment 100 years before when he sparked air, dissolved the resulting oxides of nitrogen, added more oxygen, sparked and removed the oxides again. He continued until he had a residual gas, about 1 per cent of the original volume, which would not combine with oxygen. He made no comment on the gas. Rayleigh and Ramsay repeated the experiment and then subjected the gas to spectroscopic analysis. The emitted light (here are those quantized photons again) convinced them there was more than one element present. They named the principal constituent argon.

Careful purification and distillation experiments over the years led them to the isolation of neon, argon, krypton and xenon. They had added a whole new family to the periodic table by being concerned with a spread in analytical results of about 0.5 per cent. They also provided valuable insights into atomic structure and bonding once it was noted how many atoms in nature react to achieve the same number of electrons as their noble gas atoms had. Knowledge of acceptable and unacceptable interpretations of the spread in values of analytical data pays off! And it is probably true that poor analytical data are responsible for more errors, and more serious errors, in chemistry than any other single cause.

1.6 Energy and change

The second law of thermodynamics has played a fundamental role in the history of science far beyond its original scope. Suffice it to mention Boltzmann's work on kinetic theory, Planck's discovery of quantum theory, and Einstein's theory of spontaneous emission, all of which were based on the second law of thermodynamics.

Thus, in his speech, when accepting the Nobel Prize in 1977, Ilya Prigogine was underlining the importance of energetics [4].

One of the great chemical problems of our time involves the availability of energy sources. This highlights the controversy as to how much thermodynamics (to use a term which frightens many) should be in school chemistry. Surely it is a great dis-service to avoid the energy problem in chemistry courses. Let us see if some new approaches might make the subject more interesting, more understandable and of more long-term use.

One of the most powerful generalizations discovered by scientists is the law of conservation of energy: the total energy of the universe is a constant. How then do we account for, or interpret, the increasing concern throughout the world that we are running out of energy? It is not possible to run out of something which is present in a constant amount. Perhaps the earth is losing energy faster than energy flows in; is this therefore a local problem?

But the chemical evidence from tree rings, fossils, isotope distribution in carbonates and other data on ancient times shows this not to be the case, at least as far as temperature is concerned. The temperature of the earth has fluctuated about $\pm$ 5 °C around an average of 12 °C for a million years, but no long trend is observable. Of course, temperature is an insufficient measure of energy content, but it is true that energy gain in a system often raises the temperature while energy loss lowers it. At least at this level, we have evidence that there is little net change in the energy on earth.

The reason for the constant energy level on earth is not hard to find. By far our biggest in-flow of energy is from the sun, and our biggest out-flow of energy is infra-red radiation to space. The two are almost equal so the net energy content of earth remains unchanged. There is a contribution (of less than 1 per cent) of radioactive energy from within the earth to the energy radiated out and there is about an identical amount of photosynthetic capture of solar energy in. The two almost equal rates have kept the total energy balance unchanged to a very good approximation,

$$\text{solar energy in} = \text{infra-red radiation out}, \Delta E_{\text{earth}} = 0.$$

We can think of this change in terms of photons of visible light in, photons of infra-red light out. The average energy of the photons in is about twenty times the average energy of the infra-red photons out, so the energy flow can also be described as

$$n \text{ (solar photons in)} \rightarrow 20n \text{ (infra-red photons out)}, \Delta E_{\text{earth}} = 0.$$

there has been no change in the total energy, but there has been a big change in the number of photons in existence, that is in the distribution of the energy to twenty times as many particles (photons). The energy becomes more spread out and less concentrated. Its packets of energy (energy per particle) are much smaller. Ever since the work of Carnot and Clausius in the nineteenth century it has become increasingly clear that it is this spreading out of energy to which we refer when we complain of running out of energy. The amount of energy remains constant, but its availability (measured in energy content per photon, per particle, or per molecule) decreases.

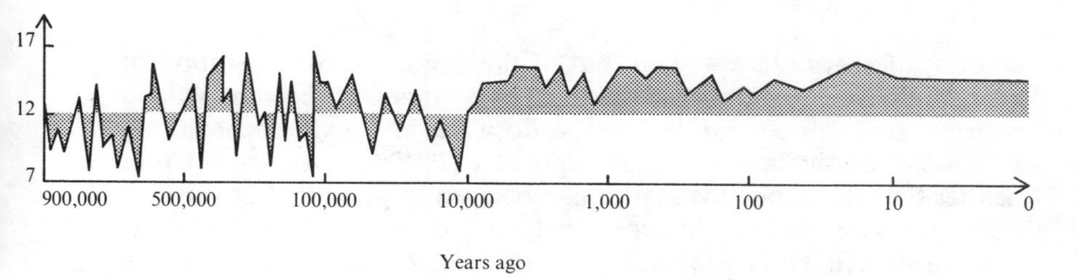

FIG. 4. World temperature as a function of time. (Data from *Understanding Climatic Change*, National Academy of Science, Washington, D.C., 1975.)

Explosives are substances which have a high energy per molecule, as are fuels. Good explosives and good fuels release their energy under readily controlled conditions to form a larger number of molecules, each with a much lower energy content. The released energy spreads out into the surroundings doing work, as in excavation or powering an engine, and/or raising the temperature as in a furnace. Typically, at the same time the energy spreads out, the number of molecules increases so that the matter also spreads out into more particles (molecules). This spreading out can also do work and/or provide heat. Here are two typical examples in terms of chemical equations for the explosion of glyceryl trinitrate (nitroglycerin) and for the burning of gasoline (which is here represented by the single compound, octane):

$$H_5C_3(ONO_2)_3(l) = \frac{3}{2}N_2(g) + \frac{5}{2}H_2O(g) + 3CO_2(g) + \frac{1}{4}O_2(g),$$
$$\triangle H^{\ominus}_{298} = -1540\,\text{kJ mol}^{-1} \text{ of } H_5C_3(ONO_2)_3$$

$$C_8H_{18}(l) + 12.5O_2(g) = 8CO_2(g) + 9H_2O(g),\ \triangle H^{\ominus}_{298}$$
$$= -5450\,\text{kJ mol}^{-1} \text{ of } C_8H_{18}$$

Each reaction liberates considerable energy to the surroundings and also produces a large increase in numbers of molecules. Both energy and matter have spread out. Study of *all* changes shows that *all* total changes are characterized by the spreading out of energy and/or matter.

It is interesting to note that cell growth in living systems operates under the same principle: a cell takes in nutrients, then produces two new cells plus waste materials. Energy is liberated to the surroundings of the process, and the waste materials are simpler and smaller molecules than the nutrients. The net change, because the original cell can be considered a catalyst for the process, can be represented as:

cell + nutrients = 2 cells + waste molecules + energy to surroundings
Net: nutrients = cell + waste molecules + energy to surroundings

Energy is spread out and so has much of the matter.

So we see that energy problems are not primarily a question of the amount of energy in a system, but of the degree of concentration of energy into a small number of photons, or particles, or molecules. Such concentrated sources can be used to do work or provide heat during which the energy and/or matter spread out. But, once spread out, neither the energy nor the matter can be used to provide that work or that heat again. In fact, to get the products of a change back to their original state (if it is possible at all) always requires more energy than the earlier spreading out process produced. Cyclical processes always result in a net spreading out of energy and/or matter, and so lead to a decrease in the ability to work or provide heat.

EXPERIMENT 4: THE TENDENCY TO SPREAD OUT

When heat spreads out from a system we say the change was exothermic. It is easy to detect exothermic reactions by noting the corresponding temperature rise during reaction in the system before the heat has a chance to escape to the surroundings. Many common reactions are exothermic, consistent with the suggestion that energy does tend to spread out during a change. In the late nineteenth century it was believed by some that all spontaneous reactions were exothermic. But it was quickly discovered that this was not so. Many spontaneous changes are endothermic, cooling off during the change and absorbing energy from the surroundings. In all endothermic changes, however, it is observed that mass (that is atoms and/or molecules) spreads out: gases form, solids dissolve in liquids, crystals melt to liquids, molecules dissociate into smaller molecules. Test this for yourself to see if you can find any change which does not involve the spreading out of energy (that is, it is not exothermic), nor does it involve the spreading out of mass (that is, it does not give a less condensed form of matter after the change).

In each of the following decide whether energy and/or mass spreads out. Use only a thermometer, or your finger touching the outside of the container, to check energy flow.

Burn a match, treat iron with an acid, add salt to water, add solid ammonium chloride to water, boil water, allow a liquefied wax to solidify, mix about 1 *M* solutions of any two chemicals that give a safely observed reaction.

Observe any change in terms of energy flow in and out and increased or decreased spreading out of matter.

Let us pause for a moment to note that the above has not invoked the term 'entropy', a term which has become frightening to many, and mystically incomprehensible to most. We have used only work and heat, both terms all too familiar to students throughout the world, plus energy, a term much used by teachers in exhorting students to greater efforts. Furthermore, we have used examples the students understand, not pistons and gases they have never thought about. We have also used chemical examples, rather than terms such as adiabatic and isothermal expansions, because thereby a course based on molecular behaviour is more likely to seem 'chemical' to chemistry students. And finally, we have avoided mathematical expressions without compromising accuracy. The resulting loss in precision is more than offset by increased clarity for most students at school level. At higher grades, precision and clarity become more and more closely associated, but at lower grades they are often confused. Clarity of understanding must come before precision of description.

If we return for a moment to the law of conservation of energy we can see that were all properties conserved there could be no change. The fact that change is constant and necessary to life suggests we should also try to describe it. Of course, it is dangerous to generalize from a few examples (energy gain and loss with photons, one explosive, burning gasoline, cell division). But it is true that all total changes which have been studied (from falling apples to explosion of a supernova) fit the generalization: all total changes are accompanied by a net spreading out of energy and/or matter. Just as the conservation of energy has been formalized in the first law of thermodynamics, so the spreading out of energy and/or matter has been formalized in the second law. And the second law is just as important as the first if one is to understand change and what limits the direction in which change can occur.

Two Nobel Prizes in chemistry, to Lars Onsager (in 1968) and to Ilya Prigogine (in 1977), have been given for extending the power of thermodynamics. Both the contributions are couched in exact and rather abstruse mathematical language and neither has yet been included to any great extent in tertiary courses. But it is interesting to note that similarly slow 'trickle-down' has occurred throughout our history. The Nobel Prizes were started in 1901 and have since been awarded to ninety-four chemists. Ideas from about twenty-four of the first sixty winners are now part of most school courses. This means that some 40 per cent of the work before 1960 now directly influences school chemistry. But only about 15 per cent of the more recent is included. There is a lag of about twenty years. And, of course, some 60 per cent of the work is never found suitable for introductory courses. With the increasing use of thermodynamic ideas in all realms of human thought, I predict that these recent contributions will find more use. I am encouraged in this by Paul Samuelson, Nobel Prizewinner in economics in 1970, who said he would not have got his prize if he had not taken an undergraduate course in thermodynamics!

I believe we will shortly find a basic understanding of energy and change almost essential in reaching reasonable decisions. We will also discover much more effective new teaching methods, based on these new needs. As a result, understanding energetics and change will have a place at least as important as the gas laws in introductory chemistry courses. In fact, the emphasis will probably exceed that on the gas laws which are much more restricted in scope and power to handle common human questions. Let me continue with some recent ideas on irreversible thermodynamics, the field addressed by Onsager and Prigogine.

It was pointed out above that cyclical processes always result in a net decrease in the availability of energy. You never get back where you started. 'You can't go home again.' There is no 100 per cent recovery. Well, if not 100 per cent, what is the best one can do? I suggest that before long Le Chatelier's principle will be joined by Prigogine's principle.

Le Chatelier pointed out that when a stress is applied to a system at equilibrium, the system reacts to relieve the stress. Prigogine points out that systems relieve small stresses by following a path which produces the minimum rate of spreading out of energy and/or matter. He also points out that this is the most efficient path in terms of loss of availability of energy. And he further points out that if the stress becomes great the most efficient path is not followed so that the availability of energy decreases much more rapidly. With energy availability a major human problem it seems clear that we may learn much from the Prigogine principle:

> When given boundary conditions prevent the system from reaching thermodynamic equilibrium . . . , the system settles down to the state of 'least dissipation' . . . this property is strictly valid only in the neighbourhood of equilibrium [4].

One reason for awarding the Nobel Prize to Prigogine is the strong indication that this principle with many others from science (for example, the first two laws of thermodynamics) will apply strongly to most, if not all, human activities. If so, surely introductory chemistry is a good place to become aware of the principle. There are many chemical examples of systems close to but not at equilibrium. Some readily available examples are the gain in crystal size when precipitates form slowly, the evenness in boiling if low heat is used, the lengthening of total power of an electric cell if used at low current either on charge or discharge, the increased efficiency of distillations at slow throughput.

The changing face of chemistry at school level may well be twenty years behind the Nobel Committee (and much farther behind the time of discovery), but change it does and quite possibly in accord with the Prigogine principle of minimum energy dissipation! But, if there is no energy dissipation for a time, Prigogine predicts even more energy dissipation will enventually be required. Maximum efficiency in use of educational energy requires steady and continual change, not just periodic spurts of innovation.

1.7 Ammonia—a case history in chemistry

Life does depend on accurate replication of molecules, and its complexity often requires that an enzyme should accept one molecular species or type and transform it to equally specific products . . . [but] many examples are now available in which one enzyme can accept more than one molecular species as substrate but still transforms each of them with absolute, though hidden, control of the stereochemistry of reactions [5].

Almost every introductory chemistry course uses the synthesis of ammonia from its elements as an example of equilibrium, of catalysis, and of the applications of Le Chatelier's principle. Some point out the importance of ammonia as an intermediate raw material in the synthesis of nitric acid, nitrogen-containing fertilizers, explosives and even polymers. A few show the interrelationships of ammonia to the problem of human hunger and military uses. Almost none point out that it is third in tonnage rank among synthetic chemicals, nor do they discuss present and future trends in its synthesis and use. Yet the study of ammonia provides a wealth of examples of the application and implications of synthetic chemistry, as well as the changes which have occurred in the last 100 years. It is a field in which future trends will be of great interest as well.

Ammonia has been known by its characteristic smell since ancient times, and was analysed in terms of its elementary composition in 1777. But it was not until the mid-nineteenth century that the extent of its importance to humans began to be appreciated. One clear statement was by Sir William Crookes in 1898, who pointed out that the food reserve in England was only a two-week supply and in great need of enhancement. One of the then limits to agricultural production was known to be nitrogenous fertilizer. The average yield of wheat in England on unfertilized land was 12 bushels per acre (0.12 kg m^{-2}), which increased to 36 bushels if sodium nitrate was provided at the rate of 500 pounds per acre ($\sim$ 0.06 kg m^{-2}). Many farms worldwide now produce as much as 100 bushels per acre. The change is partly due to new varieties of wheat, but largely to more available fertilizers, especially those containing nitrogen from synthesized ammonia.

The earliest sources of soil nitrogen were lightning in air producing about 10 pounds per acre ($\sim 1 \times 10^{-4}$ kg m^{-2}) (if in a thunderstorm area), leguminous plants producing up to 200 pounds per acre if properly tended and animal fertilizers of variable, but often high, effectiveness. Many early people, for example, routinely put a fish in each hillock when planting corn or other such crops.

EXPERIMENT 5: SIMPLE FERTILIZER RICH IN NITROGEN

Plant a seed crop, such as corn (maize), one sample in unfertilized local soil, another in the same soil but with a small fish near the seed. Use other fertilizers in various amounts to study the desirable range at which to fertilize. Finally, try ammonium nitrate. Use several seeds for each experiment.

A human may consume about 400 kg of food per year of which about 20 kg should be protein containing about 4 kg of combined nitrogen. So the 4 billion (4×10^9) people currently alive require a total of about 10^{10} kg of nitrogen fertilizer if it were all 100 per cent converted into food. The actual current worldwide rate of use of nitrogen fertilizer is about 10^{11} kg of nitrogen, just barely enough to prevent massive starvation. Almost 100 per cent of this fertilizer comes by way of the chemical synthesis of ammonia from air, water and natural gas. One possible net reaction, based on the 1 : 4 ratio of oxygen : nitrogen in air is:

$$10\,H_2O(g) + 2\,O_2(g) + 8\,N_2(g) + 7\,CH_4(g) = 16\,NH_3(g) + 7\,CO_2(g).$$

The actual process is complicated and often is done in five or six steps. First methane and steam are reacted:

(a) $$CH^4(g) + H_2O(g) \rightarrow 3\,H_2(g) + CO(g).$$

Air and more methane are then added to remove oxygen and generate more hydrogen mixed with nitrogen:

(b) $$O_2(g) + 4\,N_2(g) + CH_4(g) \rightarrow 2\,H_2(g) + CO_2(g) + 4\,N_2(g).$$

Then these gases are mixed with more steam to convert most of the carbon monoxide to carbon dioxide and generate further hydrogen:

(c) $$CO(g) + H_2O(g) \rightarrow H_2(g) + CO_2(g).$$

Carbon dioxide and any unreacted carbon monoxide are removed from the resulting gas to give a gas containing mainly nitrogen and hydrogen in a ratio somewhat greater than 1 : 3.

The final synthesis of ammonia uses a modification of the original Harber-Bosch process developed between about 1910 and 1914. Haber, a physical chemist, found a catalyst and conditions of temperature and pressure that produced good yields of ammonia rapidly enough to be commercial. The reaction gives ammonia:

$$N_2(g) + 3\,H_2(g) \rightleftarrows 2\,NH_3(g).$$

Bosch, a chemical engineer, designed equipment which could stand the high pressures, about 200–400 atm, at the high temperatures, 400–500 °C, that Haber had found necessary. The reason for the catalyst (still iron suspended on an aluminate pellet) and the high pressure and temperature are so well known they are omitted here, so we can get on with the implications and possible futures.

The most dramatic implication of the Haber-Bosch process today is that, in its absence, world population could hardly have exceeded the 3 billions actually reached in 1960, and certainly could not reach the 7 billions figure widely predicted for about 2010. It is to fertilizers and feeding humans that about 75 per cent of the current ammonia production goes. The other single big use is explosives, about 10 per cent of the total. It is interesting to note that one of the original drives to perfect the process in Germany in the early 1910s was not fertilizer, but the need for explosives in the event of war. In this and most other cases, a single chemical can be used for a large variety of ends. It is up to humans to decide for which purposes any chemical is to be used.

The efficiency of the Haber-Bosch process for producing ammonia for fertilizer has been so high and the manufacturing costs so low there has been little change in the process. But two factors over the last ten years have changed the picture considerably. First, costs of shipping have risen until they have become an appreciable part of the total price. Second, costs of natural gas (or the alternate petroleum) are so high and growing so fast as to begin to make fertilizers very expensive. Nor is there any hope in the immediate future that either the cost of transportation or of sources of hydrogen will cease their rapid rise. For the first time since 1915 there is very great interest in producing ammonia locally to diminish transport costs, and to using water as the major source of hydrogen as petroleum products increase further in price. Both air and water are, of course, available in all agricultural communities. A fine net reaction would be:

$$3H_2O + N_2 = 2NH_3 + \frac{3}{2}O_2 \qquad \Delta H^{\ominus}_{298} = -325 \text{ kJ mol}^{-1} \text{ of } NH_3$$

We know that the energy cost of this reaction is about 325 kilojoules per mole of ammonia. The present Haber-Bosch process has almost no chemical energy cost overall, but the high temperature and pressure it requires make the real energy cost comparable to the one from water and nitrogen. Furthermore, we know a system that makes the water–nitrogen process work using sunlight as the energy source at normal temperature and pressures and with no capital investment! The system, of course, is the bacterial one operating at the roots of leguminous plants.

Thus we come back Cornforth's statement, which was reproduced at the beginning of the section, 'Life does depend on accurate replication of molecules . . .'.

No one knows yet how the bacteria convert atmospheric nitrogen gas to ammonium compounds. But thousands of chemists are now trying to find out. It seems clear that a series of enzymatic catalysts, some containing molybdenum, are synthesized and used by the plants. The feeling among chemists is that what can be done by a bacterium on a

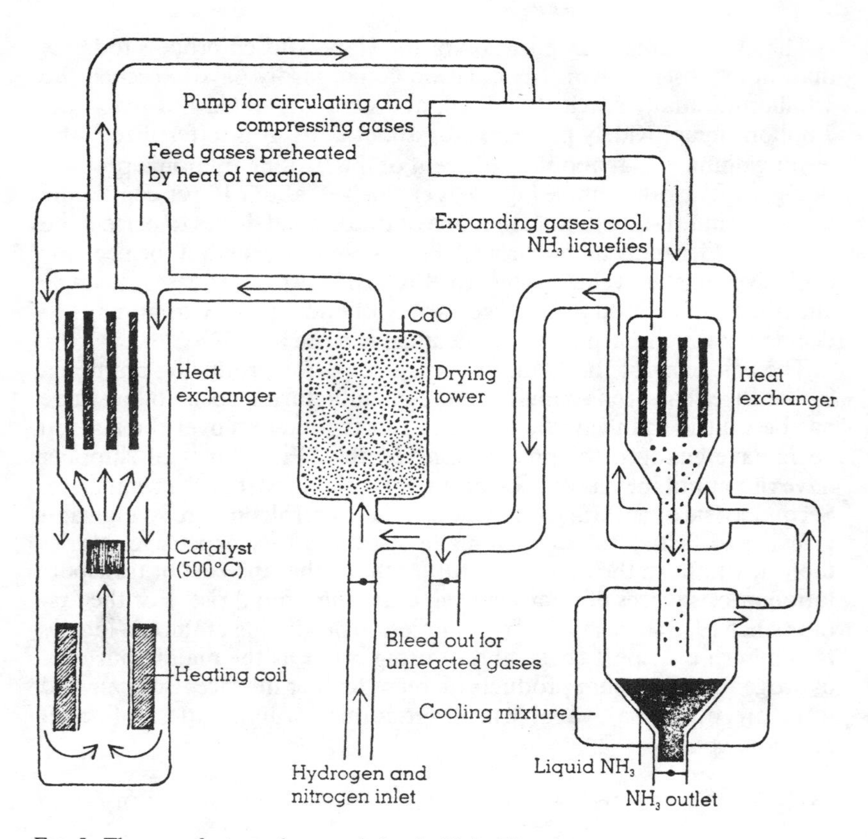

FIG. 5. The manufacture of ammonia by the Haber-Bosch process.

root should be possible on a much larger scale in an industrial plant, so research continues. If it succeeds, one limit to human populations will have been relaxed. If it fails, it is unlikely the world can support more than about 10 billion people at a reasonable standard of living just in terms of protein needs as petroleum gets scarce. It is not that we will run out of the necessary elements to make fertilizers; matter is conserved. It is low-cost energy and chemically feasible raw materials such as petroleum which may be limiting.

The transitions in ammonia synthesis are an excellent example of an early discovery of need, followed by a practical method to fill the need leading to inquiries as to how it worked, then why it worked. The why is still not understood, even for the Haber-Bosch process, at the molecular level. But quantitative measurements have provided such good data that industrial plants are designed with confidence they will work, and they do. Plant costs get lower compared to raw materials, energy and transport, until the latter become the principal items of cost to the user, and in our present state these latter costs are rising so rapidly that there is great incentive for chemists to find a process which is local (to minimize transport), uses sunlight (to minimize energy), and uses water and air (to minimize raw material problems). The search will probably be successful. After all, bacteria can do it already. Why not just do it their way, or modify the bacteria so they could grow on roots of crop plants not just on legumes? Then each plant would be self-fertilizing. This would synthesize fertilizer right where it is needed—on the roots of the growing plant! If population projections of 10 billion and more people early in the twenty-first century are to be fulfilled, one reason will be life-enhancing discoveries by chemists and other scientists, discoveries emphasizing comprehension and control of change at the level of molecular behaviour.

We know the detailed structure of a growing number of enzymes, we can synthesize reasonably efficient substitutes for several of them already, and we are also rapidly learning to modify bacteria to synthesize needed chemicals. If population growth is controlled, there is every likelihood that increased food production based on these techniques will banish mass hunger.

The changing faces of chemistry more and more emphasize and use comprehension of bulk behaviour in terms of quantized changes at the molecular level. Further successes here, coupled with serious consideration and action on human social problems, may also provide a rewarding trend to the changing faces of humanity.

References

1. Kerr, R. A. *Science*, Vol. 210, 1980, p. 514.
2. See, for example, B. Gutte and R. B. Merrifield, *Journal of the American Chemical Society*, Vol. 91, 1969, p. 501. For a helpful introduction to the subject, see R. Barker, *Organic Chemistry of Biological Compounds*, Chapter 4, Englewood Cliffs, N.J., Prentice-Hall, Inc., 1971.
4. L. Pauling, *Journal of Chemical Education*, Vol. 57, 1980, p. 38.
5. I. Prigogine, *Science*, Vol. 201, 1978, p. 777.
6. J. W. Cornforth, *Science*, Vol. 193, 1976, p. 121.

2 Curriculum innovation in school chemistry

R. B. Ingle and A. M. Ranaweera

The period since 1960 has seen a remarkable outpouring of curriculum innovation activity. New courses were first produced in a few industrialized countries, since when many countries all over the world have established curriculum development projects. It would not be an exaggeration to say that the changes in school chemistry that have occurred on a worldwide scale during the 1960s and 1970s have greatly exceeded those of the previous five decades.

The earliest curriculum projects were conceived essentially as national projects; that is, they were a response to educational and national needs in their countries of origin, although, in the event, several of them exerted wide influence in many other countries. But early in the curriculum development era, Unesco took the initiative in establishing a number of projects that were international in scope in that they sought to serve the curriculum needs of a number of countries belonging to a particular geographical region. These projects have served a valuable function in a number of ways, especially by putting curriculum developers from one country into contact with those of another.

During the 1960s and early 1970s most curriculum projects were produced centrally and, as a result, were somewhat prescriptive in tone. With the passage of time has come the realization that such innovations do not necessarily bring about the hoped-for changes in schools. As a result, there has been an increased interest in locally based curriculum development work. Such developments hold out considerable promise, although, for the foreseeable future, the need for national-based projects will continue.

Perhaps the most significant change that has occurred in the 1960s and 1970s is that the bounds of the subject have radically changed. Chemistry is now widely taught within a broad context, especially in the early secondary school years, in which it is now most commonly taught as part of an integrated science course, although other regroupings of science subjects have also found favour. Science is increasingly seen as comprising not only chemistry, physics and biology, but also geology or earth science. And the human, social and moral perspectives of chemistry are being regarded more and more as integral to the subject instead of as optional extras.

One of the main emphases in science nowadays is its practical orientation brought about by new aims in science education which placed more emphasis on science as a process. However, the practical work in today's science courses provides not only a means of stretching the intelligence, but also for encouraging the young to work harmoniously and collaboratively with each other. The benefits of well-organized practical work are very real for all young people—both for those who are in some way physically handicapped and for those who are not. Unfortunately, until recently, chemistry was often considered to be unsuitable for the physically handicapped because of its practical component. It is now realized that, possibly because of its practical work, chemistry is a particularly valuable subject for the handicapped, providing them with the opportunity to work normally with other young people.

Finally, it is necessary not only to develop new curricula but to evaluate them. In this respect much progress has yet to be made.

This chapter therefore looks at some of the principal curricula of the 1960s, 1970s and early 1980s to illustrate these ideas. It is of course not possible to be encyclopaedic; the curricula discussed are intended to be illustrations to the discussion.

Future trends are forecast and discussed in this chapter and in Chapter 8.

2.1 Introduction

The production of chemistry curricula is a challenging task as a variety of needs have to be met. Some students, albeit a very small proportion, will eventually pursue chemistry or some other scientific or technological study at university level, often to take up a science-based career. But schools also, for example, must help students to prepare for courses leading to technical and related careers as well as satisfying the needs of those whose formal education ceases when they leave school to enter a world in which both work and leisure are profoundly affected by science. And, as teachers know only too well, they must provide for a wide spectrum of students of varying aptitudes and interests in science.

It is no easy task to meet all those needs, especially when there are serious shortages of well-qualified teachers and material resources. The construction of a suitable school chemistry curriculum is also complicated by the fact that science itself is dynamic. Provision has to be made for the incorporation of new knowledge and principles into the curriculum which, if the syllabus is not to become overcrowded, means that some parts of it must be discarded or contracted. Many doubt whether it is desirable to use the same curriculum and methodology of teaching for all students beyond the age of about 14. For this reason, it is common to provide more than one science course for use in the later secondary-school years.

There cannot be a general prescription for school chemistry curricula, as each country has its own needs and priorities. Some must consider first and foremost the production of scientists and technologists; some others are deeply concerned with providing an education for the whole ability range, while others are simply trying to provide enough teachers to continue a science programme. The boundary conditions are set by political, social and educational aims, as well as by the cultural milieu and by economic realities [1, 2, 3].

This chapter contains a description of some of the chemistry curricula that have been developed over the last twenty or so years, the reasons why they were introduced and, in some cases, how they were written and implemented. This will be done for two principal reasons. First, because we can learn from the successes as well as the failures of past endeavours and, second, because these curricula contain a wealth of interesting ideas, some of which may be of value in building new courses.

No attempt can be made in the following pages to describe or even to mention all the curriculum development projects containing chemical material that emerged during the 1960s and 1970s. Examples will be taken to illustrate the diversity of approaches that have been used and the trends that are taking place. Some sources of information about chemistry curricula in various countries that may be of value to those involved in the development and implementation of chemical curricula are given in Appendix 1.

Many curriculum innovation projects in science have sent reports to the International Clearinghouse on Science and Mathematics Education at the University of Maryland, United States, which are available in published form, thus forming an invaluable collection of information about curriculum development worldwide up to the date of the last report in 1977. The reports given the salient features of each project, including the name and address of the person or agency from whom further particulars may be obtained. The *International Newsletter on Chemical Education* is also a most useful source of current information.

The various appendices at the end of this chapter list some of the projects in the secondary-school age range that have a chemical component, from which it will be seen that curriculum development in science has become a worldwide activity. Some projects are directed towards the teaching of chemistry as a separate subject, whereas others treat chemistry as part of a more generalized or integrated form of science education. In no way are these appendices comprehensive; they provide examples of curricula.

2.2 Three early influential chemistry projects

Awareness that the science curriculum is in need of reform is almost as old as science teaching itself. However, until about 1950, changes in the curriculum were usually made gradually, as a result of occasional revisions of examination syllabi. A new climate of opinion then came about reflecting the increased importance that was attached both to science and to education. Major changes were seen to be necessary in the content of science courses which badly needed bringing up to date, as well as in the method of teaching. Progress on a number of fronts was clearly needed; in the improvement of laboratories and apparatus, in teacher training and in the curriculum. However, at the time renewal of the curriculum itself was seen as the first priority. Curriculum development as a large-scale organized activity started in the United States in the late 1950s. Similar activity followed in the United Kingdom in the early 1960s, and within a few years many countries all over the world had started to set up curriculum development exercises.

In chemistry, three projects led the way. In the United States, Chemical Bond Approach (CBA) and Chemical Education Material Study (CHEM Study) were the forerunners, while in the United Kingdom, Nuffield O-level Chemistry was the first chemistry project to be published. These projects have made a significant impact in their countries of origin, and also influenced the course of curriculum development in many other countries. As they have all been fully described elsewhere, the main points about each of these projects, together with references to which the reader can go for full details, are summarized in Appendices 2, 3 and 4.

Although there are significant differences between these projects, they all arose at a time when there was widespread concern about the production of scientific manpower. Each of them involved leading scientists and educators, and had as their aim the production of improved curriculum materials. Both in the United States and the United Kingdom it was felt that in the past chemical education at the school level had placed undue reliance on the acquisition of facts for their own sake, and was out of touch with modern developments in chemistry. Furthermore, the subject was invariably taught as a number of 'truths'. The many applications of chemistry in everyday life and in industry were only touched upon. Because of these shortcomings, all these projects emphasized the following:

Updating chemistry in the light of modern knowledge of the subject.

Giving the students a good understanding of the subject. Factual knowledge would of course still be required, as a basis for theoretical understanding. Many of the great themes of chemistry such as periodicity, the mole, structure, kinetics and energetics were introduced even in these introductory courses. The projects were particularly concerned with physical chemistry (see Appendix 5 for the content of CHEM Study), and are sometimes now referred to as 'concept-based' curricula.

Providing some insight into modes of scientific thinking. Although the revolution in scientific thinking brought about by the work of Popper and Kuhn had yet to occur (Section 2.12), students would be expected to acquire a much more profound understanding of how scientists think and work by reading about the work of great chemists and in the course of discussion resulting from their own laboratory investigations.

Emphasis was also placed on the following points, especially in the United Kingdom:

Practical work was seen as having a dual role. As in the past, it would illustrate and 'make real' the chemistry which was being taught. But equally important, it would help to stimulate students in exercising scientific and critical modes of thinking in grappling with laboratory investigations. Laboratory-based inquiry teaching became the order of the day.

Giving students an awareness of the applications of chemistry in everyday life and in industry through a study of such topics as fibres, plastics, elastomers, detergents, drugs and insecticides.

One of the most significant contributions to the success of these projects was the participation of leading scientists such as Nobel Laureate Glenn Seaborg in the United States and Sir Ronald Nyholm in the United Kingdom. They gave their services as advisers and mentors, not in the form of 'tablets of stone'. Their encouragement was a key reason for the ready acceptance of the projects by both the scientific and educational communities. This is perhaps a lesson that has not always been learnt by some curriculum developers.

2.3 Regional chemistry projects

Most curriculum projects in the world have been set up on a national basis. However, in the 1960s the idea of establishing projects to serve a large regional area consisting of a number of countries appeared particularly attractive as scarce skills could be located in one place and made available to other countries in the region. Perhaps the best known example of a regional chemistry project is the Unesco Pilot Project for Chemistry Teaching in Asia which was one of a number of projects sponsored by Unesco in the field of science teaching in the 1960s [4, 5]. The aim of the Chemistry Pilot Project was to put chemical educators from many Asian countries into touch with one another and with experienced leaders in chemical education from other countries as well as to provide training in curriculum development. It was hoped that this would be of benefit in their work when they returned home. Unesco had a well-equipped laboratory at Bangkok in Thailand, which served as a regional meeting and working centre for chemistry educators from Asia and to provide information and consultancy services on innovations in chemistry teaching through a network of 'study groups' located in each Asian country. Set up initially for one year, the Pilot Project continued throughout the period from 1964 to 1970. The idea was that the project would contribute towards both the regional needs of Asia and the national needs of Thailand.

Twenty teachers of chemistry from universities, the Ministry of Education and science curriculum units and schools representing fifteen Asian nations came together in 1965 for a year's work as the International Working Group of the project. The participants and staff studied together, carrying out laboratory investigations of certain chemical systems and developing prototype instruction materials on three topics which planning meetings held earlier in Moscow had identified as 'connecting threads' in much of the chemistry taught in schools. They were: (a) mass relationship among reacting substances; (b) energy in chemical reactions; (c) the role of structure in chemistry.

The International Working Group pointed out that

> many attempts to improve chemistry teaching are frustrated by insufficient attention to the nature of chemistry itself. The basic question is: what could be accomplished through a careful examination by competent chemists and experienced teachers of the principles and practice that comprise chemistry? The Pilot Project asked the teachers to take this question seriously and has arranged for them to devote considerable time to a careful examination of selected areas of chemistry content. This examination, referred to as content analysis, had already indicated a variety of gaps and ambiguities in conventional presentations of chemistry.

Content analysis was carried out in the following way:

> (i) the analysis of each topic began with participants questioning the adequacy of definitions of terms commonly used in textbooks; (ii) participants then searched for 'operational' ways of relating these terms with student experience by carrying out a considerable amount of laboratory research on suitable experiments for student use; (iii) finally, participants tried to link each topic to other parts of chemistry.

The group felt that there were many merits in analysing topics in this manner. It helped to clarify the logical structure of the topic and enabled the participants to formulate precise questions about it. It also enabled subject-matter specialists (chemists) and curriculum specialists (educationalists) to work together in a fruitful way. Furthermore, it was valuable in identifying specific points at which operational research would be needed in order to provide experimental evidence for the usual textbook assertions. But above all, content analysis made the task of designing teaching and evaluating schemes more clear cut and direct.

Perhaps one of the most valuable insights that arose from the project was the realization of the value of a dialogue between chemists and curriculum specialists. Nevertheless, the impact that the project has had in bringing about changes in the chemistry curricula of Asian countries, although considerable (for example, in Thailand and the Philippines), was not as great as might have been hoped, perhaps because the importance of the following questions was not fully realized at the time:

Was the project too ambitious in terms of the number and variety of countries covered?

Were the staff involved in sufficiently close contact with ordinary classroom teachers in Asian countries?

Were the units which were produced difficult to incorporate into national curricula, possibly because of their fragmentary nature?

Did the project over-emphasize difficult concepts in theoretical and physical chemistry?

Another well-known regional project which came to be known as the School Science Project in the East African countries of Kenya, Uganda and the United Republic of Tanzania developed along rather different lines. The initial driving force for reform came not so much from an external agency but from science teachers in these countries whose interest had been aroused by the recent publication of curriculum innovations in other countries. A British organization then known as the Centre for Curriculum Renewal and Educational Development Overseas (CREDO) was asked to help, and G. Van Praagh [6] ran courses for chemistry teachers in all three countries. A conference was held in Nairobi in 1968 which was attended by the Chief Education Officer for Kenya and by the Chief Inspectors of Uganda and the United Republic of Tanzania, with representatives of the institutes of education, leading teachers and representatives from Zambia, Malawi and CREDO. The need for new science curricula and for more appropriate examinations was discussed, and it was agreed that the three ministries of education would work together in producing new four-year courses in biology, chemistry and physics. The aim was to develop modern courses relevant to the needs of the countries which would stress understanding rather than rote learning. The courses would contain much practical work and involve an investigational approach to science teaching.

The work of writing was done by groups of teacher educators in the three countries who made extensive use of ideas found in Nuffield Chemistry. The draft materials were tried out in a dozen or so schools in each country. The teachers in these schools were urged to send feedback to the writing teams. In the early years, most of the work was done in

Uganda and the United Republic of Tanzania. The United Republic of Tanzania withdrew from the scheme in the early 1970s because of changing emphases in national educational objectives and the use of Swahili as the language of instruction. Kenya and Uganda continued to use the project material and to submit a sizeable minority of schools for the special examination set by the East African Schools Examination Council. Now, Kenya is revising the curriculum devised in the School Science Project (SSP) [7] and both Kenya and Uganda have taken the interesting step of deciding to fuse traditional and new courses into a single unified course. This is discussed later in the chapter (Section 2.8).

One of the products of this curriculum development was a set of background readers written in a simple language having titles such as *Salt in East Africa* [8] and *Fermentation and Distillation* [9] which provided examples of chemistry in the home and in industry.

The project undoubtedly helped to raise the standard of awareness and interest in modern chemistry curricula in East Africa. Its strength lay principally in the commitment and enthusiasm of chemistry educators in these countries. CREDO rendered valuable help by giving professional assistance and timely small grants and by trying to exert a co-ordinating role between the three countries. There was some valuable cross-fertilization of ideas between the participating countries, and the setting up of a special examination by the East African Schools Examination Council was a major achievement. But one of the lessons to be learnt from the project is the importance of obtaining early agreement for a suitable means of publication when the materials are at the end of the development phase. As a result of a failure to agree on this, publication was delayed. It fell to a small number of individuals to bear the burden of printing, duplicating and distributing materials. By contrast, Malaysia, Nigeria and many other countries used experienced publishers to produce the curriculum materials, and involved them in discussions from the inception. Another serious weakness was the lack of well-laid plans of implementation; it was left to schools individually to decide whether or not to use the materials and enter their pupils for the SSP examination. On the other hand, to allow choice between 'traditional' and 'new' curricula may be thought to be an imaginative and sensible decision, given the difficulties of giving intensive preparation to all the teachers.

2.4 National chemistry projects

2.4.1 *For students up to the age of about 16*

The establishment of a curriculum innovation project on a national basis obviously avoids the complexities involved in planning and execution that are encountered by regional projects which serve countries having different priorities and policies. In countries that have a centralized education system, the development of new curricula is usually under the control of the Ministry of Education. A good example of a project that had well-conceived plans for curriculum development is the Modern Chemistry project in Malaysia (Appendix 6). These plans took account of the need not only to produce a modern curriculum, but also to help teachers to use it effectively, to improve the provision of laboratories and equipment and to produce a more appropriate form of examination for students who had completed the course.

The drafting of the modern chemistry course (and also of courses in biology, physics and general science) in Malaysia was carried out in writing workshops which were convened for a few weeks each year to prepare the following year's materials for students and teachers. In the early years, the work of writing was carried out by groups of Malaysian science teachers under the guidance of a few experienced science teachers from other countries. Courses took place each year which were attended by teachers, inspectors and ministry officials. Their purpose was twofold. First, to familiarize everyone involved with the work for the forthcoming year that had just been written, and second, to consider critically the previous year's work in the light of experience of teaching it in trial schools. The material was extensively revised before publication of the final version, and this was the responsibility of a group of key Malaysian teachers who had taught the course over the two previous years. The curriculum development was linked to a restructuring of the examination, and training courses were provided to help examiners to write questions which would test both the aims and the content of the new curriculum. The implementation of the project throughout the country was carefully planned by the Ministry. Each year it directed additional schools to take up the project, and in so doing it did its best to ensure that teachers in the schools concerned had attended familiarization courses (Chapter 5, page 236).

Cuba provides another example of a nationally based chemistry project [10]. During the 1960s, as in most countries, there was an enormous increase in the number of students. A national committee, in a diagnosis of the shortcomings of the education system, identified the weaknesses in the teaching of science. Among them were:

(a) that curricula were full of ideas and concepts which were not adequately developed or made use of at later stages; (b) that there was a large amount of duplication in the curricula; (c) that there was little co-ordination between the various science curricula being studied by a student; and (d) that the curricula were out of date scientifically.

A team consisting of representatives from the Ministry of Education, scientists, university teachers and schoolteachers came together. The team was assisted by specialists from the Union of Soviet Socialist Republics and the German Democratic Republic, and has worked since 1975 on the project. Their guiding principles included:

(a) the importance of matching the curriculum to the stage of the student's mental development; (b) ensuring that, as far as possible, each topic is introduced through tasks at the *concrete* level; (c) emphasizing experimental work; (d) excluding concepts of doubtful educational value and including only those which are fundamental to the curriculum; and (e) including practice at numerical problems.

The teaching is done in two cycles, the first in the eighth and ninth grades (two lessons a week) and the second in the tenth to twelfth grades (three lessons a week).

The work is structured linearly with five 'levels': (a) empirical–analytic level; (b) molecular–atomic level; (c) level of the periodic law; (d) level of atomic structure; (e) level of electrolytic dissociation.

In the first cycle, students are introduced to chemical concepts and phenomena up to the level of atomic structure and study the principal types of inorganic compounds, their genetic relationships and behaviour. The periodic law is then introduced, which leads to the electronic structure of the atom. The cycle ends with an introduction to some organic compounds.

The second cycle includes the theory of electrolytic dissociation and a deeper study of chemical relationships making use of physico-chemical principles, energetics and equilibria and kinetics. The cycle ends with the study of some more organic compounds and a review of the work covered.

Overall, the choice of concepts has been made to illustrate the importance of Cuba's developing chemical industry.

Information about a number of nationally based chemistry projects for pupils up to the age of about 16 is given in Appendix 6. It should be noted that chemistry is now being increasingly taught as a component of some form of unified science course, for the reasons discussed in Section 2.5.

2.4.2 *Advanced projects (beyond age of 16)*

Turning now to projects for students in the 16 to 19 age range, these are less numerous. Two early examples, CBA and CHEM Study have already been outlined in Section 2.2. Another influential project was the Nuffield Advanced Chemistry Project, the salient points of which are given in Appendix 7.

The organizer of Nuffield Advanced Chemistry describes it [11]:

The Nuffield Advanced Chemistry Project began in October 1965, following the completion of the Nuffield O-level Chemistry Project. It therefore had the benefit of the considerable experience of curriculum development gained by the team responsible for the latter project.

The Project was set up to produce a teaching scheme, and an appropriate method of assessment, for schools entering candidates for the Advanced (A) Level Examination for the General Certificate of Education (GCE), which is normally taken at age 18, at the end of a two-year 'sixth-form' course, by students in roughly the top 10 % of the ability range. As with the O-level project the impetus for the curriculum changes which were proposed by the Advanced Chemistry Project came largely from chemistry teachers in schools.

The strength of the Project lies mainly in its encouragement of understanding by the student, which is done largely by fostering class discussion. The success of such an approach depends largely on cooperation between teachers and students. It is undoubtedly true that this course suits some teachers better than others, which must be the case for any teaching scheme. The course has a greater factual content, most of which should be learned, than some commentators realise, but every fact has a purpose in the general development of the subject. Also, since practical work has an integral part to play in the scheme, it is true that students do not perform a large amount of repetitive work involving a small number of techniques, such as the use of balance, pipette and burette in volumetric analysis, and the separation techniques of qualitative analysis. They are, however, exposed to a much wider range of techniques than was the case in the past, including calorimetry, pH measurement, photocolorimetry, spectroscopy, polarimetry and the measurement of cell emf. Novel features of the course include the Special Studies and the use of a measure of internal assessment, especially of practical work.

As with the O-level project, the scheme has been widely adopted: in 1979, approximately 9000 students [took] the Nuffield A-level examination from some 600 schools.

During the ten years of the project, a great deal of experience in teaching the course was gained and adaptations of the course have been made to suit local conditions. The decision to prepare a second edition of the Nuffield publications was made for a number of reasons. There was a need to update industrial applications, nomenclature, units and other factors that change with time, but, more important, the opinions of teachers and students showed that some improvements could profitably be made.

One of the unique features of the course is the meetings that take place each year following the examinations, in which teachers discuss the examination with the examiners and with each other, and exchange ideas about the teaching of the course.

The opinions of teachers using the course on what would be desirable changes were obtained from these meetings and from the detailed questionnaire. Since the revision exercise began, a second questionnaire was sent and the broad aims of the revision as well as points of detail have been discussed at teachers' meetings.

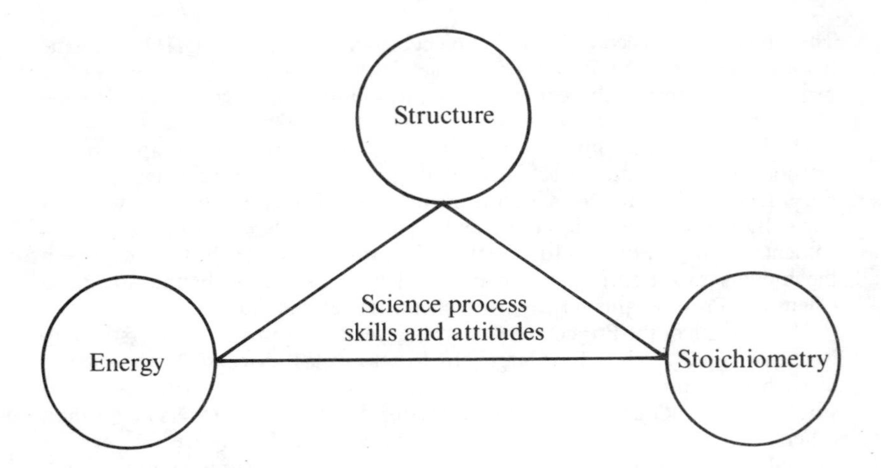

FIG. 1. Themes for chemistry curricula, grades 11 and 12 (Thailand).

A new approach to energetics has been developed, in which the basic idea behind the Second Law of Thermodynamics is introduced during the first year of the course. When the student has gained some familiarity with the idea, it will be used as a criterion on which to decide the direction of spontaneous change [12].

Among many other advanced projects that deserve mention are those developed in Thailand and in India. The following accounts illustrate varying approaches to the development of chemistry curricula, as well as some of the difficulties encountered. The Director of the Institute for the Promotion of Teaching Science and Technology (IPST) in Thailand describes the new advanced chemistry curriculum, which is now in use in all the country's secondary schools [13]:

The Chemistry Team of IPST formulated the following broad aims which they felt were valid for any science course:

1. To develop an understanding of the basic principles and theories of science.
2. To develop an understanding of the nature, scope and limitation of science.
3. To develop a scientific attitude.
4. To develop skills important for scientific investigation.
5. To develop an understanding of the consequences of science on man and his physical and biological environment.

The IPST Chemistry Team considered it necessary to formulate criteria for the selection of content which will give some rationale to support the choice of such content. It was decided that the content selected should: (a) lead to modern science and reflect ideas and structure of the most recently accepted advances; (b) show continuity and follow a logical conceptual scheme; (c) have a capacity for unifying and explaining the widest variety of phenomena and data; (d) be teachable within the time allocated and suitable for students' intellectual

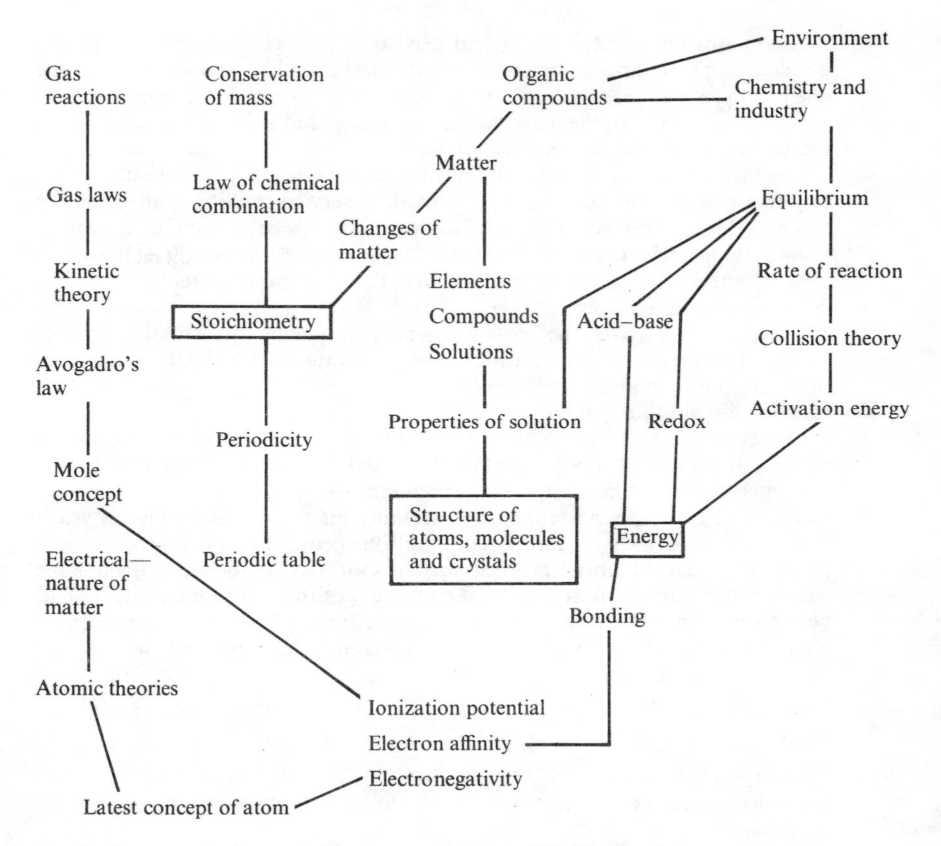

FIG. 2. Conceptual scheme of chemistry, grades 11 and 12 (Thailand).

maturity and interest; (e) be within the capability and experience of the teachers' adequacy of resources; (f) contain examples of important applications to technology, agriculture, medicine, industry and social sciences in Thailand; (g) bring out the relationship to other fields of study, especially to the natural sciences and mathematics; and (h) contribute to the growth and development of the individual as well as society and stimulate the proper utilization of natural resources and the preservation of the environment. The chemical themes chosen are illustrated in Figure 1.

A conceptual framework was then built around these basic themes. A great deal of shifting of ideas and topics from one chapter to another continued through the period of development. The final conceptual scheme and the interrelationship between topics are shown in Figure 2.

The chemistry course integrates text and laboratory manual in an effort to encourage teachers to teach chemistry as a learner-centred inquiry involving both theory and practice. Experiments were designed and an inexpensive kit that contains most of the equipment needed to perform the experiments was developed. The course includes a study of Thai chemical industries and some chemical problems relevant to Thai society. An IPST-produced film, *The Salt Farm*, is an example of the materials developed to Thai chemical concepts, Thai industry and their impact on Thai society together.

The Institute and the Ministry planned to have the chemistry curriculum fully implemented in 1976 and intended to provide in-service-training to all chemistry teachers during 1975, one year prior to the full implementation. Due to restrictions in budgetary provision, only some 70 per cent of the schoolteachers were able to attend the in-service courses and the training of teachers therefore continued for several years.

The budget to cover the cost of science equipment for schools was also limited. This created problems which were aggravated further by the inadequate production of equipment, both in quantity and quality, and by the lack of an effective distribution system.

The staffing of the curriculum team in the early years depended almost entirely on secondment; some members worked full time and others part time. The membership consisted of a representative cross-section of chemistry teachers, science supervisors, teacher-training instructors and university lecturers. It was not easy to recruit many full-time people to serve on the curriculum team. The most serious problem was the short term of the secondment. One year was too short in most cases and continuity of the Institute's activities had been jeopardized by the departments of the Ministry recalling key personnel after one or even two years. In most cases a minimum of three years would have been more satisfactory.

Despite these difficulties which are common to most curriculum development projects, the programme has really been implemented across the whole country. It has been well received and has not required much revision.

In India, a senior secondary-level curriculum has been developed by the National Council of Educational Research and Training (NCERT) as a model for adoption or adaption by the states according to their needs and resources. The course is described thus [14]:

The major objectives of this curriculum at senior secondary level are:

To strengthen the concept developed at a secondary level and further develop new concepts to provide a sound background for higher studies.

To develop a competence in students to offer professional courses like engineering, medicine, etc., as their future career.

To acquaint the students with different aspects of chemistry used in daily life and enable them to recognize that chemistry plays an important role in the service of man.

To expose the students to different processes used in industries and their technological applications.

To provide relevant content materials useful for vocational courses.

To develop an interest in students to study chemistry as a discipline.

Teaching of chemistry at higher secondary level (+2 stage) is a major change from the traditional teaching in India. For example, chemistry is presented as a unified subject. There is no traditional classification as physical, inorganic or organic chemistry. Basic chemical concepts are developed in the beginning and applied to the study of elements and their compounds provided in the later units. Further, there is an effort to relate macroscopic to microscopic behaviour (atomic and molecular structure). There are two sets of the textbook developed separately for each academic year.

The laboratory part of the course in chemistry includes some traditional, open-ended and environmental investigations. This course is designed to develop skill, scientific attitudes and future training for research. Major emphasis is placed on proper selection and use of experimental tools, improvisation of laboratory apparatus with the help of readily available materials and critical observations of various phenomena involved in various experiments.

Students taking professional and vocational courses may not be interested in abstract theory; instead they would like to know how the principles of chemistry can be applied. With this in mind, blending of facts and theory has been done by emphasizing the application of principles rather than their derivations. Thus, equal emphasis has been given to chemical principles and descriptive chemistry in the text, always correlating the two. Examples of units which bridge the academic and the vocational are electroplating, electrorefining, purification of water (using zeolytes) for industries and drinking, corrosion, photography, agriculture (fertilizers and pesticides), nutrition and food preparation. Nevertheless the course provides sufficient background for professional courses such as medicine and engineering.

In planning this course, we faced many problems. In particular, the course was expected to provide a base for advanced academic courses in science and for professional courses (like engineering, medicine). At the same time, some of the units of the course were expected to serve as bridges between academic and vocational courses as described. In the absence of relevant research studies, it was difficult to decide the proportion in which various content areas should be presented. Experts invited from academic and professional areas during the planning of the course emphasized their own areas, and the curriculum became too long and complex. The schoolteachers involved in the planning were not able to contribute as effectively as hoped. This was due to a lack of knowledge at a higher level, or because they felt inhibited in a team containing university teachers. Thus, in the writing of the textbooks, they did not have enough influence on the depth of the treatment of the subject-matter that would be appropriate.

Implementation also had its difficulties. Due to the major changes from the traditional course, there was resistance from practising teachers towards accepting the new course since they anticipated a lot of preparation for effective handling of it. Motivating the teachers, therefore, became a key issue in implementation strategies.

There was also a communication gap between the curriculum developers, textbook writers and the practising teachers. The schoolteachers were oriented in summer institute programmes by teachers mostly drawn from college and university departments. A survey of the summer institute programme revealed that the lack of full success was due to three factors: (a) reluctance of the teachers to learn new things; (b) short duration of the summer institute (fourteen days); and (c) inadequate preparation on the part of instructors in the institutes.

Moreover, other problems had to be faced. In the initial stages, only pupils' textbooks were supplied. A teacher's guide and other instructional material, handbook for teachers and supplementary readers, which the teachers should have been provided with at the start, were delayed. The approach to teaching at +2 stage is based on experimentation and inquiry (as envisaged in the curriculum). But laboratory facilities in schools are not adequate, particularly for secondary and higher secondary levels.

In many cases, problems were purely organizational and administrative. In the initial stages, rigid attitudes of the principals and administrators offered resistance to the new course. In many cases, it was felt that this resistance was due to a communication gap.

In spite of all these difficulties, the courses at secondary and higher secondary level were initially implemented in nearly 1,000 schools, located in different parts of the country, which were affiliated to the Central Board of Secondary Education, New Delhi, an autonomous body under the Ministry of Education. Only recently some of the states have adopted NCERT courses and textbooks at secondary level. Most of the states have adopted the 10 +2 pattern of education and some are modifying their courses in the light of higher secondary courses prepared by NCERT.

The chemistry course at secondary level has been revised in the light of feedback received from the teachers. Subsequently a review committee, set up by the Ministry of Education to reduce the work load, made drastic changes in the course recommending two alternative courses based on a 'disciplinary approach' and 'a combined science approach'. Evaluation of the textbooks was based on teachers' opinions received through questionnaires and by interview, experts' opinion and comments received from the teachers during summer institute programmes. Based on this short-term evaluation, necessary modifications have been made in the textbooks. Now a long-term evaluation has been planned.

In some countries, physics is more dominant in the curriculum than chemistry, the latter becoming a relatively poor relation. This occurs, for example in Tunisia [15] and in the Upper Volta [16] among others; physics and chemistry are taken by the same teacher and, as the division of teaching hours is frequently in favour of physics, often at the expense of chemistry.

Some examples of chemistry projects for senior secondary-school students are summarized in Appendix 8.

2.5 Locally based projects

Most of the curriculum projects have proposed changes not only in the content of what is to be taught but also in the manner in which the developers thought that it should be taught. While it is clear that curriculum development projects have often brought about valuable changes in what is taught, there is little evidence to suggest that it has fundamentally changed the manner of teaching. In-service education of teachers can perhaps help. Yet teachers can attend courses to be exposed to 'the new teaching philosophy', but often find it extraordinarily difficult to put the new ideas into practice on their return to school. As Eggleston has said, 'we cannot carry out successful development aimed at altering styles of teaching as well as content from a far away position remote from the classroom [17]'.

One strategy which has met with some success is to develop projects locally—among a small group of schools that are close together, or even within one school. Teachers then become active curriculum developers themselves and find themselves having to think through their own aims, objectives and style of teaching.

Locally based curriculum development has flourished best in those countries that do not have centralized systems of education, particularly in parts of North America and some European countries. Three examples will first be described; from the Federal Republic of Germany, Canada and the United Kingdom. In the fourth example, from the United States, a local curriculum was adopted more widely.

In the Federal Republic of Germany, the *Länder* (the individual states) have autonomy in curriculum matters. The following report describes the curriculum in Hesse [18]:

The Hessean Curriculum does not give a prescription of definite topics that must be taught. There is complete freedom for teachers to teach any topic provided the following four criteria are fulfilled: an experimental approach is to be used; instruction should be based on problem finding and problem solving; general chemistry should be taught by observing and interpreting the properties of substances; the importance of the facts and principles taught for the individual and society should be stressed.

Separate committees of schoolteachers designed new school chemistry curricula for the lower and upper secondary stages, that is for the fifth to tenth and eleventh to thirteenth years of school education. The resulting curricula replace the 'lists of subject matter' that had formerly served as the curriculum. Stress has been given to learning objectives rather than to content matter.

Two major groups of general aims were identified by the committees of teachers: general aims referring to chemical science; and general aims referring to knowledge, skills and attitudes to be developed in the students with respect to their future role as educated citizens.

Among the aims relating to chemical science, the following were stressed:

The scientific method. The methods used by scientists to carry out activities and investigations in science and the ways and means of articulating scientific results.

Chemistry as an experimental science.

Chemistry as a complex science. The order within this complex science, patterns and generalizations. Relationship between material changes and energy changes.

Under the more general aims relating to the future role of students as educated citizens the following were identified: (a) extrapolation of values and attitudes learnt in chemistry to the outside world and everyday life; (b) chemistry and the environment, developing a critical attitude towards environmental problems.

In Canada, the Alberta Chemistry project, known as ALCHEM, is described by its leader [19]:

The project, for students in the 15 to 18 age range, was started by teachers in Alberta because it was felt that Chemical Education Material Study was not pedagogically suitable for 80 per cent of students who took chemistry in Alberta, owing to its particularly strong theoretical framework. To the aims accepted up till 1980, other aims have been added which feature clearly in ALCHEM:

Pre-1980: to present chemistry rigorously; to make chemistry easy and enjoyable to learn; to provide applied and descriptive chemistry contexts;

Post-1980: to present a modern (twentieth century) view of the nature of science; to integrate a science and society theme into both the core and elective materials; to compare and contrast science as a way of knowing with other ways of knowing.

Application of these guiding principles to the construction of a chemistry curriculum resulted in the elimination of many long-established topics such as the gas laws, electronic configurations, rate laws and value of equilibrium constants. These were replaced by applied chemistry topics. A most important point is that the materials were *written, revised and edited entirely by practising high school teachers*. One of the main difficulties encountered in getting the project officially accepted was simply the widespread assumption that something produced *locally* must be second-rate!

Now that the materials have been published [20], it can be seen that they form an excellent and coherent chemistry curriculum for use with a wide age and ability spectrum. The *slow* development of the project, due in part to the part-time participation of its staff of school teachers, has been a source of *strength*, since it has made it possible to incorporate many aspects of curriculum approaches that have emerged over the last decade or so: social relevance, use of local chemistry industry; technological thrust, and a conceptual level which matches the group of the students for whom it was devised. The core and elective units provide a number of ways of matching the varying learning patterns of students [21].

School-based projects are often given little publicity, but it sometimes happens that a group of teachers in a school who start by talking through their aims finish up by producing a new curriculum which is eventually published. An example of this is from the United Kingdom—the Wreake Valley Integrated Science project for use in the 11 to 13 age range which, like the ALCHEM project, started from a belief that existing curriculum packages were not entirely suitable. The working method adopted by the science staff of Wreake Valley Comprehensive School was to discuss each week's work of their science course in a meeting some weeks before teaching it in order to pool ideas and decide on a general approach [22]. One member of the department would then write the resources for the topic, which would be reproduced and tried out by all the members of the group with their classes. After this had been done, the group would convene again, normally a week or so later, to discuss the way the work had gone, so that it could be revised or rewritten, before being put aside for trial again the following year. After proceeding for some years in this sort of way, their materials were eventually published and are now used by a number of other schools in the United Kingdom. Once again professional standards comparable with national projects were achieved, but even if they had not been, it would still have been worth while as an exercise in staff development.

The Wreake Valley course is but one example of many; all these courses show the shift from the centralized to localized curriculum development. For example, the Independent Learning Project for Advanced Chemistry was produced by the Inner London Education Authority especially in London schools for use particularly in schools that are short of well-qualified teachers of chemistry [23]. The material provides a great deal of useful guidance for both teachers and students. Inquiries from many parts of the world suggest that it may find extensive use elsewhere.

In 1970, about 200 chemistry teachers came to the University of Maryland in the United States to participate in a regional symposium. In group discussions and casual conversations, teachers emphasized the need for a new approach to teaching high-school chemistry that would be more relevant and interesting to the 80 per cent or so of students who are not preparing to be scientists. As a result, more than 100 local high-school teachers, science supervisors and university professors, engaged in the development of the modular Interdisciplinary Approaches to Chemistry (IAC) programme [24]. It was agreed that the new high-school chemistry should have a broader perspective, that it should relate chemistry to other natural sciences, that it should be investigative and that the programme should be flexible.

After the instructional characteristics had been defined, the proposed laboratory activities were developed and tested with experienced class-room teachers in a way reminiscent of the developments for the Nuffield projects. Individual authors wrote the full narrative for each module only after developing the appropriate laboratory activities. As a result, it is a student-centred course.

The modules include an introductory core, discussing basic concepts and skills of chemistry and a series of additional modules that continuously reinforce and extend these concepts and skills by drawing on areas of biochemistry, inorganic chemistry, organic chemistry, nuclear chemistry, environmental chemistry and physical chemistry. Further details of this influential course are given in Appendix 13. Although it began as a 'local' course, it is used nationally in different ways. Its modules can be strung together to form a course. Alternatively, single modules can be inserted into other courses. As a result of this type of development, a school in the United States may offer several different courses in chemistry, depending on the interest and career goals of the student. It is not unusual for a teacher to be teaching three or more different types of chemistry courses in a single day.

IAC has been translated and has influenced courses in other countries; for example, in Thailand. It is truly a 'local' course that has 'gone' international.

2.6 Reappraisal of the specialized teaching of chemistry

It may be noted that, with the exception of the locally based projects discussed in the last section, most of the chemistry projects described so far were intended for the minority of students who remain at school for an extended period of secondary education. In the 1970s, a number of questions were beginning to be asked about such curricula, of which the following were of particular concern to many educators: Were these chemistry curricula too concerned with theory? Did they pay enough attention to the needs of less academically gifted or motivated school students? Should increasing attention be paid to the other sciences when developing chemistry curricula? Do they take sufficient account of the role of science in our everyday lives and thus help future citizens to take advantage of and face the problems of living in an increasingly technologically orientated society?

The early projects were intended to replace the old descriptive method by a more modern and coherent conceptual approach. But in many parts of the world it was felt that the early concept-based projects went too far in stressing theoretical aspects of chemistry. As Power has pointed out:

> ... as chemistry has progressed, the models we use in chemistry have become more and more abstract. The use of highly abstract concepts and mathematical models which are connected to experimental observations by complex logical relationships has enabled chemistry to achieve great advances, particularly in the area of structural chemistry. Yet the very steps which have increased the power of chemistry as a mode of intellectual inquiry have generated formidable problems associated with its teaching and learning at the secondary school level [25].

In the United Kingdom, doubts began to be expressed, partly as a result of the work of Shayer [26] that the conceptual demands made by 'concept-based' courses may be too difficult for all but a minority of students. And there was concern, too, from the work of Ormerod and others [27] that the strong theoretical basis of many science curricula, coupled with a disregard for the human and social dimensions of the subject, may be partly responsible for causing some students, and perhaps particularly girls, to give up the study of chemistry as soon as they are allowed to do so.

In Sri Lanka, Gunatillake [28] has emphasized that the realization that for 90 per cent of pupils senior secondary-level education is terminal has

> necessitated a reappraisal of the curriculum at this level. As a result the chemistry curriculum was redesigned to meet the needs of the majority. Every attempt was made to make it satisfying and challenging to pupils.... The traditional division of the subject into different compartments like inorganic, organic and physical was done away with and these areas were integrated into units so as to facilitate the understanding of unifying principles and concepts in chemistry. Practical work was designed to help pupils understand principles and concepts, so that it would no longer be a series of exercises confined to developing skills in qualitative and quantitative analysis. The descriptive chemistry component was reduced but non-traditional areas like resources with respect to Sri Lanka, plant-based chemicals, chemicals used in agriculture, medicine, industry and pollution were included in the syllabus. Every attempt was made to make the chemistry that the pupils learnt relevant to their needs.

In Canada, too, similar concerns have been expressed. Gillespie [29] has spoken of the need to develop courses that will present 'a realistic and balanced view of modern chemistry so as to provide students with a useful knowledge of factual chemistry, an appreciation of the scope of the subject, and a critical attitude towards theories'. Newbold [30] explains how this has been achieved:

> The high-school chemistry curricula in some provinces have recently undergone revision with a view to including more descriptive chemistry, since that aspect had been previously neglected. For instance, the grade 11 and 12 chemistry curricula in British Columbia have been redesigned to give more attention to descriptive areas of chemistry (including laboratory experiments). These new curricula have more student 'hands-on experiments', and define a 'core' of material comprising states of matter, stoichiometry, chemical periodicity, chemical kinetics, thermochemistry, redox chemistry and acid-base reactions. Topics such as: bonding beyond ionic bonding and sharing of electron pairs, crystal structures and determination of molecular shapes, quantum chemistry and the shapes of orbitals, spectroscopy, biochemistry, and transition elements, have been eliminated or designated 'for enrichment only'.

And in Australia, Power [25] wrote:

> A chemistry textbook by Hunter, Simpson and Stranks entitled, *Chemical Science* has been published [31]. Both the new syllabus (produced by the

Australian Academy of Science, which is to form the basis of new syllabi in several Australian states) and this textbook represent a reaction against what is seen as an overemphasis on theory. It is claimed that the students of first generation reforms emerge quite 'unaware of the properties of the commonest and simplest chemical substances' although they can talk about 'abstract and difficult concepts such as bond orbitals and free energy'. It has been said that the old 'uncomprehending learning of facts has been largely replaced by the uncomprehending learning of theories'.... Happily the new syllabus has given some recognition of this, in that concepts like free energy and entropy are jettisoned along with the quantum mechanics model of the atom.

There is a short but inclusive introduction to the project entitled 'The Future of the Australian Academy of Science School Chemistry Project' by Watts, one of the two Project Officers [32], which traces some of the seminal ideas to the International Conference on Introductory Chemistry held at McMaster University [33] and which led to an impassionate cry on the relationship of theory to evidence in school syllabuses [34].

In New Zealand, Hitchings argued it was not just that some syllabuses have overemphasized theory, but rather they have failed to show the prime importance of observation. He wrote [35]:

Recently revised examination prescriptions in New Zealand contain general aims worded to reinforce the importance of observations. In part, it is stated that 'the purpose of the prescription is that it should enable the pupil: (a) to gain knowledge of the specific major patterns of chemical behaviour, and of the manner in which these patterns have been interpreted according to the kinetic theory of matter; (b) to understand the relationship between experimental observations and creative thinking on the progress of science and so on.'

These aims were interpreted in the prescription as follows:

States of matter. '*Gases*. A descriptive treatment of the kinetic theory of the gaseous state.... General equation of state for ideal gases.'

became

'A knowledge of the relationships between pressure, temperature and volume of a gas and of the equation of state for an ideal gas. An explanation of this behaviour in terms of a descriptive treatment of the kinetic molecular theory of the gaseous state.'

In the original statement, the explanatory theory is mentioned before the equation of state which summarises observable behaviour. This is reversed in the revised statement:

Rates of reaction '*Collision Theory of Chemical Reactions.*' 'Simple collision theory; activation energy; effect of temperature and concentration on rates; the role of catalysts.'

became

'Rates of Reaction': Effect of temperature and concentration on reaction rates. The role of catalysts. A qualitative explanation of these effects in terms of the simple collision theory including the concept of activation energy.

The same reversal is made with regard to the topic Chemical Periodicity.

Chemical periodicity 'The electron configurations and valency relationships of the elements in the two short periods . . .'

became

'The periodic pattern of the chemical properties of the elements from hydrogen to argon. . . . An explanation of this chemical behaviour in terms of the electron configuration of the atoms.'

In the case of reaction rates explanatory concepts were listed before observation.

As can be seen from this discussion, there is widespread concern that some chemistry curricula may be too narrowly theoretical, and, as a result certain trends have begun to emerge. These include:

Chemistry is now more often taught as a part of a broader group of subjects, instead of as a separate subject. This is particularly so in the early secondary-school years (Section 2.7).

Attempts to relate chemistry more fully to the surroundings of the pupil by, for example, including examples of local chemical industry (Section 2.8) and the environment (Section 2.9).

Adopting, in some cases, a topic-based or thematic approach to science teaching (Section 2.7.4 on integrated science).

2.7 Chemistry in a broader context

In many countries, curriculum innovation in the 1970s became increasingly involved with chemistry, not so much as a single subject, but as part of a larger whole. In this way it was believed that it could be made more interesting and relevant to the needs of the majority of pupils up to the age of about 16. A number of regroupings emerged which may be represented: (a) physical science (chemistry and physics) plus biological science (biology); (b) general science—sometimes called combined science (chemistry, physics and biology loosely combined together); (c) integrated science (as general science, but more closely integrated, usually with an element of the social and environmental sciences).

It is worth noting that, in most countries of the world, science is taught in some general form in the early school years, and that it invariably becomes specialized in the 16 to 19 age range. There is, of course, a variation from country to country in the age of changing from a general to a specialized form of science education, and some countries make different arrangements for their arts and science sections. The patterns for the Arab countries shown in Appendix 9 illustrate the differences in practice followed by a number of countries; a similar variety of practice is to be found in other regions of the world. This diversity provides, incidentally, an illustration of one difficulty which may be encountered whenever curriculum materials are developed on a regional basis for use in a variety of countries.

2.7.1 *Chemistry as a component of physical science*

The physical–biological split recommends itself on a number of grounds. Chemistry and physics are closer to each other than to biology, and the reduction of three science subjects to two results in some lessening in the load on the student. Many countries combine their physics and chemistry courses together with the omission of some subject-matter to teach the subject referred to as 'physical science'.

It is hard to point to many physical science courses in which the elements of physics and chemistry are significantly integrated below the age of 16, perhaps because it is mainly at more advanced school and university levels that opportunities of achieving a conceptual integration are greatest. Some examples of physical science courses developed in several countries are given in Appendix 10.

2.7.2 *Chemistry as a component of general science*

Concern was first expressed that science courses were too specialized early in this century. The view then gained ground that science courses in schools should be broadened and humanized in order to make them suitable, not only for those going on to study science at university, but also for those who would not. In the United Kingdom, for example, several influential reports, including *Science for All* [36], were published in the early years of the century which led, in the 1920s and 1930s, to the development of general science courses. General science was based on the idea that it should form an essential part of the general education of all young people, and was seen by its proponents as humanistic as well as scientific in scope, broad rather than deep in content. It is interesting to note that there are similarities between this thinking and that of the 1970s which was discussed in Section 2.6.

It is unfortunate that general science often failed to meet these ideals, and that it became, more often than not, a scrappy mixture of physics, chemistry and biology. Perhaps partly because of this shortcoming, general science never really ousted the teaching of the specialist subjects during the 1930s and 1940s, and it became regarded as a weak subject suitable for those who did not have much aptitude or interest in science.

Further, in the United States, the main criticism was that general science was often taught as 'read about talk about' science without a practical component. This, too, led to a second-class reputation for general science.

There is some difficulty in terminology in talking about general science. 'General science', 'science', 'combined science' all mean more or less the same thing, although some would attach different shades of meaning to each of these terms. Furthermore, it should not be assumed that a course described as 'integrated science' necessarily has a higher degree of integration than one described as 'science' or 'general science'.

2.7.3 *Chemistry as a component of integrated science*

Numerous reasons have been put forward for teaching integrated science [37] during the last twenty years. It has been widely argued that:

If the student is to obtain a coherent view of science, it is necessary to make numerous links between its various branches. This is difficult to do when science is taught as three separate subjects, especially when each of them is taught by a different teacher.

Most of the applications and social implications of science (e.g. the motor car, or DDT) involve more than one science. For example, the motor car could be more easily linked to aspects of biology, chemistry and physics within a well-integrated science course, than within three separate specialized courses.

Teaching three separate subjects may require too much school time. For example, instead of spending two school periods on each subject per week in the first two years of secondary school, four per week might be allocated to integrated science—a saving of two periods per week. This reduction may be necessary to prevent too much of the pupils' curriculum time being spent on science, and it may also find favour because it results in some reduction of the need for science teachers, laboratories and equipment.

Integrated science is widely associated with the movement to teach science for the majority. In some cases, integrated science has been developed from the older and more academically oriented courses that were already in existence, by increasing the descriptive element at the expense of the mathematical or quantitative aspects. In this way the conceptual structure is left largely unchanged. This approach was adopted for example, in the 14 to 16 age range in Malaysia because it was felt that there would be many practical advantages in not allowing too great a difference to develop between the general science course for 'arts' students and the specialist science courses in biology, chemistry and physics for 'science' students. This policy helped to prevent imposing too many changes on teachers simultaneously, and permitted the same apparatus to be used for both the general and the specialist courses.

A more radical approach towards the development of integrated science courses is to make use of some integrating factor. In the United Kingdom, the Schools Council Integrated Science Project (SCISP) has done this by using the idea of 'patterns' as a way of organizing and using information. By way of an example, the notion of a 'population' can be studied in systems of molecules, of bacteria and in human populations. Another approach is to use the well-known idea of a topic approach, e.g. the subject 'water' may be studied from physical, chemical and biological points of view. This approach avoids the artificial division of the subject into the various branches of science which is particularly inappropriate for those who are not specializing in chemistry, and may readily be extended to include geographical, human, social, economic and other implications of the subject (Section 2.9).

Using the integrating theme, 'Investigating the Earth' was an ambitious effort which exerted a considerable influence on the junior secondary curriculum in the United States. It includes structure, properties of matter and chemistry of the oceans, soils, minerals and atmosphere. This curriculum inspired many texts.

It is sometimes pointed out that many integrated science courses are 'short' of chemistry material. This is because the organizers of such courses have felt that the need to produce courses suitable for the majority should take priority over the notion of content balance. A great deal of chemistry, it is pointed out, is conceptually very demanding (see Section 2.6) and is not appropriate except for the more able and intellectually minded students.

Many countries are now using some form of integrated science, particularly in the early years of secondary school, as is apparent from a recent worldwide review [38]. Some examples are given in Appendix 11.

Several integrated science projects have had a particularly wide influence outside their country of origin and two of these are outlined below:

Scottish Integrated Science was developed between 1964 and 1969 for use by all pupils in the 12 to 14 age range. The development was part of the continuous process of curriculum reform in Scotland. It has had a very high level of adoption in Scotland and has been adapted for use in many countries including Malaysia, Nigeria, Lesotho and Hong Kong. Although it gives clear straightforward advice to teachers, it makes considerable demands on pupils by using the concepts of energy and particle theory as organizers for much of the work. The course has undergone a major evaluation and revision.

The Australian Science Education Project (ASEP) was developed during the period from 1969 to 1973 for use by all pupils in the 12 to 15 age range with the aim of instilling in pupils an understanding of man, his physical and biological environment, and the importance of interpersonal relationship, skills and attitudes as well as imparting some

Search 6.1

Breakfast energy

Discussion

You probably often feel hungry and tired after a hard game of football or running round. After a good meal, you feel better but not ready for another game of football. As you probably know, it takes a while for the food to be digested and reach your whole body. So after a few hours you feel energetic again. You need sleep, too, but sleep does not give you energy—it just gives the mind and body a rest.

If you do not have good food for breakfast you may find yourself short of energy in the afternoon.

Scientists can measure the amount of energy you can get from food. Energy is measured in kilojoule (kJ) and a boy of your age needs about 10 000 kJ a day while a girl would need about 9 000. About one-third of this should come from your breakfast, so you should eat enough breakfast to give you about 3 000 kJ.

Some adequate breakfasts

Bread and butter, honey, orange, Milo and liver

Cocoa, condensed milk, porridge and mango

Pawpaw, milk, fish and plantain

The Problem

How much energy did you get from your breakfast today?

Suggestions

Look up pages near the beginning of this book to find the energy value of each thing you ate. If you cannot find the food listed there ask your teacher for help.

The Results

Things I ate	How much I ate (cups, pieces, teaspoons)	How much energy it gave me (kJ)
	Total	

Put your peg in the right place on the teacher's board. When everyone's pegs are in, copy the results onto the graph on the next page.

FIG. 3. Example of practical work in *New World Science* (see Appendix 11, ref. 8 for details). Reproduced with permission.

understanding of the nature, scope and limitations of science. It is a flexible course that may be used in an independent learning situation. Among the many interesting integrated science projects developed in the Third World, mention may be made of one published under the title *New World Science* (Appendix 11). The example from the project in Figure 3 gives some idea of the freshness of the approach, the use of science in the student's own environment, and the careful choice of language. Although it contains a comparatively small proportion of specifically chemical material, the project exploits a wealth of worthwhile scientific ideas that can be developed using comparatively simple apparatus.

2.7.4 *Chemistry in relation to other subjects in the school curriculum*

In many countries of the world, the primary-school curriculum bears little relation to that of fifty or so years ago. Then, the subjects were reading, writing and arithmetic. Now the curriculum is conceived much more as a whole. Although there is still a need for deliberate and carefully planned work in communication and number skills, many primary-school activities are now broadly based. The primary-school curriculum has, to a considerable extent, become integrated, and many good primary-school teachers possess a broad background which enables them to guide their pupils' learning on a variety of topics as often based on the surroundings of the school. This sort of work involves not only communication and number skills, but also aspects of history, geography and science.

By contrast, the secondary-school curriculum generally consists of a number of separate 'subjects'. All too often, there is little or no co-ordination between them. The reason for this state of affairs lies in the way many secondary-school teachers have been trained (Chapter 6), and the public examination system which is also strongly subject-bounded. The move towards an integrated science curriculum may be seen as a small but significant step towards achieving a more unified secondary-school curriculum. But although it has brought the science subjects somewhat closer together, it is still possible to teach science without considering other areas of the curriculum, such as language, mathematics and social studies.

The creation of a fully integrated secondary-school curriculum may appear a distant or even an unobtainable goal but, as a first step, it is clearly desirable to try to co-ordinate the existing school subjects. This is so seldom done that it is surprising that Sir Percy Nunn [39] declared as long ago as 1918 that:

> the curricula in science, geography and mathematics should be thought out as

a whole. There is only one way of bringing about effective co-ordination of this kind: the teachers of the several subjects must meet from time to time in conclave to adjust their frontiers and regulate their policies. It should be regarded as unthinkable that any one of them should go his own way without constantly taking account of the needs and proceedings of his colleagues.

There are two school 'subjects' which relate particularly closely to chemistry, and to the other sciences. These are language [40–44] and mathematics. Language provides not simply a way of communicating with others, but quintessentially the vehicle for thought [45]. It is virtually impossible for a student to grapple with a scientific problem without the use of words. Hence the importance of encouraging the student to engage in both spoken and written forms of discourse. Practical work in science provides a particularly rich opportunity for discussion between teacher and student (and also between students). Writing skills, too, can be fostered by encouraging students to make their own record of practical work, perhaps in the form of a diary, rather than simply copying from the board.

Mathematics is another 'subject' which needs to be carefully coordinated with science [46–49]. Mathematics and science teachers need to be broadly aware of their students' progress in each other's subjects. The chemistry teacher should know, for example, precisely when students are introduced to ratio, proportion, percentage, algebraic transformations and the use of graphs in mathematics lessons, as well as their depth of treatment and the form of mathematical language used in mathematics lessons. It is all too easy for the student to be disadvantaged if the chemistry teacher is insufficiently aware of what has been taught in mathematics as well as how it has been taught [50].

2.8 The chemical industry and applied chemistry

The inclusion of material on the chemical industry is a valuable way of broadening and extending the teaching of chemistry outside the confines of the school laboratory. Even before the curriculum development era, chemistry curricula often made reference to such processes as the manufacture of lime, the smelting of iron, and the Haber process for the synthesis of ammonia. However, some curriculum innovation projects have gone much further by presenting a much more up-to-date picture of the chemical industry, and in some cases have sought to show the contribution it makes to the enrichment of our material way of life. A number of developing countries have included a study of traditional and local forms of industry which obviously relate more closely to pupils' lives than examples from far-away industrialized countries. Some notable examples are given in Appendix 12. In one example, that of Papua

New Guinea, there is a unit on chemical technology, for Grade 10 (about age 16) [51]. The *Teachers' Guide* describes four sections of work: copper, fuels, lime and water. There is a wealth of experiments for pupils to do, all related to the life of the country. For example, there is a section on obtaining copper from the Bougainville Concentrate; questions are asked which relate to the student's work in social-science lessons. Under lime, there is a delightful experiment on examining coal, various shells and limestone. There are statistics and data for the teacher, carefully documented. This trend may be seen in proposals for new curricula in Japan where topics include photography, food additives, the chemistry of aspirin, natural resources—distribution and utility [52]. The *Shimozawa Report* stresses the need not only to reduce the quantity of material in the curriculum but also to emphasize everyday phenomena rather than theoretical areas, the report building on the equally significant *Oki Report* [53].

The former East African Chemistry Syllabus (O-level, examination at 16), emphasizes applied aspects of chemistry. Its objectives speak for themselves:

At the end of this topic students should be able to:

1. Explain the industrial isolation of nitrogen, oxygen and the inert gases.
2. Recognize the advantages and disadvantages of different fuels.
3. Recall the industrial and domestic uses of water and explain the sources of pollution and their treatment.
4. State the specific uses of chemical compounds both locally and world wide.
5. Name the natural resources available locally and illustrate their use in local industry.
6. Illustrate the operation of chemical principles already encountered in both the natural environment and specific industrial processes.
7. Explain the synthesis and breakdown of natural and synthetic materials.
8. Recognize the relative advantages of synthetic materials over those of natural origin in terms of both structure and properties.

The content is divided into four sections: the atmosphere, water resources; natural and synthetic materials; mineral resources and industrial processes. The emphasis was not on new principles but on application. Thus topics such as the internal combustion engine, steam engines and fuel cells, fertilizers, pollution from fertilizers, insecticides and herbicides, plastics, availability of raw materials in East Africa, found their way into the syllabus [54].

The hope behind such work is not only that it will form an interesting and worthwhile study, but that it will help students to live a little more easily in today's society.

By contrast, we can look at the curriculum of another country. Tunisia is the third most important producer of phosphate in the world; it is a powerful member of the Arab world with its abundance of oil; it has mineral wealth, of magnesium, iodine and potash; and it has a growing chemical industry. However, Bouguerra points out that these are given scant treatment in curricula yet the chemical industry is a provider of the country's wealth and development as well as a potential source of pollution and other hazards. The curriculum says nothing about the challenges of industrialization [55]. His comments agree with those made earlier in an Alecso Conference [56]. Bouguerra's strictures on his own country's curriculum are candid and helpful, and are echoed throughout the world.

It is worth noting that individuals in universities and in the chemical industry are invariably pleased to collaborate with chemical educators in producing educational resources on a nation's chemical industry. Many of the examples given in Appendix 12 have resulted from this type of collaboration.

2.9 Social aspects of chemistry teaching: chemistry and society

The study of theoretical chemistry allows little play for human emotions and values. In the past, chemical education involved essentially the factual and theoretical sides of the subject, although some reference was usually made to the 'applications' of chemistry in the home or in industry. However, these applications were usually mentioned in a manner that precluded any discussion of human, social or economic issues. But as a result of the enormous impact that science and technology have made on our way of life, there is now a growing body of opinion that science education must become increasingly involved with these wider issues of science. It is now widely felt that ways of sensitizing students to their importance must be found. In a recent international seminar, some criteria for the inclusion of topics on science and society have been proposed [57].

A difficulty at the present time is that although many science teachers would like to broaden their teaching of science, including social and other dimensions, they often lack the knowledge of how to do so. The problem is made more acute because so many science courses at university level omit totally any discussion of the nature of science, or its social implications. Two particularly helpful texts for teachers are available [58, 59]. Examples of projects containing socially relevant material are given in Appendix 12 from which it will be seen that various techniques are being used. Some projects incorporate materials on the nature of science; others make a study of the chemical industry in the light of its social and economic consequences, while yet others use interactive exercises to simulate the sorts of decision-making that are involved in, for example, deciding where to site a refinery or nuclear power station.

Some of the projects are essentially chemistry courses containing a small element of social relevance, while others are basically science and society projects.

The topic approach to science teaching was described in Section 2.7 on integrated science (page 69), but a new generation of topic courses is now emerging in some countries that provide a new basis for including chemical studies in school curricula. These involve carefully chosen, and often isolated topics relating to aspects of practical living that impinge on most citizens today. The choice needs to vary from country to country, and even within countries depending on such factors as the predominant employment. Many countries now require the composition of products to be displayed, and warning of potentially hazardous substances given. 'Do-it-yourself' kits flourish in developed countries as both the price of labour and the amount of leisure time increases. There is scope to exploit these opportunities in chemical education, but it must be admitted that some of the attempts to do so in school science have not been particularly successful. If they are to contribute effectively to the leisure pursuits of future citizens they will need to be tightened up, and thought given to the acquisition of skills. Deliberate attempts must be made to link science and technology. Glass fibre patching and stain removing are two examples that attract many students and enable chemical understanding to be developed at a number of levels in association with skill learning. More needs to be done to develop this sort of technological learning of chemistry, but it will not be an easy task, as so few teachers have these skills, or even value them.

Another approach to science education should be mentioned which involves the 'environmental movement' and was seen originally as an important educational component in the United States (Appendix 11, note 28) and other industrialized countries. This has as its aim the development of an awareness in the quality of the environment and a concern for its intelligent management. One approach is to provide courses on the environment in schools, while another is to try to introduce environmental aspects of education into many subjects of the curriculum. There is scope for the latter approach with chemistry courses in dealing with topics such as resources and pollution, which contain an important chemical component, for example 'acid rain', which is an international problem, foams from detergents in the rivers, the disposal of polymers. But these are not the problems solely of the highly industrialized nations. The Asian Programme of Educational Innovation for Development has now encouraged several curriculum projects to include an environmental emphasis. Some of these bring meaningful chemical education within the range of primary education involving topics such as 'choosing the right fuel'. Appendices 6 and 11 indicate an interesting curriculum in Brazil, 'Unidades Modulares de Química' in which the introductory chemistry is taught in part in terms of issues important to the Brazilian people—far-reaching issues such as self-prescription in medicines and the application of nuclear energy. And in Venezuela, a project—'The Environment: A Learning Resource for Chemistry'—has been developed for 14 to 18 year olds, involving the study of chemistry in the environment and including a simple environmental research project (Appendix 13).

2.10 Teaching chemistry to the handicapped

The curriculum of young people who are handicapped, whether orthopaedically or as a result of impaired vision or hearing, has come under scrutiny in recent years, and it is being increasingly questioned as to whether practical subjects such as physics, chemistry and technology should be omitted from their curriculum. In the past, it was often felt that such practical subjects would be unsuitable and even hazardous. However, it is becoming more widely recognized that this need not be the case. Omission of these subjects is most unfortunate, as they can help to motivate handicapped children. As Thompson has said, 'science can become an unsuspected ally in the struggle to provide success for the handicapped child' [60].

In the past, the education of young handicapped people was more often carried out in special schools than is the case today, when increasing numbers of them are taking their place in ordinary schools. As laboratory work is commonly carried out in pairs or in small groups, the incorporation of a handicapped student into such a group does not normally present any particular difficulty [61]. Indeed, science, as a practical subject, provides a good opportunity for getting handicapped and ordinary children to work together so that they come to regard such partnership as a perfectly normal state of affairs.

There are a number of modifications which may be made to give handicapped students maximum access to chemistry as an experimental science:

In direct teaching, it is desirable to use large visual aids and individual diagrams for visually impaired members of the class. The orthopaedically handicapped or wheelchair-bound student can have a 'reserved' spot near the door for ease of access, although care will have to be taken to ensure that this does not impair escape of the class in an emergency.

In teaching chemistry in special schools containing students with several types of handicaps, it is often possible to arrange working in small groups in such a way that there is a 'balance' of disabilities in each group. There is, of course, no reason to assume that handicapped students are any less careful or mentally alert than other students. But it is important to ensure that adequate safety arrangements are made for handicapped students in case of fire. *Teaching Chemistry to Physically Handicapped Students* [62] and a special issue of the *Journal of Chemical Education* [63] are most useful sources of information.

A number of experiments may be modified for use by mobility handicapped in a manner analogous to that in which cars are adapted for 'disabled drivers'. Some adaptations are available for appliances such as gas and water taps, and science benches can be adapted for pupils in wheelchairs [64]. When the necessary adaptations have been made, handicapped students may well be able to carry out practical work in chemistry without difficulty.

The use of microcomputers (Chapter 4) would appear to merit special consideration. There is undoubtedly value in all pupils carrying out computer simulations from time to time and possibly the handicapped can do so a little more often: trials in some British schools have successfully used commercial programs for this purpose.

Several curriculum innovation projects have been specially adapted for use by the handicapped. A notable example is the Science Curriculum Development Study project in the United States which has been adapted for use with those who suffer impairment of sight or hearing [65].

A recently published book *Science for Handicapped Children* [66] contains many experiments on elementary chemistry and provides valuable advice for parents and teachers of handicapped children.

Useful tips and special devices appear from time to time in the literature and some addresses are given where inquiries can be made [67].

2.11 Evaluation and revision

It is highly desirable that science curricula should be evaluated. It is necessary to find out, for example, whether a particular curriculum in chemistry is engaging the interest of the pupils for whom it is intended. Are there any topics in a course which appear to be beyond the capabilities of the pupils, perhaps because they are introduced at too early an age? Is the practical work appropriate? Is any of it too complicated to perform, too time consuming, or too dangerous? Are the pupils able to handle the mathematics required in the course of their science with reasonable confidence? Are the pupils' materials written in suitable language, and are they of appropriate level of sophistication? Do the diagrams and illustrations of the pupils' texts communicate effectively with the pupils? Are cross-sectional diagrams, for example, 'read' by the pupils? Would perspective drawings be more suitable with some age groups?

These are but a few of the issues that will need to be considered in evaluating any course in science. Such evaluation may be carried out during, immediately or some years after the production of the curriculum materials themselves. It is sometimes argued that it is preferable to carry out an evaluation during the course of development of a curriculum development project. The school trials that were a feature of many curriculum innovation projects could be regarded as a form of evaluation—their purpose was to throw light on the suitability of the materials, principally by obtaining the views of teachers on them. But although teachers must certainly be consulted, the views of other informed persons need also to be considered. It is possible that this may be achieved more effectively by having an evaluator attached to the development team.

The idea of carrying out an evaluation some years after the publication of a project has the merit that the views of a much larger and more representative group of teachers and pupils may be probed. The suitability of a project, not just for the innovative teachers who are usually the first to take up a new curriculum, but of a wider group can then be judged. This was done, for example, in the evaluation of Nuffield Chemistry which was carried out four years after the publication of the first edition. The evaluation considered the sorts of issues raised in the first paragraph of this section [68] and led to a completely new set of publications. By contrast, the evaluation of Scottish Integrated Science centred on the suitability of the worksheets for use with mixed-ability groups of pupils [69].

In Scotland, Johnstone [70–73] has employed a simple but valuable method of obtaining information on the difficulties encountered by students, which does not depend on the subjective views of teachers. Students who have completed the course are invited to criticize it and classify each of the topics of the course under one of four headings: Easy to Understand; Difficult to Understand; Never Understood; Never Studied.

Samples, each of 1,000 pupils, were taken in consecutive years, and the results were found to be reproducible. The areas of greatest difficulty were found to be:

(a) those based on chemical formulae (for example, the difference between covalent and ionic bonding, writing chemical equations, ion-electron half-equations);
(b) practical topics involving multi-stage thought (e.g. acidic and basic oxides, precipitation reactions);
(c) organic topics with a common link of hydrolysis, esterification or condensation (hydrolysis of carbohydrates, formation of esters, condensation polymerization).

The latter was found not to be due to a visual problem associated with complex formulae, but simply to a lack of understanding of functional groups and their role [73].

It is probably fair to observe that a serious shortcoming of curriculum development in almost every country of the world has been the failure to evaluate the new curricula, or to evaluate them in too specialized or narrow a manner. It seems probable that in the future both the setting up and the evaluation of curriculum development projects in science and mathematics will, like many other aspects of education, become matters of increasing public concern. In Malaysia, for example, some parents in the 1970s took the unusual step of writing to the press to express disquiet about certain aspects of curriculum innovation in science. In Kenya, not only have repeated questions been asked by parents about 'new mathematics', but Parliament took the extreme step in 1981 of calling for its abolition. In the United Kingdom, the Assessment of Performance Unit has been set up as a result of social and political pressures to monitor the performance of representative groups of pupils throughout the country in language, mathematics and science [74]. Likewise in the United States, a monitoring programme, the National Assessment of Educational Progress, has been set up [75].

The development and evaluation of curriculum packages in chemistry may be expected to be increasingly subject to public scrutiny. As a step towards public accountability, it is desirable that a much larger and more representative group of individuals should be involved in any broad review of the science curriculum. Such a group might comprise not only university and school chemistry teachers, but also industrialists, educators who are not exclusively concerned with teaching science, professional communicators in science (such as broadcasters), parents and a variety of lay persons.

Two particularly interesting examples of ways in which the views of the community have been sought on educational issues deserve mention here. The first is the Sierra Leone education review [76] which provided a forum for considering many issues that faced the country in which individuals from many walks of life put forward their ideas. The second, from the Federal Republic of Germany, which dealt more specifically with the science curriculum, is an article entitled 'Physics Education for Everyone in an Industrial Country' [77]. The techniques described in the paper for collecting the views of many groups of well-informed persons could readily be adapted to other subjects, including chemistry, and for use in predominantly agricultural countries.

2.12 Concluding remarks

As has been illustrated, the 1960s and 1970s have witnessed an astonishing growth of curriculum activity throughout the world. Not only has the content of school curricula been brought more up to date, but its aims have become increasingly broad. A great deal of variety among the many attempts made to renovate chemistry teaching is apparent, but one trend has been remarkably constant, that towards laboratory work. Indeed, in many countries, the school chemistry curriculum is now essentially laboratory-based, and a large proportion of the students should be occupied with practical work. The financial cost in terms of laboratories, apparatus and chemicals has been considerable. It is widely believed that practical work is inherently valuable and the following proverb has often been quoted in justification [78]:

> I hear and I forget,
> I see and I remember,
> I do and I understand.

It must, however, be emphatically stated that 'doing' does not necessarily produce understanding. This requires much more—above all, the student needs to be intellectually engaged in thinking about science. But how should this be brought about? What form should this engagement take? It is here that the insights afforded by a thorough study of the history and philosophy [58, 79–83] of science may be particularly valuable in the years to come. Philosophies of, for example, Karl Popper [84] and Thomas Kuhn [85] could have profound implications for the way in which science is taught and could shed light on the interplay between 'theory' and practical work. If the limited financial resources available to science education are to be put to the best use, there needs to be a continual evaluation of chemical education. We hope that this chapter will help chemistry educators to be more aware of what has been achieved in making school chemistry more lively and relevant in other countries and, as a result, stimulate further questioning, not only about what is being done today but also about what needs to be done in the future.

References

1. B. Wilson, *Cultural Contexts of Science and Mathematics Education: A Bibliographic Guide*, Centre for Studies in Science Education, University of Leeds, 1981; B. Wilson, *Studies in Science Education*, Vol. 8, 1981, p. 27.
2. R. B. Ingle and A. D. Turner, 'Science Curricula as Cultural Misfits', *European Journal of Science Education*, Vol. 3, 1981, p. 357.
3. R. Driver, 'Cultural Diversity and the Teaching of Science', in T. Trueba and C. Barnett-Mizrahi (eds.), *Bilingual Multicultural Education and the Professional: From Theory to Practice*, Rowley, Mass., Newbury House, 1979.
4. *First Report and Evaluation of the Unesco Pilot Project for Chemistry Teaching in Asia, 1964–70*, Paris, Unesco, 1972.
5. *New Trends in Chemistry Teaching*, Vol. I, p. 335, Paris, Unesco, 1967.
6. G. Van Praagh, private communication.
7. M. Sinclair, *International Newsletter on Chemical Education*, Vol. 15, 1981, p. 17.

8. R. T. Allsop, *Chemistry Readers for East Africa: Salt in East Africa*, Kampala (Uganda), Ministry of Education, 1968.
9. R. B. Ingle, *Chemistry Readers for East Africa: Fermentation and Distillation*, Kampala (Uganda), Ministry of Education, 1968.
10. L. C. Perez, in C. P. McFadden (ed.), *World Trends in Science Education*, p. 42, Halifax, Nova Scotia (Canada), Atlantic Institute of Education, 1980.
11. E. H. Coulson, private communication.
12. B. J. Stokes, private communication.
13. Nida Sapianchai, private communication.
14. R. D. Shulka, private communication.
15. M. L. Bouguerra, in A. Kornhauser, C. N. R. Rao and D. J. Waddington (eds.), *Chemical Education in the Seventies*, 2nd ed., p. 250, Oxford, Pergamon Press, 1982.
16. S. Sib, *International Newsletter on Chemical Education*, Vol. 13, 1980, p. 5.
17. J. S. Eggleston, M. J. Galton and M. E. Jones, *Processes and Products of Science Teaching*, p. 122, London, Macmillan Educational, 1976. (Schools Council Research Studies.)
18. G. Keller, private communication.
19. F. Jenkins, private communication.
20. For a descriptive brochure on the publications write to J. M. Le Bel Enterprises Ltd, 10372 60 Avenue, Edmonton, Alberta T6H 1G9 (Canada).
21. F. Jenkins, 'Custom Tailoring Chemistry Curricula to the Culture', in W. T. Lippincott (ed.), *Proceedings of the Sixth International Conference on Chemical Education*, College Park, Md., University of Maryland, 1982.
22. See Appendix 11.
23. C. T. Robertson and L. Bulman, *Education in Chemistry*, Vol. 17, 1980, p. 133. Twenty textbooks and teachers' notes, 1983, 1984. Inquiries to Independent Learning Project for Advanced Chemistry (ILPAC), 94–100 Leigham Court Road, London SW16 2QD (United Kingdom).
24. M. H. Gardner (ed.), *Interdisciplinary Approaches to Chemistry*, London, Harper & Row, 1973. Revised 1977–79.
25. C. N. Power, *The Australian Science Teachers Journal*, Vol. 26, 1980, p. 15.
26. R. B. Ingle and M. Shayer, *Education in Chemistry*, Vol. 8, 1971, p. 182; M. Shayer and H. Wylan, *School Science Review*, Vol. 59, 1977, p. 377; M. Shayer and P. Adey, *Towards a Science of Science Teaching: Cognitive Development and Curriculum Demand*, London, Heinemann Educational, 1981.
27. M. B. Ormerod and D. Duckworth, *Pupils' Attitudes to Science—A Review of Research*, Windsor (United Kingdom), NFER Publishing Company, 1975. See Section E: 'Early Age of Arousal'.
28. N. A. C. Gunatillake, in A. Kornhauser, C. N. R. Rao and D. J. Waddington (eds.), *Chemical Education in the Seventies*, 2nd ed., p. 236, Oxford, Pergamon Press, 1982.
29. R. J. Gillespie, *Chemistry in Canada*, Vol. 28, 1976, p. 23.
30. B. T. Newbold, in A. Kornhauser, C. N. R. Rao and D. J. Waddington (eds.), *Chemical Education in the Seventies*, 2nd ed., p. 64, Oxford, Pergamon Press, 1982.
31. R. Hunter, P. Simpson and D. Stranks, *Chemical Science*, Sydney (Australia), Sydney University Press, 1978.

32. D. Watts and N. Bayliss, *School Chemistry Project: A Draft of a Secondary School Chemistry Syllabus for Comment*, Canberra City, Australian Academy of Science, 1978.
33. International Conference on Introductory Chemistry, McMaster University, Ontario (Canada), 19–23 June, 1978.
34. H. Bent et al., *International Newsletter on Chemical Education*, Vol. 10, 1978, p. 17.
35. T. R. Hitchings, in P. Childs and J. E. Gowan (eds.), *Proceedings of the International Conference on Chemical Education*, p. 25, Dublin, 1980.
36. See, for example, E. W. Jenkins, *From Armstrong to Nuffield*, p. 70, London, John Murray, 1979.
37. S. A. Brown, *Studies in Science Education*, Vol. 4, 1977, p. 31.
38. S. Haggis and P. Adey, 'A Review of Integrated Science Education Worldwide', *Studies in Science Education*, Vol. 6, 1979, p. 69.
39. P. Nunn, in J. Adams (ed.), *The New Teaching*, Chapter 5, London, Hodder & Stoughton, 1918.
40. J. Britton, T. Burgess, N. Martin, A. McLeod and H. Rosen, *The Development of Writing Abilities, 11–18*, London, Macmillan Educational, 1975. (Schools Council Research Studies.)
41. *A Language for Life: Report of the Committee of Enquiry. Chairman, A. Bullock*, London, HMSO, 1975.
42. A. Bullock, *School Science Review*, Vol. 57, 1976, p. 621.
43. *Science Teacher Education Project*, Chapter 4 and Topic 6, New York, McGraw-Hill, 1974.
44. C. Sutton, *School Science Review*, Vol. 62, 1980, p. 47.
45. R. Farrar and J. Richmond (eds.), *How Talking is Learning* (Inner London Education Authority Oracy Project 1971–1977). Available from ILEA Learning Materials Service, Highbury Station Road, London N1 1SB (United Kingdom).
46. *Co-operation between Science Teachers and Mathematics Teachers* (proceedings of a Conference jointly organized and sponsored by Unesco, the Committee on Teaching of Science (CTS) of the International Council of Scientific Unions (ICSU), the International Commission on Physics Education (ICPE), the International Commission on Mathematical Instruction (ICMI) and the Institute for the Didactics of Mathematics (IDM)), Bielefeld, 17–23 September 1978, Institut für Didaktik der Mathematik der Universität Bielefeld. See also P. J. Fensham, *European Journal of Science Education*, Vol. 1, 1979, p. 347.
47. G. Woods, in A. Rogerson (ed.), *Mathematics and Chemistry: The Classroom Interface*, in the series Co-operation between Science Teachers and Mathematics Teachers, ICSU Committee on the Teaching of Science, 1981. Available from the Association for Science Education, College Lane, Hatfield, Hertfordshire, AL10 9AA (United Kingdom). See also, J. Ling, *Mathematics across the Curriculum*, Chapter 2, Glasgow, Blackie, 1977.
48. R. B. Ingle and A. D. Turner, *Education in Science*, No. 93, 1981, p. 31; R. B. Ingle and A. D. Turner, *Education in Chemistry*, Vol. 18, 1981, p. 48; J. N. Lazonby, J. E. Morris and D. J. Waddington, *Education in Chemistry*, Vol. 19, 1982, p. 109.
49. G. Mathews, 'Science as Handmaiden of Mathematics', *Proceedings of the*

International Congress of Mathematics, p. 571, Canadian Mathematical Congress, 1974.
50. A. D. Turner, *Mathematics in School*, Vol. 8, 1979, p. 14; *School Science Review*, Vol. 60, 1979, p. 773.
51. B. Deutrom and M. Tapo (eds.), *Science: Chemical Technology, Teachers' Guide, Grade 10*, Boroko (Papua New Guinea), Department of Education, 1980.
52. J. T. Shimozawa, *Journal of Science Education of Japan*, Vol. 2, 1978, p. 181. (In English.)
53. M. Oki, *The Oki-Report*, Tokyo, Ministry of Education. The Committee began work in 1967. For details, see ref. 52.
54. Kenyan Advanced Certificate of Education (KACE) in Chemistry. For further details, Ministry of Higher Education, P.O. Box 30426, Nairobi (Kenya).
55. M. Bouguerra, private communication.
56. *Proceedings of the ALECSO Conference on Chemical Education, Alexandria (Egypt), 1976.*
57. 'Malvern Seminar on Science in Society', *ICSU Newsletter from the Committee on Teaching of Science*, Vol. 5, 1980, p. 6. See also the *International Newsletter on Chemical Education*, Vol. 15, 1981, p. 1 and subsequent articles.
58. J. L. Lewis (ed.), *Science and Society; Teachers' Guide*, London, Assocation of Science Education (ASE)/Heinemann, 1981.
59. J. Ziman, *Teaching and Learning about Science and Society*, New York, Cambridge University Press, 1981.
60. B. E. Thompson, *Science and Children*, Vol. 13, 1976, p. 5.
61. A. V. Jones and A. Barnett, *Special Education: Forward Trends*, Vol. 7, 1980, 25.
62. *Teaching Chemistry to Physically Handicapped Students*, Washington, D.C., American Chemical Society, 1981.
63. D. Tombaugh, *Journal of Chemical Education*, Vol. 58, 1981, p. 222; D. Smith, *Journal of Chemical Education*, Vol. 58, 1981, p. 226; D. Lunney and R. C. Morrison, *Journal of Chemical Education*, Vol. 58, 1981, p. 228; I. D. Brindle, J. M. Miller, M. F. Richardson, W. Balenovich, M. Benkel and T. Biernacki, *Journal of Chemical Education*, Vol. 58, 1981, p. 232; A. B. Swanson and N. V. Steere, *Journal of Chemical Education*, Vol. 58, 1981, p. 234.
64. J. T. Moore 'The Mobility Handicapped Student in the Science Laboratory', *Journal of College Science Teaching*, Vol. 11, 1981, p. 362.
65. *Science Curriculum Improvement Study* is published by Rand McNally, Chicago, Ill. (United States). Inquiries to the Director, Science Curriculum Improvement Study, Lawrence Hall of Science, University of California, Berkeley, Calif. 94720 (United States).
66. A. V. Jones, *Science for Handicapped Children*, London, Souvenir Press, 1983 (simultaneously published in Canada).
67. The following addresses may be helpful: The Secretary, The Association for Science Education, College Road, Hatfield, Hertfordshire, AL10 9AA (United Kingdom); A. V. Jones, Department of Physical Science, Trent Polytechnic, Clifton, Nottingham (United Kingdom); National Science Teachers Association, 1742 Connecticut Avenue, Washington, D.C.

(United States); S. Malcolm, American Association for the Advancement of Science, 1776 Massachusetts Avenue, N.W., Washington, D.C. (United States); Centre for Multisensory Learning, Lawrence Hall of Science, University of California, Berkeley, Calif. 94720 (United States); B. Thompson, Science for the Handicapped Association, University of Wisconsin-Eau Claire, Eau-Claire, Wis. 54701 (United States) from whom a comprehensive bibliography is obtainable.

68. R. B. Ingle and E. H. Coulson, *Education in Science*, Vol. 61, 1975, p. 15.
69. S. K. Kellington and A. C. Mitchell, *An Evaluation of New Science Worksheets for Scottish Integrated Science*, London, Heinemann, 1978.
70. A. H. Johnstone, *Studies in Science Education*, Vol. 1, 1974, p. 21.
71. A. H. Johnstone and N. C. Kellet, *European Journal of Science Education*, Vol. 2, 1980, p. 175.
72. A. H. Johnstone, *Chemical Society Review*, Vol. 9, 1980, p. 365.
73. N. C. Kellet and A. H. Johnstone, *Education in Chemistry*, Vol. 11, 1974, p. 111.
74. Assessment of Performance Unit, Department of Education and Science (United Kingdom), (a) *Science Working Group Assessment of Scientific Development: A Consultative Document*, London, HMSO 1977; (b) *Mathematical Development Secondary Survey Report No. 1*, London HMSO, 1980; (c) *Science in Schools, Age 11, Report No. 1*, London, HMSO, 1981.
75. 'National Assessment of Educational Progress', *Science Technical Report: Summary Volume, Science Report—04-5-21*, Denver, Colo. (United States).
76. A. T. Porter and V. O. Younge, *Sierra Leone Review: All Our Future*, Freetown, University of Sierra Leone, 1976.
77. K. Frey, P. Haüssler, L. Hoffmann, J. Rost and H. Spada, 'Physics Education for Everyone in an Industrial Country,' in C. P. McFadden (ed.), *World Trends in Science Education*, p. 116, Halifax, Nova Scotia (Canada), Atlantic Institute of Education, 1980.
78. R. F. Kempa, *Research for the Classroom and Beyond: A Report of a Symposium*, p. 1, London, Education Division, The Chemical Society, 1978.
79. P. Stevens, *Journal of Philosophy of Education*, Vol. 12, 1978, p. 99.
80. J. Donnelly, *School Science Review*, Vol. 60, 1979, p. 489.
81. G. F. Williams, *Education in Chemistry*, Vol. 16, 1979, p. 102.
82. M. Richardson and C. Boyle, *What is Science? An Introduction to the Philosophy and Sociology of Science*, Hatfield, (United Kingdom), Association for Science Education, 1979. (Study Series, 15.)
83. N. J. Selley, *School Science Review*, Vol. 63, 1981, p. 252.
84. K. Popper, *The Logic of Scientific Discovery*, London, Hutchinson, 1959; *Conjectures and Refutations*, London, Routledge & Kegan Paul, 1963.
85. T. Kuhn, *The Structure of Scientific Revolutions*, Chicago, Ill., University of Chicago Press, 1970.

Appendices to Chapter 2

Appendix 1. Some useful sources of information about chemistry curricula throughout the world

International Newsletter on Chemical Education;[1*] *British Council Science Education Newsletter*;[2] *Annotated Bibliography for Science Education*;[3] *Reports of the International Clearinghouse*;[4] *Chemical Education in the Seventies*;[5] *European Journal of Science Education*;[6] *Bibliography of Chemistry Education Journals*;[7] various international conference reports, such as *International Conference on Chemical Education: Chemical Education in the Coming Decades* (held at Ljubljana, Yugoslavia, 1977);[8] *South East Asian Workshop on New Trends in Chemistry Teaching in Schools* (held at RECSAM, Penang, Malaysia, 1978);[9] *Chemical Education in Developing Countries* (held in Penang, Malaysia, 1979);[10] *International Conference on Chemical Education: Interaction between Secondary and Tertiary Levels* (held in Dublin, Ireland, 1979);[11] *International Symposium on World Trends in Science Education* (held in Halifax, Nova Scotia, Canada, 1980);[12] *Sixth International Conference on Chemical Education: Chemistry in a Diverse World* (held in Maryland, United States, 1981);[13] *New Trends in Chemistry Teaching*.[14]

* Notes to the appendices appear on pages 103–113.

Appendix 2. Chemical Bond Approach (CBA) in brief

Country of origin	United States
Age range	16–18
Date of inception	1959
Project Director	L. E. Strong
Most important texts	*Chemical Systems* and *Investigating Chemical Systems*[1]
Adoption in the United States	Actual adoption limited, but influence considerable
Influence in other countries	Well respected, but implementation inhibited, possibly because of its strong orientation towards physical chemistry
Further information	See below[2]
Revision	Revised 1964–77 with aim of decreasing the reliance on physics, and lowering the reading level of the materials.[3]

Appendix 3. Chemical Education Material Study (CHEM Study) in brief

Country of origin	United States
Age range	16–18. Also used as an introductory course at university level
Project Director	J. A. Campbell (to 1963); G. C. Pimental
Date of inception	1960
Most important texts	*Chemistry: An Experimental Science* (textbook, laboratory manual, teachers' guide)[1]
Adoption	Considerable in the United States[2]
Influence in other countries	Widespread and important,[3] particularly in Canada and Latin America. Books translated into 19 languages and films into 8: Chinese, Danish, French, German, Greek, Gujarati, Hebrew, Hindi, Icelandic, Italian, Japanese, Korean, Persian, Portuguese (Brazil), Portuguese (Portugal), Russian, Spanish (Latin America), Spanish (Spain), Swedish, Thai, Turkish
Further information	A very full account of the project is given in the book *The CHEM Study Story*[4]
Revision	No official revision as such, but three groups of authors were authorized to write texts based on CHEM Study in order to bring it up to date and to prevent it from becoming stereotyped.[5]

Appendix 4. Nuffield Chemistry (O-level) in brief

Country of origin	United Kingdom (England, Wales and Northern Ireland)
Age range	11–16
Project organizer	H. F. Halliwell
Date of inception	1961
Most important texts	*Introduction and Guide: The Sample Scheme Stages I and II: the Basic Course*; numerous background books[1]
Adoption	Adopted by many schools in the United Kingdom
Influence in other countries	Has had considerable influence in many countries of the world especially those of the British Commonwealth. Modified forms used in Kenya, Uganda and Malaysia. Also carefully studied by curriculum developers in a number of European countries. Translated into Japanese, Spanish, Italian and German. Some Background Books were translated into Farsi and Polish. Stage I of the Sample Scheme has been translated into French.
Further information	See below[2]
Revision	Revised Nuffield Chemistry is now published. Main texts: *Teachers' Guides*, I, II and III (3 volumes), *Handbook for Pupils*, *Chemists in the World*, and ten options. *Teachers Guides*, I and II are also published in German.[3]

Appendix 5. Content of the CHEM study course

Section	Chapter titles	Experiments	Films
An introduction to scientific activity	1. Chemistry: an experimental science	1. Scientific observation and description 2. Behaviour of solids on warming 3. Melting temperature 4. Combustion of a candle 5. Heat effects	
Some fundamental concepts of chemistry	2. A scientific model: the atomic theory	6. The weights of equal volumes of gases 7. Copper immersed in silver nitrate	*Gases and How They Combine*
	3. Chemical reactions	8. Mass relationships in chemical change	
	4. Gas phase: kinetic theory	9. Reaction of a metal with hydrochloric acid	*Gas Pressure and Molecular Collisions*
	5. Liquids and solids	10. Reacting volumes of solutions 11. Reactions between ions	*Electric Interactions in Chemistry*
	6. Structure of the atom and the periodic table		*Chemical Families*
Macroscopic view of chemical reactions: energy effects	7. Energy effects 8. Rates of chemical reactions	13. Heat of a reaction 14. A study of reaction rates	*Molecular Motion* *Introduction to Reaction Kinetics* *Catalysis*

reaction rates			
equilibrium	9. Equilibrium in reactions	15. Chemical equilibrium	*Equilibrium*
solubility	10. Solubility equilibria	16. Solubility product measurement	
acids and bases	11. Acids and bases		
oxidation–reduction calculations		17. Heat of acid–base reactions	
		18. Hydrogen ion concentration from indicators	*Acid–base Indicators*
		19. Le Chatelier's principle	
	12. Oxidation–reduction reactions	20. Introduction to oxidation–reduction	
		21. Electrochemical cells	*Electrochemical Cells*
	13. Chemical calculations	22. Reactions between ions	
		23. Quantitative titration	
A microscopic view of substances:	14. Why we believe in atoms	24. Black box experiment	
		25. Electrolysis: weight relations	
atoms			
periodic table	15. Electrons and the periodic table		*Hydrogen Atom as Viewed by Quantum Mechanics*
molecules			
solids and liquids			*Ionization Energy*
	16. Molecules		*Chemical Bonding*
		26. Properties of *cis-trans* isomers	*Shapes and Polarities of Molecules*
	17. Solids and liquids	27. Packing of atoms or ions in a crystal	*Crystals and Their Structures*
Descriptive chemistry	18. Carbon compounds	28. Reactions of hydrocarbons and alcohols	*Mechanism of an Organic Reaction*
		29. Preparation of organic acid derivatives	*Synthesis of an Organic Compound*
	19. Halogens	30. Electrolysis of KI	*Bromine–Element from the Sea*
		31. Chemistry of iodine	

Section	Chapter titles	Experiments	Films
	20. Third row	32. Chemistry of third row elements	
	21. Second column	33. Qualitative analysis	
	22. Fourth row transition elements	39. Preparation of a complex salt	*Vanadium, A Transition Element*
	23. Sixth and seventh row elements		*Transuranium Elements*
	24. Biochemistry		*Biochemistry and Molecular Structure*
	25. Chemistry of earth, planets, and the stars		

Supplementary experiments: 12. A study of reaction; 34, 35, 36. Qualitative analysis; 37. Anion exchange resin separations; 38. Corrosion of iron; 40. Preparation of potassium dichromate; 41. Preparation of chrome atom.

Supplementary films: Vibrations of Molecules; *Molecular Spectroscopy*; *Molecular Motions*; *High Temperature Research*; *Nitric Acid*; *A Research Problem: Inert(?) Gas Compounds*. See also 'MARS: Chemistry Looks for Life in a New CHEM Study Film', described in *European Journal of Science Education*, Vol 2, 1980, p. 328. (Versions with English and German commentaries available.)

Source: G. C. Pimental and D. W. Ridgeway, 'CHEM Study: Knowledge of Chemistry'. *Science Activities*, Vol. 43, 1972.

Appendix 6. Examples of nationally based chemistry projects for students up to the age of about 16

Country	Project	Age range
Brazil	Unidades Modulares de Química	15–17[1]
Cuba	New chemistry programme	13–15[2]
Federal Republic of Germany	Institute of Pedagogy and Science:	
	IPN Chemistry 5/6	10–11[3]
	IPN Chemistry 8/13	14–19
Israel	Chemistry in Modern Society	14–16[4]
Japan	Modernization of Secondary School Chemistry	15–17[5]
Kenya	School Science Project in Chemistry	12–16[6]
Malaysia	Modern Chemistry for Malaysian Schools	15–17[7]
Netherlands	Experimental Chemistry for the Secondary Schools	14–17
Nigeria	Nigerian Secondary Schools Science Project in Chemistry	13–16[8,9]
Sri Lanka	Chemistry Curriculum Development Project	14–15+[10]
Uganda	School Science Project: Chemistry	12–16[11]
United Kingdom (England, Wales and Northern Ireland)	Nuffield O-level Chemistry	11–16[12]
(Scotland)	Alternative Syllabus in Chemistry for Secondary Schools	12–17[13]

Appendix 7. Nuffield Chemistry (A-level) in brief

Country of origin	United Kingdom (England, Wales and Northern Ireland)
Age range	16–18
Project Director	E. H. Coulson
Date of inception	1965
Most important texts	*Students' Books 1 and 2, Teachers' guides 1 and 2, The Chemist in Action*, and the Special Studies[1]
Adoption	Widespread in the United Kingdom
Influence in other countries	Influential, although not so widely adapted as the O-level project
Further information	See below[2]
Revision	Revised edition will be published in 1984[3]

Appendix 8. Examples of advanced chemistry projects for students

Country	Project	Age range
Argentina	Enseñanza Actualizada de la Química	17[1]
Australia	Australian Academy of Science School Chemistry Project	16–18[2]
Austria	New Chemistry Curriculum (Project Group Chemistry)	15–18[3]
Canada	Alberta Chemistry Education Project	15–18[4]
	Turn On Chemistry Program for Non-Science Major Students	16–17[5]
Colombia	Improvement of Science Teaching in Colombia	14–18[6]
Cuba	New chemistry programme	15–18[7]
Federal Republic of Germany	Institute of Pedagogy and Science (IPN): *Curriculum Chemistry 8/13*	14–18[8]
	Hesse chemistry curriculum	16–19[9]
India	National Council of Educational Research and Training	16–18[10]
Indonesia	New chemistry programme	15–18[11]
Israel	Experimental Chemistry Programme for Secondary Schools	16–18[12]
Republic of Korea	Korean Science Education Project	6–18[13]
Thailand	The Institute for Promotion of Teaching Science and Technology (Grades 11 and 12 Chemistry)	16–18[14]
United Kingdom (England, Wales and Northern Ireland)	Nuffield Advanced Chemistry	16–18[15]
	Independent Learning project for Advanced Chemistry	16–18[16]
United States of America	Chemical Bond Approach	16–18[17]
	Chemical Education Material Study	16–18[18]
	Interdisciplinary Approaches to Chemistry	16–18[19]
Unesco	Pilot Project for Chemistry Teaching in Asia	16–18[20]

Appendix 9. Patterns of school science programmes in Arab countries (1973)

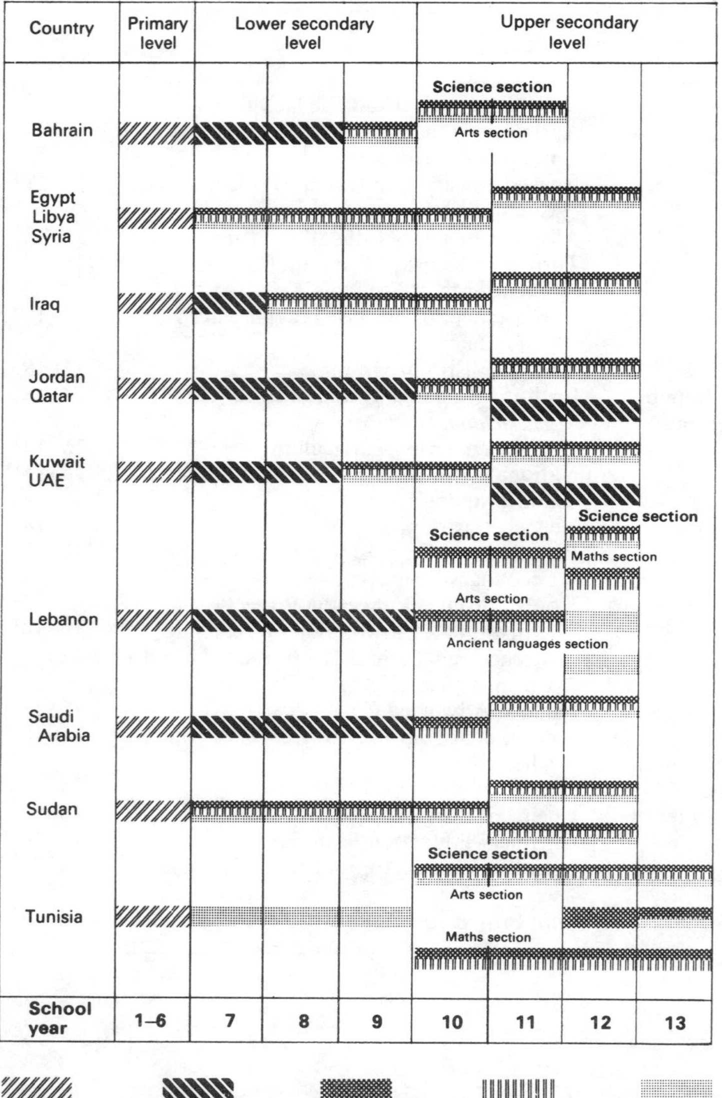

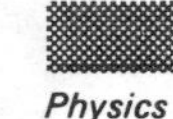

Primary science *General science* *Physics* *Chemistry* *Natural history (or biology)*

After J. E. Arrayed, *A Critical Analysis of School Science Teaching in Arab Countries*, Beirut, Longman Arab World Centre, 1980. (© Librairie du Liban 1980, reproduced with permission.)

Appendix 10. Examples of physical science projects

Country	Project	Age range
Argentina	Introductory Physical Science	15[1]
Australia	Physical Science—Man and the Physical World	16–17[2]
Denmark	*Ask Nature*	12–15[3]
Fiji	Physical Science	11–16[4]
France	Initiation aux science et techniques	11–16[5,6]
Indonesia	Physical Science	
Israel	Physics Chemistry Project	12–17[7]
Morocco	Material and Documentation Centre	12–14[8]
Norway	Active Physics Chemistry Project	13–16[9]
Pakistan	Science Education Project at Institute of Education and Research	14–16[10]
Papua New Guinea	Physical Science[11]	
Thailand	Physical Science[12]	
United Kingdom (England, Wales and Northern Ireland)	Nuffield Physical Science	16–18[13]
United States of America	Introductory Physical Science	12–14[14]
	Physical Science II	12–17[15]
Africa and United Kingdom	*Beginning Physical Science*	11–13[16]

Appendix 11. Examples of integrated science projects

Country	Project	Age range
Arab States	Integrated Sciences at the Intermediate Schools in the Arab States	13–15[1]
Australia	Australian Science Education Project	12–15[2]
Bostswana	Science by Investigation in Botswana	12–15[3]
Botswana, Lesotho and Swaziland	Boleswa Integrated Science	12–15[4]
Brazil	Integrated Science	15–16[5]
Canada	*Science: A Way of Knowing*	15–16[6]
Caribbean	Integrated Science for the Caribbean	12–15[7]
	MONA Integrated Science Project	12–15[8]
Ethiopia	New science curriculum	14–15[9]
Federal Republic of Germany	Integrated Natural Science Curriculum	12–15[10]
Ghana	Project for Science Integration	12–14[11]
India	Integrated Science	13–15[12]
Italy	The Structure of Matter	14–15[13]
Jamaica (see Caribbean)		
Japan	Studies on the Integrated Science Curriculum	12–17[14]
Lesotho	Science Curriculum Development Programme	12–15[15]
Malaysia	Integrated Science	11–14[16]
Nigeria	Science Teachers' Association of Nigeria (STAN) Curriculum Development Project	12–14[17]
Sierra Leone	Core Course Integrated Science	13–15[18]
Swaziland	Integrated Science Project	12–15[19]
Thailand	The Vocational Science and Mathematics Curriculum Project	16–19[20]
Trinidad (see Caribbean)		
United Kingdom (England, Wales and Northern Ireland)	Schools Council Integrated Science Project	13–16[21]
	Wreake Valley School Science Department	11–14[22]
(Scotland)	Scottish Integrated Science	12–14[23]
	Science in S1 and S2	12–14[24]
United States of America	Intermediate Science Curriculum Study	12–15[25]
	Integrated Science for Independent Study	16–18[26]
	Foundation Approaches to Science Teaching	11–15[27]
Unesco	Programme in Integrated Science Teaching	6–16[29]

Appendix 12. Examples of resources on the chemical industry and applied chemistry

Africa	The reference volume *Africa*[1] *69/70* contains accounts of mining, energy and petrochemical developments in various parts of the continent. See also *Chemical Technology in Africa*[2]
Australia	
Australian Science Education Project	Units such as 'Petroleum', 'Sticking together' and 'Metals'[3]
Canada	
Action Chemistry Series	A series including *Chemistry of Man and Molecules*, *Chemistry of Photography*, *Chemistry of the Car*, *Water*, *Chemistry Ecology and Chemistry of Metallurgy*[4]
Ghana	Soap, caustic soda, stabilization of rubber latex, fireproofing of hemp fibres,
Guinea-Bissau	Soap[5]
Israel	Contains work on fertilizers, pesticides,
Chemistry for the High School	petroleum and rocks and minerals.[6] Various examples[7]
Kenya	Extraction and uses of tin[8]
Malaysia	Soap[5]
Mali	*Chemical Processes in New Zealand* is a com-
New Zealand	prehensive volume (409 pp.) describing most industrial processes in the country.[9]
Papua New Guinea	
Physical Science Project	Interesting scheme for grade 10 on copper, fuels, lime, water, soaps and detergents with well-integrated practical work[10]
Philippines	Self-learning materials on applied topics in chemistry including desalination, biogas, tanning[11]
Sierra Leone	Soap[5]
United Republic of Tanzania	'Salt Production in Tanzania'[12]
Thailand	
(a) Physical Science	Units on medical chemistry, synthetic materials, agricultural chemistry, natural resources, energy[13]
(b) Vocational Science and Mathematics Curriculum Project	Units on metals and corrosion, fuels and lubricants, food additives[14]
Tonga	Ethanol from cassava, charcoal from timber, soap from coconut oil, biogas[15]
Uganda	'Chemical Industry in Developing Countries'[16] 'Iron and Education in Uganda'[17]

United Kingdom (England, Wales and Northern Ireland)	
(a) Nuffield O-level Chemistry	Background Books: *Petroleum*, *Plastics, Fertilisers and Farm Chemicals*, and many others[18]
(b) Revised Nuffield Chemistry	*Chemists in the World* and *Handbook for Pupils*, especially Part 2, 'Chemistry and our lives'[19]
(c) Nuffield Advanced Chemistry	*The Chemist in Action* and Special Studies *Food Science*, *Biochemistry*, *Chemical Engineering*, *Ion Exchange* and *Metallurgy*[20]
(d) School Science Review	'The Chemistry of a Mini' See diagram below[21]
(Scotland)	Interactive Teaching Packages[22]
Zambia	
Curriculum Development Centre	Eleven guides for teachers form a major resource on the country's chemical industry. Titles include the production of lime, industrial gas, ammonium nitrate, cement, treated water, pyrites, soap, detergents, edible fats and beer[23]
General	Unesco, 'Chemistry and Industry'[24] Third World Science Group[25]

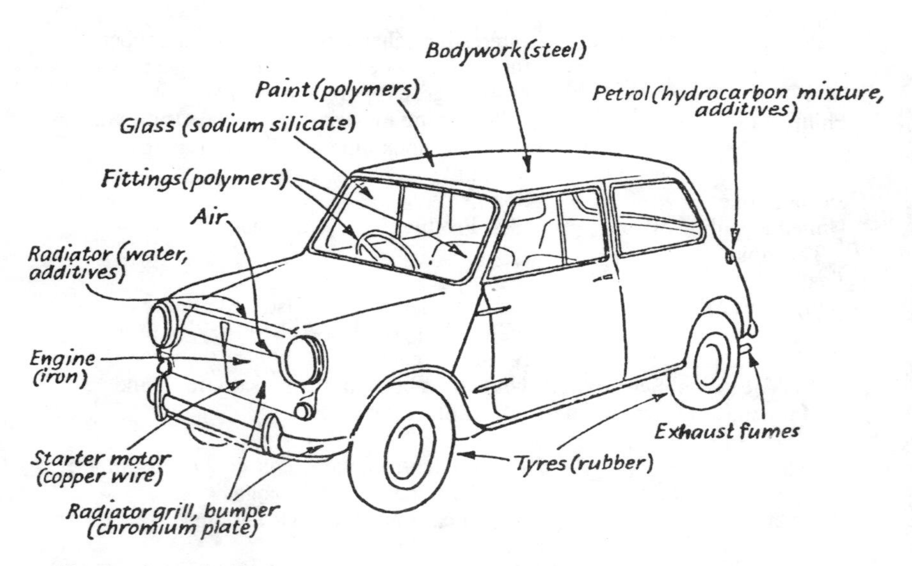

The chemistry of a Mini.

Appendix 13. Examples of projects and texts containing material of social relevance

Country/Project	Age range	Notes
Australia		
Physical Science—*Man and Physical World*	16–18	See unit 'Science and Society'[1]
Chemistry and the Market Place	14–18	Chapters on chemistry in the laundry, kitchen, boudoir, garden, etc.[2]
Canada		
Science: A Way of Knowing	16–18[3]	
Caribbean		
Science and People in the Caribbean	14–16[3]	
Israel		
Chemistry for the High School[4]		
Chemistry in Modern Society	14–16[5]	
United Kingdom (England, Wales and Northern Ireland)		
Revised Nuffield Chemistry	11–13	See Study Sheets: *Fresh Air?* and *Water* for examples of a simple approach to social relevance[6]
	13–16	*Handbook for Pupils*, chapter 'Man, Chemistry and Society'[6]
Schools Council Integrated Science Project	13–16[3]	
Nuffield Advanced Chemistry	16–18	Special Studies[3]
Joint Matriculation Board, Manchester (JMB)	16–18	Alternative Advanced Level Syllabus. Chemistry Case Studies[7]
Working with Science	16–18	A set of some twenty-five booklets on such topics as the motor car, beer making, dealing with prejudice and supersition, interpreting statistics, and so on. Intended for use with students who are not academically minded[8]
Association for Science Education: Science in Society	16–17	Material consists of a teachers' guide, student readers and decision-making simulation exercises[9]
Science in a Social Context (SISCON) SISCON—in schools	16–18	Devised first for use in universities, but now adapted for use in schools[10]

Country/Project	Age range	Notes
United Kingdom (Scotland)		
Interactive Teaching Packages	14–18	Intended for use with existing chemistry courses[11]
United States of America		
Interdisciplinary Approaches to Chemistry (IAC)	16–18	Modules can be used together as a course or singly to enrich other courses[12]
Intermediate Science Curriculum Study	12–15	
Integrated Science for Independent Study	16–18[3]	
Chemistry in the Community	16–18	New programme under development to link chemistry closely to everyday life[13]
Combating the Hydra	13–15	Four teaching units—fibres pesticides, energy, food additives, written from the point of view of their impact on society (cf. Appendix 12)[13]
Chemistry, Man and Society	16–18	Topics include air pollution, consumer chemistry, nuclear chemistry and energy conservation[14]
Chemistry, Man and Environmental Change	16–18	Topics include global air pollution, water pollution, environmental contamination with heavy metals, radioactivity and the environment[15]
Venezuela		
The Environment: a living resource for chemistry	16–18	Involves a study of chemistry in the environment, with project work[16]
General		
Social content in chemistry courses[17]		

Notes to Appendices

Appendix 1

1. Published by the Committee on the Teaching of Chemistry of the International Union of Pure and Applied Chemistry (IUPAC/CTC) and available from the IUPAC Secretariat, Bank Court Chambers, 2–3 Pound Way, Cowley Centre, Oxford OX4 3YF (United Kingdom).
2. Available from Education, Medicine and Science Division, The British Council, 10 Spring Gardens, London SW1A 2BN (United Kingdom), and also through their representatives in many countries of the world.
3. Published at the School of Education, University of the West Indies. The first number was published in January 1980. Editor: J. Reay.
4. Published by International Clearinghouse, Science Teaching Centre, University of Maryland, College Park, Md. (United States); 4th report, 1966; 5th, 1967: 6th, 1968; 7th, 1970; 8th, 1972; 9th, 1975; 10th, 1977. Editor: J. D. Lockard.
5. Available from Pergamon Press Ltd, Oxford OX3 OBW (United Kingdom), 2nd ed. 1982. Editors, A. Kornhauser, C. N. R. Rao and D. J. Waddington.
6. Published by Taylor and Francis Ltd, 10–14 Macklin Street, London WC2B 5BF, in collaboration with Institute for Science Education (IPN), Kiel (Federal Republic of Germany). Editors: K. Frey, R. Kempa, G. Delacôte, P. Guidoni, K. Keohane, G. Schaefer and R. Sexl.
7. Available from IUPAC (see Note 1 for address). Editors: J. N. Lazonby and D. J. Waddington.
8. Published by DDU Univerzum, Ljubljana (Yugoslavia). Editor: A. Kornhauser.
9. The final report may be obtained from the Regional Centre for Education in Science and Mathematics (RECSAM), Penang (Malaysia). It contains articles on chemistry curricula in Indonesia, Malaysia, Philippines, Singapore, Thailand.
10. Published by the Committee on Science and Technology in Developing Countries, 1979. Inquiries should be addressed to the International Council of Scientific Unions (ICSU), 51 Bd de Montmorency, 75016 Paris (France). Editors: C. N. R. Rao and S. Radhakrishna.
11. Available from P. Start, Department of Chemistry, University College, Dublin (Ireland). Editors: P. E. Childs and J. E. Gowan.
12. *World Trends in Science Education*. Halifax, Nova Scotia (Canada), Atlantic Institute of Education, 1980. Editor: C. P. McFadden.
13. Published by the Department of Chemistry, University of Maryland, College Park, Md. (United States). Editor: W. T. Lippincott.
14. *New Trends in Chemistry Teaching/Tendances nouvelles de l'enseignement de la chimie*. Vol. I, 1967; Vol. II, 1969; Vol. III, 1972; Vol. IV, 1975 containing an article by M. J. Frazer, 'Up-to-date and Precise Learning Objectives in Chemistry', pp. 43–53, which makes many points of comparison between chemistry projects at both secondary and tertiary levels in a number of countries; Vol. V, 1981. Published by Unesco, Paris (France).

Appendix 2

1. Text, teachers' guide, laboratory guide and examinations published by Webster Division of McGraw-Hill, New York.
2. See following articles: A. H. Livermore and F. L. Ferris, 'The Chemical Bond Approach Course in the Classroom', *Science*, Vol. 138, 1962, p. 1077; R. W. Heath and D. W. Stickell, 'CHEM and CBA Effects on Achievement in Chemistry', *The Science Teacher*, Vol. 30, 1963, p. 45; S. B. Oyanedel, *Revista Chilena de Educación Quimica*, Vol. 3, 1978, p. 196.
3. See *Tenth International Clearinghouse Report*, p. 253. New texts: *Chemical Systems II*, 1976, and *Investigating Chemical Systems II*, 1977.

Appendix 3

1. Published by W. H. Freeman, San Francisco, Calif., who are still distributing the original materials.
2. D. W. Ridgway and G. C. Pimental, 'CHEM Study—Its Impact and Influence', *High School Journal*, Vol. 53, 1940, p. 216.
3. R. J. Merrill and D. W. Ridgway, *The CHEM Study Story*, San Francisco, Calif., Freeman; 1969.
4. See, for example, V. Mondler and J. Silberstein, 'The CHEM Study Adaptation in Israel', in P. Tamir et al. (eds.), *Curriculum Implementation and its Relationship to Curriculum Development in Science*, pp. 301–3, Jerusalem, Israel Science Teaching Centre, 1979; G. C. Pimental and D. W. Ridgway, 'CHEM Study: Knowledge of Chemistry', *Science Activities*, Vol. 43, 1972, pp. 40–5.
5. F. A. Cotton and L. D. Lynch, *Chemistry: An Investigative Approach*, Boston, Mass., Houghton Mifflin, 1968, 1970; and R. W. Parry, P. M. Dietz, R. L. Tellefsen and L. E. Steiner, *Chemistry: Experimental Foundations*, Englewood Cliffs, N.J., Prentice-Hall, 1970, 1975 with supplementary materials; and P. R. O'Connor, J. E. Davis, E. L. Haenisch, W. K. MacNab and A. L. McClellan, *Chemistry: Experiments and Principles*, Lexington, Mass., D. C. Heath, 1977.

Appendix 4

1. Published by Longman/Penguin, but out of print since the publication of *Revised Nuffield Chemistry*.
2. G. Van Praagh, 'The Nuffield Chemistry Project: Education through Chemistry', *New Trends in Chemistry Teaching*, Vol. 1, p. 328, Paris, Unesco, 1967. G. Van Praagh, 'Relevance of Nuffield Chemistry in Developing Countries', *New Trends in Chemistry Teaching*, Vol. III, p. 270, Paris, Unesco, 1972. For a description of the first edition of *Nuffield Chemistry* see Longman/Penguin Books, 1966. For a discussion and comprehensive references see R. B. Ingle and A. J. Jennings, *Science in Schools: Which Way Now*?, pp. 21–30, London, University of London, Institute of Education/Heinemann Educational, 1981.
3. *Revised Nuffield Chemistry* is published by Longman for the Nuffield Foundation, London (United Kingdom). The publications are listed in Chapter 1 of *Teachers' Guide I*, 1975, and the background to the work of the revision is explained in Chapter 5.

Appendix 6

1. Unidades Modulares de Química is based at the Centro de Treinamento para Professôres de Ciências de São Paulo and is supported by the Ministry of Education. The course is designed as a modern approach to the teaching of chemistry avoiding the division of 'general', 'organic' and 'inorganic'. An attempt has been made to link chemistry with cultural, economic and social factors. The project has the following units: *Unit I—The chemist's tools:* chemical reactions are sources of energy and sources of materials; predictions about chemical reactions are possible through the knowledge of atomic structure and chemical bonding; *Unit II—Chemical reactions—sources of energy:* most of the energy we use comes from chemical reactions; main combustibles employed; combustibles and environment; alternative sources of energy; *Unit III—Chemical reactions—a compromise between reactants and products:* how and why we change the rate of chemical reactions; the importance of the knowledge of the mechanism of chemical reaction; chemical equilibrium; *Unit IV—Chemistry—some fields of action:* chemistry in agriculture; chemistry in pharmacy; soap and synthetic detergents; electrochemical cells; nuclear reactions application.

   The published material consists of four students' units and one teachers' guide: A. Ambrogi, *Unidades Modulares de Química*, Brasília, MEC/SEPS, 1980. *Química—*

Unidades Modulares, I. Título; II. Versolato, Helena F.; III. Lisbôa, Julio Cézar Foshini; IV. CESISP, São Paulo.

2. L. C. Perez, 'The Renovation of Science Education in Socialist Cuba', in C. P. McFadden (ed.), *World Trends in Science Education*, p. 42, Halifax, Nova Scotia (Canada), Altantic Institute of Education, 1980.
3. Institut für de Pädagogik der Naturwissenschaften an der Universität Kiel, Olshausenstrasse 40–60, D-2300 Kiel (Federal Republic of Germany). See *Ninth International Clearinghouse Report*, p. 162, for a comprehensive list of descriptive and research references; and K. Riquarts, 'Evaluation of IPN Science Curricula for Grades 5 and 6 in the Federal Republic of Germany', in P. Tamis et al. (eds.), *Curriculum Implementation and its Relationship to Curriculum Development in Science*, Jerusalem, Israel Science Teaching Center, 1979. See also the IPN texts *Stoffe und Stoffumbildungen: ein Weg zur Atomhypothese*, Stuttgart, Ernst Klett, 1979.
4. Israel Science Teaching Center, Hebrew University, Jerusalem (Israel). See S. Novick and F. X. Sutman, 'A Socially Oriented Approach Through Carbon Compound Chemistry', *The Science Teacher*, Vol. 40, 1973, p. 50.
5. Department of Chemistry, Faculty of Science, The University of Tokyo, Tokyo 113 (Japan). See Y. Hayashi, K. Ide, T. Nishihira, Y. Takebayashi, Y. Yoshida and T. Shimozawa, *Kagaku-Kyoiku*, Vol. 19, 1970, p. 137 (in Japanese), and *Ninth International Clearinghouse Report*, p. 45.

   The textbooks have been used in seven schools and the experiments tried out in twenty-two schools. Professor Y. Tanizaki (Nagaoka University of Science and Technology) revised and published *Kagaku no Kiso* (Fundamentals of Chemistry), Tokyo, Maruzen. See also J. T. Shimozawa, *Education in Chemistry*, Vol. 19, 1982, p. 104.
6. See Section 2.3. The following materials were based on the School Science Project Chemistry course: J. V. Binns, D. A. Futcher, S. Pardham, J. S. Rank, J. W. Steward and P. J. Towse, *Introducing Chemistry* (pupils' books and teachers' guides), London, Edward Arnold, 1974.
7. See Section 2.4.1. For an overview see: Hamdan, in C. N. R. Rao and S. Radhakrishna (eds.), *Chemical Education in Developing Countries*, p. 2, COSTED, 1979; S. P. Koh and S. E. Loke, ibid., p. 120; S. P. Koh and S. E. Loke, in A. Kornhauser, C. N. R. Rao and D. J. Waddington (eds.), *Chemical Education in the Seventies*, 2nd ed., p. 191, Oxford, Pergamon Press, 1982; S. P. Koh and S. E. Loke, in P. E. Childs and J. E. Gowan (eds.), *Proceedings of the International Conference on Chemical Education, Dublin, 1979*, p. 167, 1980; and *Final Report of the South East Asian Workshop on New Trends in Chemistry Teaching in Schools*, p. 53, Penang (Malaysia), RECSAM, 1978. The curriculum materials are published under the title: *Modern Chemistry for Malaysian Schools*, two pupils' books and two teachers' guides, Kuala Lumpur (Malaysia), Longman, 1973.
8. The Nigerian Secondary Schools Science Project (NSSSP) which was developed by the Comparative Education Study and Adaptation Centre (CESAC). The project was developed over a period of ten years from 1970, for pupils in the age range 13–15 with average or above average ability. Not only was there an attempt to stress the relationship of chemistry to the students' culture and environment, but also to make it practically based. Students' and teachers' books have been prepared, together with a special examination. Only a relatively small fraction of the school chemistry population take the programme because most school laboratories are ill-equipped and only a few schools were able to afford the change without direct support from CESAC.
9. Comparative Education Study and Adaptation Centre, University of Lagos, Nigerian Secondary Schools Science Project, three students' books and a teachers' guide, Ibadan (Nigeria), Heinemann Educational, 1980.
10. For details: Curriculum Development Centre, 255, Banddhaloka Mawata, Colombo 7 (Sri Lanka).
11. See Section 2.3.
12. See Appendix 4.

13. Scottish Education Department, St Andrew's Avenue, Edinburgh, Scotland. The books by A. H. Johnstone and T. I. Morrison, *Chemistry Takes Shape*, London, Heinemann Educational, 1970 (a series of five students' books and five teachers' guides for the 12–17 age range) have exerted wide influence particularly in Scotland.

Appendix 7

1. Published by Longman, London (United Kingdom).
2. See, for example, E. H. Coulson, 'Nuffield Chemistry at A-level', *Education in Chemistry*, Vol. 6, 1969, p. 200, and E. H. Coulson, 'Nuffield Advanced Science—Chemistry: An Account of Stewardship', *School Science Review*, Vol. 52, 1970, p. 261. For further discussion see R. B. Ingle and A. J. Jennings, *Science in Schools: Which Way Now?*, pp. 30–1, London, University of London Institute of Education/ Heinemann Educational, 1981 and the comprehensive list of references cited therein.
3. See Section 2.4.1. For further details, B. J. Stokes, King's College School, Wimbledon, London SW 19 (United Kingdom).

Appendix 8

1. Instituto Nacional para el Mejoramiento de la Enseñanza de las Ciencias (INEC), Av. Eduardo Madero 235–7° piso, Buenos Aires (Argentina). Materials in Spanish. See *Ninth International Clearinghouse Report*, p. 171.
2. The Australian Academy of Science established a new chemistry project in November 1979 directed especially at the needs of those who will have no formal training in chemistry after leaving school. It is hoped that the new course will help the student to understand the role of chemistry and chemical technology in daily life and to understand the chemical basis of community problems such as the conservation and use of natural resources, the energy crisis and environmental problems. It will be much less directed towards some of the more complex conceptual and theoretical facets of chemistry than existing chemistry courses. Copies of an explanatory report may be obtained from the Australian Academy of Science, P.O. Box 783, Canberra City, ACT 2601. Further information from Professor A. R. H. Cole, University of Western Australia, Nedlands, Western Australia 6009.
3. E. Jarisch, *Proceedings of the Sixth International Conference on Chemical Education*, edited by W. T. Lippincott, College Park, Md., University of Maryland, 1982. Among the interesting points is the introduction of concepts of enthalpy and entropy giving the direction of chemical change, including living systems. There are sections on 'Products of Economic Importance' and 'Man and the Environment'.
4. See Section 2.5.
5. See *Ninth International Clearinghouse Report*, p. 191.
6. Director, Instituto de Ciencias, Improvement of Science Teaching in Colombia–Carrera 13, no. 38–83, Bogota (Colombia). Materials in Spanish. See *Seventh International Clearinghouse Report*, p. 68.
7. See Section 2.4. and Appendix 6.
8. Institute for Science Education (IPN), D-2300 Kiel, Olshausenstrasse 40–60 (Federal Republic of Germany). See *Ninth International Clearinghouse Report*, p. 163, for list of descriptive and research references (in German).
9. See Section 2.5.
10. See Section 2.4.2.
11. This is part of a new Indonesian school curriculum, introduced in 1975. The course is student-centred. See *Final Report of the South East Asian Workshop on New Trends in Chemistry Teaching in Schools*, p. 39, Penang (Malaysia), RECSAM, 1978. See also Section 6.4 of this book for a description of teacher in-service training.
12. See *Eighth International Clearinghouse Report*, p. 368, for a description of this project which is an adaptation of CHEM Study (Appendix 3). The group at the Weizmann Institute has been developing modules relevant to industry (Appendix 12).

13. Science Education Project, Room 1012, Unified Government Building, Seoul (Republic of Korea). Materials in Korean. See Evaluative Research on Unesco/UNICEF assisted Korean Science Education Project, 1973, The Central Education Research Institute, Seoul (Republic of Korea), and *Ninth International Clearinghouse Report*, p. 50.
14. The Institute for Promotion of Teaching Science and Technology (IPTST), 924 Sukhumvit Road, Bangkok 11 (Thailand). Materials in Thai. See Section 2.4.2, and articles in the *Final Report of the South East Asian Workshop on New Trends in Chemistry Teaching in Schools*, pp. 79 and 110, Penang (Malaysia), RECSAM, 1978; see also *Science Education News* (Washington, D.C., American Association for the Advancement of Science), October 1973; R. T. White, W. L. Butts, 'Curriculum Development in Thailand', *Australian Science Teacher*, Vol. 21, 1975, p. 47, and Institute for Promotion of Teaching Science and Technology, *A Case Study ACEID*, Bangkok (Thailand), Unesco Regional Office of Education.
15. See Section 2.4.2 and Appendix 6.
16. See Section 2.5.
17. See Section 2.2 and Appendix 2.
18. See Section 2.2.4 and Appendix 3.
19. See Section 2.5 and Appendix 13. An interesting account of the use of IAC materials for individualized learning is given by Shirley E. Richardson, *Journal of Chemical Education*, Vol. 58, 1981, p. 1007.
20. See Section 2.3.

Appendix 10

1. *Ninth International Clearinghouse Report*, p. 173.
2. Materials available from VUSEB, 437 St Kilda Road, Melbourne 3006 (Australia), under the title *Physical Science: Man and the Physical World*.
3. Students' texts: *Ask Nature I: Practical Electricity*; *Ask Nature II: Matter and the Structure of Matter*; *Ask Nature III: Chemistry of Everyday Life*; *Ask Nature IV: Motion and Energy*; Teachers' guides for each volume, all published by Gyldendal, Oslo (Norway). See P. Thomsen, 'The Role of Books in Physics Teaching', *Physics Education*, Vol. 10, 1975, p. 69.
4. Examples of materials produced by the project include Causing and Controlling Change, Radiant Energy, Inside the Atom, Rocks and Soil, Transportation, Resource Management.
5. Documents are published by LIRESPT, Université Paris VII, Tour 23, 2, Place Jussieu, 65221 Paris Cédex 05 (France). These include students' and teachers' guides on 'Module Chimie', and a teachers' guide and technical data on 'Module Polymères'.

   The origin of the project dates from 1971 and recommendations of a national commission known as the Commission Lagarrigue. Trials were carried out throughout the 1970s and in October 1981 the first pupils who have followed the whole programme left the *colleges* (schools). The curriculum is intended to provide an introduction to physics, chemistry and technology and has the following aims: to provide an experimental basis for science education to illustrate science in everyday life; to introduce pupils to the fundamentals of physics and chemistry and to give students a better 'orientation'. See, for example, R. Viovy and J.-L. Martinand, *Proceedings of the Sixth International Conference on Chemical Education*, edited by W. T. Lippincott, College Park, Md., University of Maryland, 1982.
6. The science course in Indonesia in grades 7–9 is divided into three areas: physical, biological and earth sciences. See the *Final Report of New Trends in Chemistry Teaching in Schools*, p. 39, Penang (Malaysia), RECSAM, 1978.
7. Inquiries to Curriculum Centre, Ministry of Education and Culture, Jerusalem (Israel). Students' materials are in Hebrew.
8. See *Eighth International Clearinghouse Report*, p. 15.
9. Based at University of Trondheim, Norges Laererhogskole, 700 Trondheim

(Norway). Inquiries on materials from Gyldendal Norsk Forlag, Oslo (Norway).
10. See *Eighth International Clearinghouse Report*, pp. 102 and 105.
11. See ref. 10, Appendix 12.
12. S. Wongthonglour, Ph.D. thesis, University of Monash, 1980. Also, details may be obtained from IPST, ref. 11, Appendix 8. The team leader of this three-year programme for junior secondary schools is a chemist. The discipline is strongly represented in the largest science programme at this level in the country. Practical experiments and basic concepts related to Thai culture (e.g. environmental, agricultural and food chemistry) are prominent. There are twelve modules from which teachers can choose the most appropriate. Among the most prominent chemical themes are those concerned with food, energy, medicine and synthetic materials.
13. The only physical science project in the United Kingdom at this level. For a general description of the scheme see Nuffield Advanced Physical Science, *Introduction and Guide*, Penguin, Harmondsworth, Middlesex, 1972 and J. E. Spice, 'The Nuffield Physical Science', *Education in Science*, Vol. 31, 1969, p. 38; and J. R. L. Swain, 'Project Work in Nuffield Physical Science', *School Science Review*, Vol. 58, 1977, p. 570.
14. Introductory Physical Science is intended for use with a wide range of abilities in grades 8 and 9, but it has also been used for students in grades 11 and 12 who do not plan to take further physics or chemistry. Published by Prentice-Hall, Englewood Cliffs, N.J. (United States). Translated into French (Montreal, Canada), Italian, Japanese, Korean, Philippine, Portuguese, Spanish and Turkish. This course developed in the 1960s leans more towards chemistry than to physics. There are a series of laboratory experiments and related discussion to present essential concepts.
15. Physical Science II is designed as a sequel to Introductory Physical Science for use with grades 9–12 of all abilities, except the non-readers.
16. An introductory physical science course written for use in Africa and in multi-racial schools in the United Kingdom. M. Atherton, P. Namasaka and D. Sloan, *Beginning Physical Science*, students' book and teachers' guide, London, John Murray, 1976.

Appendix 11

1. Science Department, ALECSO, Dokki Square, Cairo (Egypt). See *Tenth International Clearinghouse Report*, pp. 80–1.
2. See Section 2.7.3. Curriculum Development Centre, Australian Department of Education, 450 St Kilda Road, Melbourne, Victoria (Australia). See A. M. Lucas, 'ASEP—A National Curriculum Development Project in Australia', *Occasional Paper*, 1971, SMEAC, Eric Center, Ohio (United States); and L. G. Dale, 'Australian Science Education Project (ASEP)', *Science Education News*, Washington, D.C., American Association for the Advancement of Science.
3. Ministry of Education, P.O. Box 439, Gaborone (Botswana). Textbooks published by Heinemann, London, based on Scottish Integrated Science course 'Science for the Seventies'. *British Council Science Education Newsletter*, Vol. 38, 1979, p. 8.
4. Ministries of Education, Botswana, Lesotho and Swaziland, *Boleswa Integrated Science*, Books 1 and 2, London, Heinemann Educational, 1979.
5. Centro de Treinamento para Professôres de Ciências Exatas e Naturais de São Paulo—CECISP, Caixa Postal 11.324, São Paulo SP (Brazil). Materials in Portuguese.
6. G. S. Aikenhead, in C. P. McFadden (ed.), *World Trends in Science Education*, Halifax, Nova Scotia (Canada), Atlantic Institute of Education, 1980.
7. Curriculum materials published under the title *WISC West Indian Science Curriculum* consist of three workbooks for pupils and three teachers' guides, Heinemann Educational, London, 1974–76. For an account of the origins of the project, its content, and its underlying philosophy see *Teachers' Guide I*, pp. 1–12. The following pupils' texts are designed for use in conjunction with WISC: F. Commissiong,

F. Dalgety, C. Jackson and A. J. Mee, *Integrated Science for Caribbean Schools*, London, Heinemann Educational, 1976 (three books); P. S. Adey and R. Sorhaindo, *Science and People in the Caribbean*, Longman (Caribbean), 1975–77 (three books), which is a freer adaptation containing much socially relevant material of excellence. See also 'News from the Caribbean' in *British Council Science Education Newsletter*, Vol. 40, 1979, p. 18. The WISC materials have been adapted for use in Botswana, Lesotho and Swaziland. For an overview see W. S. King, in C. F. McFadden (ed.), *World Trends in Science Education*, p. 155, Halifax, Nova Scotia (Canada), Atlantic Institute of Education, 1980. The course is built around the limitations and character of science itself. It has a strong science and society element and does not contain much chemical material.

8. See Section 2.7.3.7. The project was based at the University of the West Indies, Mona, Kingston (Jamaica) and gave rise to published materials by J. F. Reay and A. D. Turner, *New World Science*, two pupils' books and two teachers' guides, Longman (Caribbean), 1975–76. See P. Adey, J. F. Reay and A. D. Turner, Evaluation of New Junior Secondary Science Curricula in the Caribbean Interim Report, March 1973, School of Education, University of West Indies; J. F. Reay and A. D. Turner, The Teaching of Science in Jamaica, Grades 7–9. A report based on trials of Integrated Science Curricula, January 1973, School of Education, University of West Indies.
9. Science Curriculum Development Centre, Curriculum and Supervision Department, Ministry of Education, P.O. Box 1367, Addis Ababa (Ethiopia). Materials in Amharic and English. See *Tenth International Clearinghouse Report*, pp. 3–4.
10. Curtiusstrasse 13, 1000 Berlin 45 (Federal Republic of Germany). Published materials in German. Comprehensive list of descriptive literature (all in German) in *Tenth International Clearinghouse Report*, p. 193.
11. Ghana Association of Science Teachers (GAST) Secretariat, c/o Ghana Education Service, Inspectorate Division, P.O. Box M-188, Accra (Ghana).
12. A. Vikram, Sarabhai Community Science Centre, Navrangpura, Ahmedabad 380009, Gujarat (India). Materials in Gujarati and English. See *Ninth International Clearinghouse Report*, p. 23.
13. Professor R. M. Sperandeo-Mineo, Instituto di Fisica del-l'Università, Via Archirafi 36, 90123 Palermo (Italy). See the following research studies: S. Mannino, M. B. Palma-Vittorelli, R. M. Sperandeo-Mineo, 'Reports on GIREP Conference', p. 102, Venice, 1973; M. del Vecchio, M. B. Palma-Vittorelli, R. M. Sperandeo-Mineo, M. A. Valenza, *Giornale di Fisica*, Vol. 15, 1974, p. 1; and *Tenth International Clearinghouse Report*, p. 175.
14. National Institute for Educational Research, 6-5-22, Shimomeguro, Meguro-ku, Tokyo 153 (Japan). Material in Japanese. See *Tenth International Clearinghouse Report*, p. 47.
15. *Eighth International Clearinghouse Report*, p. 12.
16. Malaysian Integrated Science has been adapted from the Scottish Integrated Science Syllabus. See A. W. Jeffrey, 'Adapting a Syllabus to a New Environment', *New Trends in Integrated Science Teaching*, Vol. I, p. 252, Paris, Unesco, 1971. Ministry of Education, Malaysia, *Integrated Science Worksheets*, three volumes, Kuala Lumpur (Malaysia), Heinemann Educational (Asia) 1970 and C. Kwai, S. S. Sekhon and A. J. Mee, *Science for the Seventies for Malaysian Schools*, Books 1, 2 and 3, Kuala Lumpur (Malaysia), Heinemann Educational (Asia), 1972–75.
17. For an account of the STAN Curriculum Development Project see S. T. Bajah, in A. Kornhauser, C. N. R. Rao and D. J. Waddington (eds.), *Chemical Education in the Seventies*, 2nd ed., p. 215, Oxford, Pergamon Press, 1982. Among the texts that have been published for use with this project are: Science Teachers' Association of Nigeria, *Nigerian Integrated Science Project*, two pupils' books and two teachers' guides, Heinemann Educational (Nigeria), 1972. New edition in full colour, 1982. See also O. J. Negede, 'An Evaluation of the Nigerian Integrated Science Project', *ICSU Newsletter from the Committee on the Teaching of Science*, Vol. 8, 1981, p. 26.

18. Curriculum Revision Unit, Science Education, Institute of Education, University of Sierra Leone, P. M. B. Freetown (Sierra Leone). Published materials: *Core Course Integrated Science Programme*, 1976, three pupils' workbooks and three teachers' guides. See *British Council Science Education Newsletter*, Vol. 34, 1977, p. 17.
19. Science Education Centre, c/o William Pitcher College, P.O. Box 87, Manzini (Swaziland). See *Swaziland Teacher's Journal*, Vol. 63, 1972, p. 25; (published by the Ministry of Education, Swaziland); *British Council Science Education Newsletter*, Vol. 21, 1973, p. 35.
20. The Vocational Science and Mathematics Curriculum Project has grown out of the recognition that the newly developed secondary school academic science courses, in chemistry for example, do not meet the needs of students in the vocational courses of agriculture, home economics, arts and crafts, commerce and industrial arts. New integrated (Chemistry, Physics, Biology) modular courses that are conceptually sound, interesting and relevant for these students are being developed and tested. Full implementation across Thailand was scheduled to be completed by 1983. Examples of modules that rely heavily on chemistry would include 'Matter and Change', 'World of Carbon' and 'Food Science', from the Home Economics/Arts and Science Curricula and 'Solutions and Colloidal Dispersions' or 'Carbon and Living Things' from the Vocational Agriculture Curriculum. For further details, IPST, ref. 14, Appendix 8.
21. See Section 2.7.3. Material published by Longman and Penguin. For a description of the aims and main features of the project see W. Hall, *Schools Council Integrated Science Project: Patterns—Teachers' Handbook*, London/Harmondsworth, Longman/Penguin, 1973. The way the learning model of Gagné is used is explained on pp. 29–33.
22. See Section 2.3. Wreake Valley School Science Department: D. Tinbergen and P. Thorburn (eds.), *Integrated Science Books 1, 2 and 3, London, Edward Arnold, 1976.*
23. Scottish Education Department, *Curriculum Paper 7: Science for General Education* (Consultative Committee on the Curriculum, HMSO, 1969) extracts from which are reproduced in 'Science for General Education: For the First Two Years and the Early School Leaver', *School Science Review*, Vol. 51, 1970, p. 692. For the actual curriculum materials see *Scottish Integrated Science Worksheets* (London, Heinemann Educational, 1971) and for a recent study of them see S. H. Kellington and A. C. Mitchell, *An Evaluation of 'New Science Worksheets' for Scottish Integrated Science*, London, Heinemann Educational Books, 1978; S. Brown, 'Scottish Integrated Science', in L. Stenhouse (ed.), *Curriculum Research and Development in Action*, Chapter 14, London, Heinemann Educational Books, 1980, which contains an extensive bibliography on Scottish Integrated Science; Scottish Integrated Science has also been widely used, with little adaptation (except of the biological content), in other countries. See R. B. Ingle and A. D. Turner, 'Science Curricula as Cultural Misfits', *European Journal of Science Education*, Vol. 3, 1981, p. 357, and I. W. Williams, 'The Implementation of Curricula adapted from Scottish Integrated Science', *Curriculum Implementation and its Relationship to Curriculum Development in Science*, pp. 295–9, Jerusalem, Israel Science Teaching Center, 1979.
24. Scottish Centre for Mathematics, Science and Technical Education, College of Education, Gardyne Road, Dundee DD51HY, Scotland. The project is designed to help science teachers who are using Scottish Integrated Science materials with classes of mixed or of low ability.
25. The ISCS materials are designed to enable students to progress at different rates and through various pathways, depending on their interest, abilities and previous experience. They are also intended to be helpful to teachers with limited special training in science education. Curriculum materials are published under the title *Probing the Natural World* by Silver Burdett, Morristown, N.J. (United States), 1970 onwards. For descriptive references see 'A Science Course Built for Junior High Students', *Research in Review*, Vol. 1, 1970, p. 1; E. Burkman, 'ISCS: An Individualised Approach to Science Instruction', *The Science Teacher*, Vol. 37, 1970, p. 27; J. S. Hathway, D. D. Redfield and W. R. Snyder, 'What Makes ISCS Tick?', *Science*

Activities, Vol. 7, 1972, p. 50. The materials have been adapted for use in Nigeria, and are published by Evans Brothers Ltd, London (United Kingdom).

26. The same group that developed ISCS for junior secondary schools later developed a series of modules for independent study for senior secondary level, with a laboratory and/or investigative flavour. The modules introduce chemical, physical, biological and earth science concepts. The modules have societal themes (see also Appendix 13).
27. This project is an interdisciplinary environmental science programme which emphasises basic concepts in the physical, biological and earth sciences. It relates them to the practical issues of man's use of the environment. For further information, The Director, FAST project, 1776 University Avenue, Honolulu, Hawaii 96822 (United States).
28. There are many environmental education projects developed in the United States that might be regarded as a form of integrated science. For a general discussion see A. M. Lucas, 'Science and Environmental Education: Pious Hopes, Self Praise and Disciplinary Chauvinism', *Studies in Science Education*, Vol. 7, 1980, p. 1; see also *New Trends in Integrated Science Teaching*, Vol. V, Paris, Unesco, 1979; V. and R. West (eds.), *Energy and Environmental Education: the European Experience*, Northwood, Middlesex (United Kingdom), Science Reviews Ltd, which contains several papers on chemical aspects of environmental education.
29. See, for example, *New Trends in Integrated Science Teaching*, Vol. I, pp. 17–28, Paris, Unesco, 1971.

Appendix 12

1. *Africa 69/70: A Reference Volume on the African Continent*, compiled and edited by the editorial staff of *Jeune Afrique*, Africana Publishing Corporation, New York (United States). See, for example, the following articles: 'Algeria: Natural Gas and the Petrochemical Industry', pp. 149–50; 'Angola: Open Cast Mining for Iron', p. 151; 'Congo-Brazzaville: The World's Biggest Potash Deposit', pp. 157–8; 'Guinea: Still More Bauxite', pp. 165–6; 'Upper Volta: Railway Unlocks the Manganese', pp. 192–3; 'Libya: Five New Oil Ports', pp. 167–9; 'Mauritania: Operation Copper Mine', pp. 177–8; 'Niger: Under the Sign of Uranus', pp. 181–3 (an account of a collaboration between the Republic of the Niger and the French Atomic Energy Commission on the exploitation of the uranium deposits); and 'Tunisia: A Regional Chemical Industry', pp. 189–91. Although they do not give chemical details, they provide a wealth of valuable background information for chemistry teachers and textbook writers. The accounts are brief, but informative and interesting.
2. J. Steward and P. Towse, *Chemical Technology in Africa*, Cambridge University Press, 1984.
3. See Appendix 11.
4. Published by Book Society of Canada, Agincourt (Canada, M1S 386), in 1974, 1979, and in the United Kingdom by John Murray, London, 1980.
5. For further details, Technology Consultancy Centre, University of Science and Technology, Kumasi (Ghana).
6. Booklet (24 pp.) entitled *Chemistry for High School*, by R. Ben-Zvi, A. Hofstein and D. Samuel, available from the Chemistry Group, Science Teaching Department, Weizmann Institute of Science, Rehovot (Israel 76100). Also the following published articles: A. Hofstein, 'Future Developments in Integrated Science Education in Relation to Technological Studies', *New Trends in Integrated Science Teaching*, Vol. V, p. 119, Paris, Unesco, 1979; A. Hofstein and N. Nae, 'Chemical Case Studies: Science-Society Bonding', *Science Teacher*, Vol. 48, 1980, p. 52; N. Nae, A. Hofstein and D. Samuel, 'Chemical Industry: A New Interdisciplinary Course for Secondary Schools', *Journal of Chemical Education*, Vol. 57, 1980, p. 366; N. Nae, A. Hofstein and D. Samuel, 'The Case for Case Studies: School Chemistry and the Chemical Industry', *Education in Chemistry*, Vol. 19, 1982, p. 20; and P. Pezara, E. Mazor,

R. Ben-Zvi and D. Samuel, 'The Chemistry of Rocks and Minerals: A New Interdisciplinary Curriculum for Secondary School', *Journal of Chemical Education*, Vol. 55, 1978, p. 383; N. Nae, A. Hofstein and D. Samuel, in W. T. Lippincott (ed.), *Proceedings of the Sixth International Conference on Chemical Education*, p. 181, College Park, Md., University of Maryland, 1982.

7. M. Sinclair, 'Secondary School Chemistry in Kenya', *International Newsletter of Chemical Education*, Vol. 15, 1981, p. 17. See Section 2.3 for a description of the East African School Science project.
8. A. F. Harmer, 'Tin Mining, Extraction and Uses; a Malaysian School Project', *School Science Review*, Vol. 61, 1980, p. 476.
9. Available from New Zealand Institute of Chemistry, P.O. Box 1926, Christchurch (New Zealand).
10. B. Deutrom and M. Tapo (eds.), *Science: Chemical Technology*, Boroko (Papua New Guinea), Department of Education.
11. See *Final Report on the South East Asian Workshop on New Trends in Chemistry Teaching in Schools*, p. 62, Penang (Malaysia), RECSAM, 1978.
12. D. R. Morgan, 'Salt Production in Tanzania', *Education in Chemistry*, Vol. 8, 1971, p. 20.
13. See Appendix 10.
14. See Appendix 11. These are written from the point of view of industry and contrast, for example, with units written for courses concerning society (Appendix 13).
15. S. Akau'ola and J. Bonato (eds.), *Small-Scale Chemical Processing*, University of the South Pacific Institute of Education, 1979. (Unesco-USP Science Series.)
16. R. T. Allsop, J. M. Freeman and R. B. Ingle, 'Chemical Industry in Developing Countries', *Education in Chemistry*, Vol. 8, 1971, p. 226.
17. J. D. Haden, 'Iron and Education in Uganda', *Education in Chemistry*, Vol. 10, 1973, p. 49.
18. See Appendix 6.
19. See Appendix 6.
20. See Appendix 8.
21. The article by C. E. Jones, 'The Chemistry of a Mini', *School Science Review*, Vol. 54, 1972, p. 35, is a chemist's description of the motor car as exemplified in the 'Mini'. It deals with the engine and bodywork, paint, tyres, battery, the petrol and the radiator. It might almost form a basis of a course in chemistry.
22. See Appendix 13.
23. The Director, Curriculum Development Centre, Lusaka (Zambia).
24. *New Trends in Chemistry Teaching*, Vol. II, p. 247, Paris, Unesco, 1969; Vol. V, p. 257, Paris, Unesco, 1981.
25. A group based at Bangor (inquiries to Professor I. W. Williams, School of Education, University College of North Wales, Lon Pobty, Bangor, Gwynedd, United Kingdom) has prepared a number of booklets on traditional crafts and technologies of scientific interest.

Appendix 13

1. Project materials available from VUSEB, 437 St Kilda Road, Melbourne 3006 (Australia). For a more recent discussion of science and society in Australian schools see ICSU, *Newsletter from the Committee on the Teaching of Science*, Vol. 8, 1981, p. 23.
2. B. Selinger, *Chemistry in the Market Place*, Canberra/London, Australian National University Press/John Murray, 1975.
3. See Appendix 11.
4. See Appendix 12.
5. See Appendix 6.
6. Published by Longman, London (United Kingdom).
7. Syllabus and the book, *Chemistry Case Studies*, available from The Secretary, Joint

Matriculation Board, Manchester M15 6EU (United Kingdom). For description, see G. Hallas and W. J. Hughes, 'The JMB Applied Chemistry Syllabus—The Place of Case Studies and Industrial Processes', *School Science Review*, Vol. 56, 1974, p. 391.

8. For accounts of the inception of the Nuffield Working with Science Project, first known as the Nuffield 16-plus Project, see M. Hurst, 'The Nuffield 16-plus Science Project', *Education in Science*, Vol. 54, 1973, p. 40 and K. Wild and J. K. Gilbert, 'A Progress Report of the Nuffield Working with Science Project', *School Science Review*, Vol. 58, 1977, p. 560.
9. Science in Society project publications are available from the Association for Science Education, College Lane, Hatfield, Hertfordshire AL10 9AA (United Kingdom). The materials are published by Heinemann Educational Books, London.
10. The SISCON materials intended for use at tertiary level are published by Butterworth and Co. Ltd, Borough Green, Sevenoaks, Kent TN15 8PH (United Kingdom), from whom a free descriptive leaflet is available. For a discussion of SISCON in schools, see J. Solomon, 'Science and Society Studies in the School Curriculum', *School Science Review*, Vol. 62, 1980, p. 213. The draft SISCON-in-schools materials are available from the Association for Science Education, College Lane, Hatfield, Herts, AL10 9AA (United Kingdom); published by Basil Blackwell, Oxford, 1983.
11. Materials available from the Scottish Council for Educational Technology, 16 Woodside Terrace, Glasgow, Scotland, from whom a price list may be obtained. For a description see N. Reid, *International Newsletter on Chemical Education*, Vol. 15, 1981, and A. H. Johnstone and N. Reid, 'Bringing Chemical Industry into the Classroom', *Chemistry and Industry*, Vol. 122, 1979.
12. See Section 2.5. Titles of modules: *Reactions and Reason: An Introductory Chemistry Module*; *Diversity and Periodicity: An Inorganic Chemistry Module*; *Form and Function: An Organic Chemistry Module*; *Molecules in Living Systems: A Biochemistry Module*; *The Heart of the Matter: A Nuclear Chemistry Module*; *The Delicate Balance: An Environmental Chemistry Module*; *Communities of Molecules: A Physical Chemistry Module*. For each module there is a teachers' guide and a students' book. Published by Harper & Row, New York. See M. Gardner, 'The Interdisciplinary Approaches to Chemistry (IAC) Program and Related Research', *Science Education Research*, Australian Science Education Research Association, Vol. 17, 1973; M. Gardner, IAC Progress Report, *REACTS*, pp. 1–6, 1972. *Proceedings of the Regional Educators Annual Chemistry Teaching Symposium*, Department of Chemistry, University of Maryland, College Park, Md. (United States). H. Heikkinen, 'Factors Influencing Student Attitudes Toward the Study of High School Chemistry' (Ph.D. Thesis, University of Maryland). See also R. D. Sherwood, 'Student Attitude and Achievement in IAC and CHEM Study', *Journal of Chemical Education*, Vol. 55, 1978, p. 733.
13. For further details, Sylvia Ware, American Chemical Society, 1155 16th Street, N.W., Washington, D.C. (United States). See also Chapter 8.
14. M. M. Jones, J. T. Netterville; D. O. Johnston and J. L. Wood, *Chemistry, Man and Society*, 2nd ed., Philadelphia, Penn., W. B. Saunders, 1976. (A laboratory manual is also available.)
15. J. C. Giddings, *Chemistry, Man and Environmental Change: an Integrated Approach*, San Francisco, Calif., Canfield Press, 1973.
16. See Section 2.9. For further details, Lilian M. de Hernández, Centro Nacional para el Mejoramiento de la Enseñanza de la Ciencia, El Marqués, Caracas 107 (Venezuela). Two books have been published: *El ambiente: un recurso para el aprendizaje de la química*, and *Sesiones de laboratorio*.
17. P. Fensham, 'Social Content in Chemistry Courses', *New Trends in Chemistry Teaching*, Vol. V, p. 31, Paris, Unesco, 1981.

3 Some methods in teaching chemistry: four studies

A. Kornhauser

This chapter is concerned, as can be seen from the title, with four different methods of teaching. Professor Kornhauser identifies four areas which are most promising in chemical education: the investigative method; problem solving; structuring of chemical knowledge; and pattern recognition.

She has selected four papers to illustrate these areas, the first by G. C. Pimental and the second by M. Vrtáčnik, G. Djokić and herself. She uses a paper by M. J. Frazer to illustrate structuring and one of her own on pattern recognition.

Professor Kornhauser points out that all can be used in any form of education; for example, whether it be in group or individualized teaching and learning, and whether or not the teacher and learner are supported by educational techniques.

Although some of the examples in the papers are found in tertiary level courses, they nevertheless illustrate and illuminate the areas for secondary school teaching.

3.1 Introduction

There is a lot of confusion in chemical education when methods are discussed. Some authors do not distinguish research findings on methods (strategies and theories) from techniques (skills); others mix both with other forms of education, such as classroom teaching, individual study, group work and so on. All this vagueness indicates not only that chemical education is still at the beginning of its development towards becoming a specific chemistry-based scientific discipline, but also draws attention to its interdisciplinary character, showing the interaction of chemistry with psychology, educational theory, sociology and even philosophy, not to speak of its sister-sciences. When so many different disciplines with diverse approaches meet, difficulties in finding a common language have to be expected. But there is also a great opportunity of creating a synthesis of the best results of all co-operating disciplines, in this way reaching much higher scientific levels.

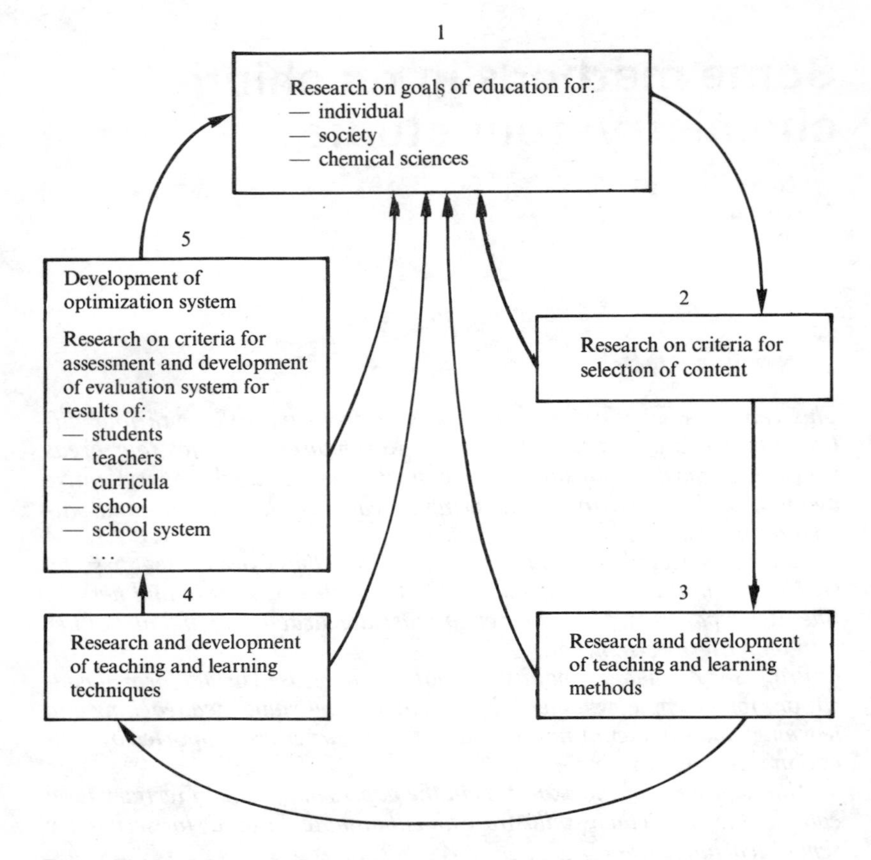

FIG. 1. Fields of chemical education as a chemical discipline.

Let me first define what is meant by chemical education as a chemical discipline (Figure 1).

Among the five fields of chemical education the most important is the third one: research and development of teaching and learning methods, which are sometimes also called strategies in educational theory. Only with the help of highly efficient methods shall we be able to orientate chemical education towards process skills, i.e. the ability to cope with the extensive and ever growing scope of chemistry data, with the increasingly abstract character of its theories and with the diversity and growing interdisciplinary character of its borders. And only via scientifically based methods can chemical education develop as a scientific discipline, for which the three main conditions are: (a) objective observation; (b) powerful methods and techniques; and (c) predictive ability.

Four methods are most promising in chemical education: the investigative method, problem solving, structuring of chemical knowledge and pattern recognition. They can be used in any form of education, i.e. in classroom teaching, group work, individual study, etc., as well as being supported by any technique, i.e. laboratory work, audio-visual techniques, the computer, etc. Method is the basic philosophy of acquiring new knowledge in its content and process.

Thus the following sections will consider:

3.2 Investigative method
3.3 Problem solving
3.4 Structuring of chemical knowledge
3.5 Pattern recognition

3.2 Investigative method

The investigative method can be met in different variants:

(a) Some authors try to follow the *historical pathway towards discovery*, showing in this way not only the main achievements, but also the errors and doubts in the process of discovering new knowledge. If we add the search for data, which, in addition to a high demand for reasoning and a critical approach, is the most essential part of this method, then its great value becomes obvious. It is therefore hard to understand the fact that examples of this beautiful and proven method are a great rarity not only in chemical education literature [1] but also in science education [2]. A clear example of the historical approach is given by Frazer later in the chapter (Section 3.4).

(b) Pimental gives a good example of the investigative learning method in a paper from the Third International Conference on Chemical Education [3]. Although he uses as his example the Second Law of Thermodynamics, he nevertheless shows that it may be understood at school level, by using this method of teaching.

Its process is outlined in Table 1. It begins with observation and description—the collection and recording of sensory perceptions of what goes on around us. This is the starting point in all science. This is where our 'scientific facts' begin. Then, experimentation and measurement lend sophistication and quantitative dimension to these facts. Next comes modelling, as we try to bring coherence to apparently disconnected observations. Likenesses between some familiar

TABLE 1. The investigative method

Observation and description
Experimentation and measurement
Modelling (theory)
Uncertainty and tentativity

behaviour and a new phenomenon enable us to transfer that familiarity–this is called 'understanding'. 'A gas behaves like a collection of billiard balls' is a comparison between a behaviour we wish to understand (the gas laws) and a behaviour that is familiar and already understood (billiard ball collisions). It is a 'model' in its qualitative formulation. It is called a 'theory' as it becomes more quantitative (the kinetic theory of gases). The role of such models and theories is to aid us in developing expectations for future events.

Finally, the investigative method intrinsically includes uncertainty and tentativity. Every observation and every measurement contains some uncertainty injected by the limitations of our observational powers and the calibrations of our instruments. Since the usefulness and validity of a model or theory depends entirely upon the accuracy with which it paraphrases observational facts, the theory can be no more certain than the observations from which it rose. Hence every theory carries some uncertainty. Hence a scientific attitude is pervaded by tentativity. As we use today's theories to predict, be it interpolation within the domain of our present experience or extrapolation outside, we must always be ready for something unexpected. We must be prepared to discard popular models as our increasing knowledge gives us new facts to encompass. Recognizing the omnipresent uncertainty and tentativity in scientific activity contributes to our goal of preparing the student for technological change. Equally important, it makes him aware of the limited ability of science to solve all problems as soon as they are recognized and to anticipate far into the future every new problem that will arise.

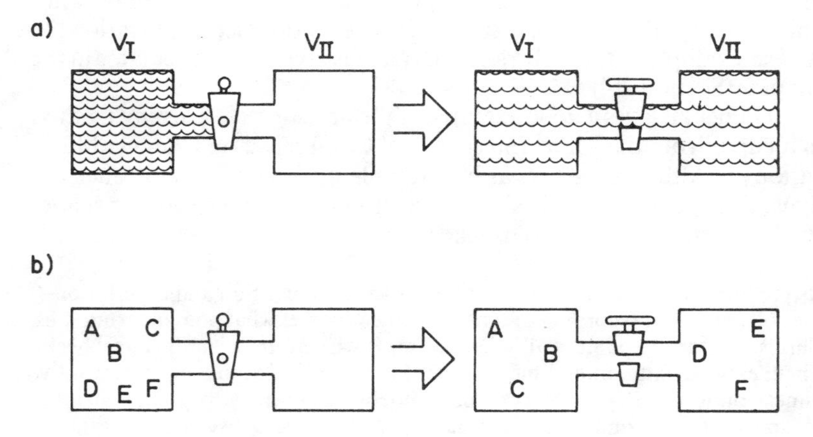

FIG. 2. (a) Expansion of a gas into a vacuum: $V_I = V_{II}$;
(b) six molecules distributing between equal volumes V_I and V_{II}.

The example I wish to develop is connected with the conceptual meaning of the Second Law, which tells us the direction of spontaneous change. The Second Law can be stated in the form 'In a spontaneous change, the entropy of the Universe increases.' But there is an equally valid statement of the Second Law that is almost self-explanatory and that immediately brings to mind familiar (and correct) analogues 'Spontaneous changes occur in the direction of increasing probability.' If the essence of the Second Law is connected with probability, we have a great reservoir of student enthusiasm and expertise waiting to be tapped. For, after all, almost all games involve some element of probability and the level of sophistication possessed by our children in their games is quite significant. This is the entry I wish to exploit and I would like to do so in terms of the simple change represented in Figure 2(a), the expansion of a gas in a bulb V_I into another evacuated bulb of equal volume, V_{II}.

Every student can predict the outcome if the stopcock is opened to permit gas movement from V_I to V_{II}. Gas will move from V_I to V_{II} until equal pressures exist in each bulb. The question is, *why* did this elementary, spontaneous change take place? If the gas is ideal, there is no energy change involved. What caused some of the molecules to move from V_I into V_{II}? Then, when the pressures were equalized, how did the rest of the molecules know that they should refrain from crossing over? Could it be that these molecules are highly intelligent and each one keeps track of every other so that they can, in concert, behave according to Boyle's Law? Not likely!

An alternate, and more appealing model is that each molecule behaves independently, ignoring quite stupidly both the behaviour of its compatriot molecules *and* the demands of Boyle's Law. Let us explore this model. But to make it tractable, we will consider how it would work with a small number of molecules so it will be easy to keep track of them. For example, the gas expansion Figure

TABLE 2. Ways to distribute six molecules between two bulbs of equal volume

Number of molecules in V_{II} (or in V_I)	Which molecules are in V_{II}	Total number of arrangements
0	None	1
1	A or B or C or D or E or F	6
2	AB BC CD DE EF AC BD CE DF AD BE CF AE BF AF $\frac{6 \times 5}{1 \times 2} = 15$	15
3	ABC BCD CDE DEF ABD BCE CDF ABE BCF CEF ABF BDE ACD BDF ACE BEF ACF ADE ADF AEF $\frac{6 \times 5 \times 4}{1 \times 2 \times 3} = 20$	20
4		15
5		6
6		1

2(a) might be considered for only six molecules, A, B, C, D, E and F initially in bulb V_I, as pictured in Figure 2(b). After this stopcock is opened, one possible outcome is pictured. Perhaps D, E and F will cross over. That would give equal pressures in the two bulbs, and Dr Boyle would be quite satisfied.

But wait! There are other possibilities! If each molecule behaves independently of all the others, perhaps D will come back into V_I. Then the pressures would not be equal. In fact there are lots of ways the molecules could be found if they behave randomly. Let us write down all of the possibilities.

Table 2 shows the number of ways we could have 0, 1, 2 or 3 molecules in bulb V_{II}. There is only one way to have no molecules in V_{II}, that is, to have them all in V_I. But there are six ways to have one molecule in V_{II} simply because there are six candidates. It could be A, or, it could be B, or, in turn, it could be any of C, D, E or F. There are even more ways to have two molecules in V_{II}, fifteen in all. But it requires no mathematical skill to write them all down and count them up. In a similar way, any high school student can decide that there are twenty ways to divide the molecules equally between the two bulbs.

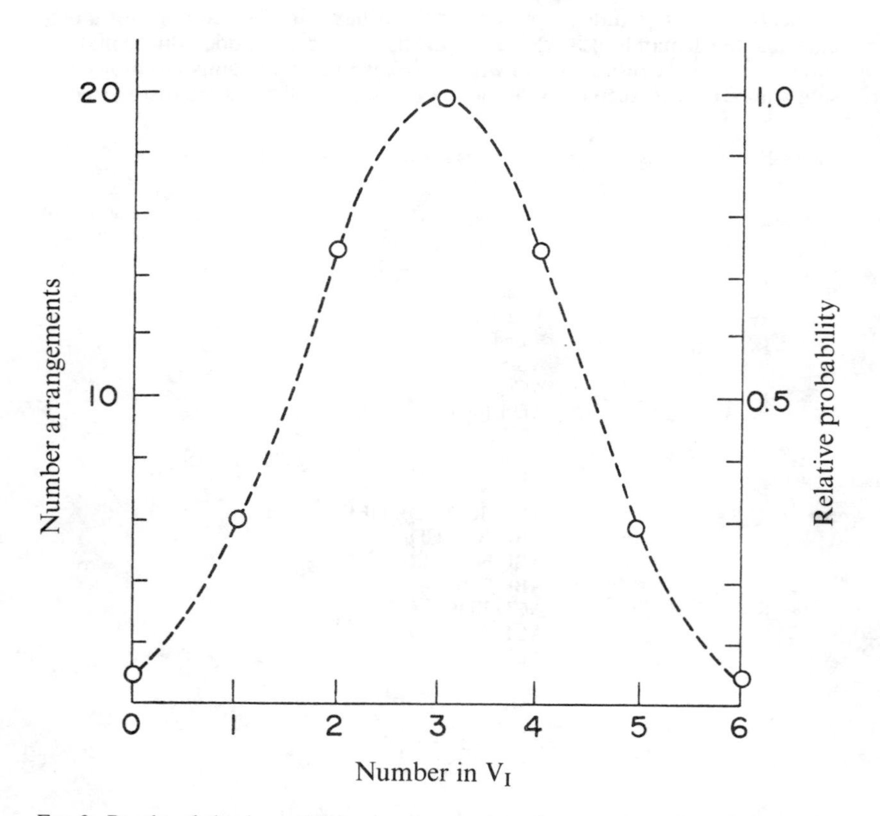

FIG. 3. Random behaviour of six molecules: number of arrangements (or relative probability) versus number of molecules in V_I.

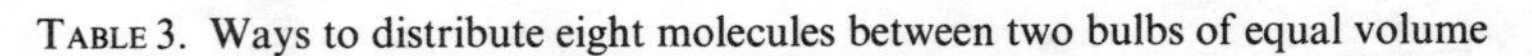

TABLE 3. Ways to distribute eight molecules between two bulbs of equal volume

Number of molecules in V_{II} (or in V_I)	Total number of arrangements	Relative number of arrangements (compared to most probable)
0	1	0.01
1	8	0.11
2	28	0.40
3	56	0.80
4	70	(1.00)
5	56	0.80
6	28	0.40
7	8	0.11
8	1	0.01

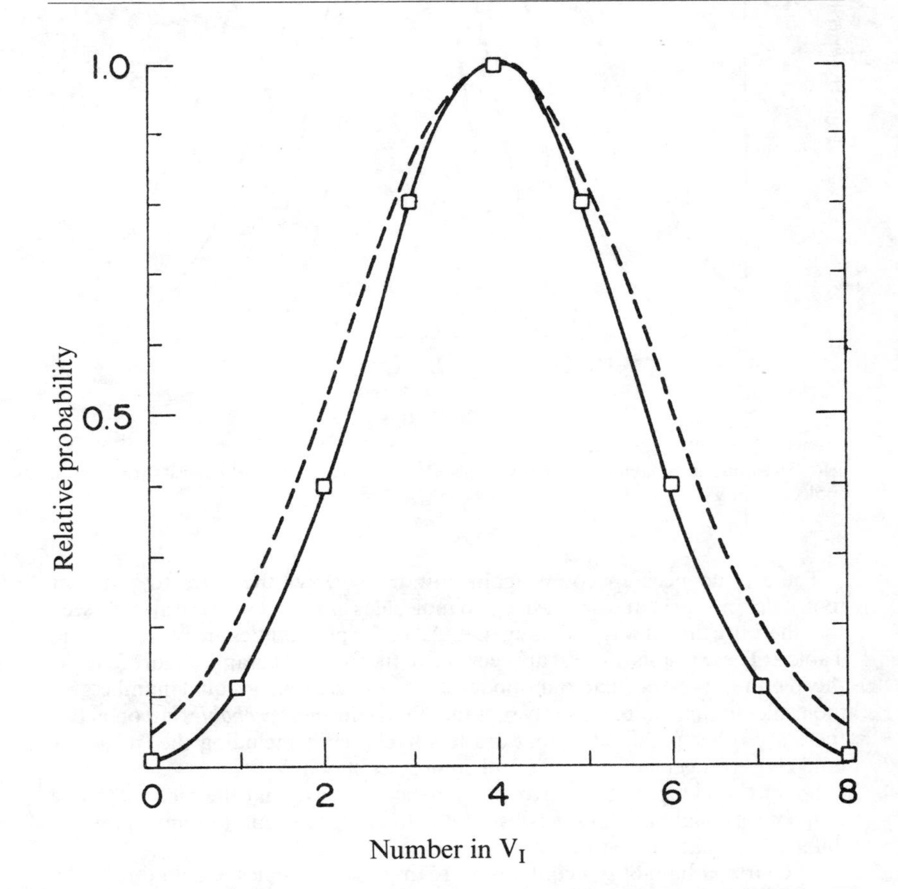

FIG. 4. Random behaviour of eight molecules: relative probability versus number of molecules in V_I (dashed curve, six-molecule behaviour for comparison).

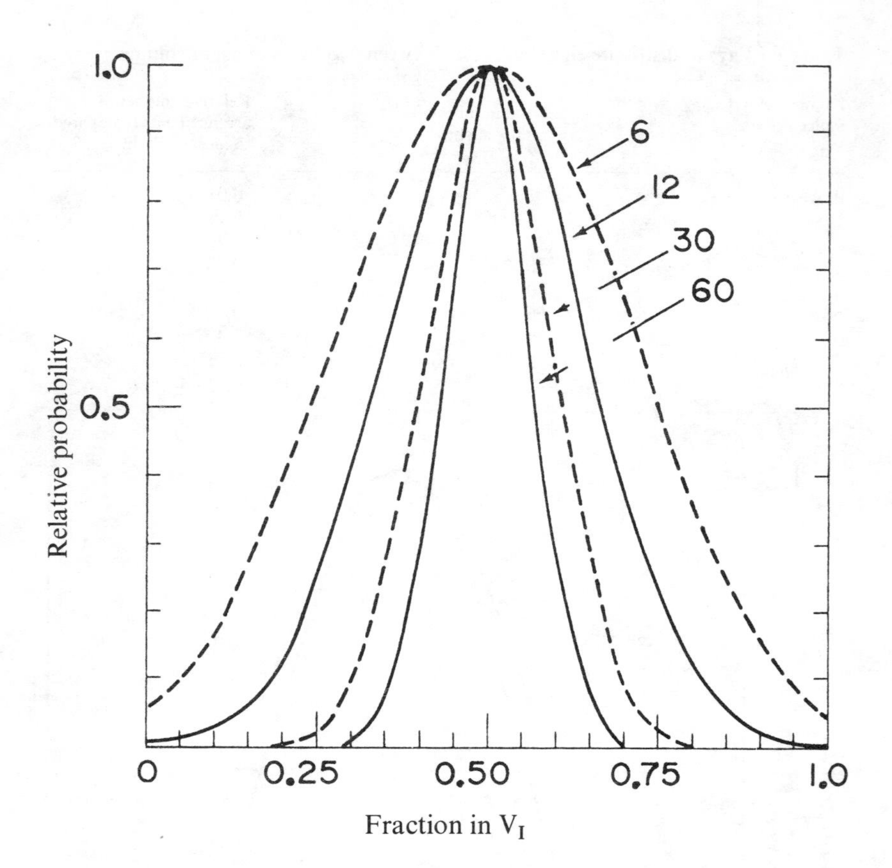

FIG. 5. Random behaviour of many molecules: relative probability versus fraction of molecules in V_I.

There is no need to count again how many ways there are to put four molecules in V_{II}. That would leave two molecules in V_I and we have already seen that there are fifteen ways of doing that. All of the possibilities are represented in Table 2. These results are best appreciated in the form of a graph. Figure 3 shows the picture. We see that our model does not *guarantee* equal numbers of molecules in each bulb, it merely says that this is the *most probable* outcome. But there are other possibilities that are less likely, even including the (relatively unlikely) arrangement that all of the molecules stay in V_I!

So, our model partially mirrors experience—it says that the most probable outcome is equal pressures, as observed with real gases, but it permits pressure differentials that we don't observe.

Of course, a bulb of gas contains more than six molecules. Could this be the trouble? That is easy to check. We merely have to construct a table like Table 2 for a larger number of molecules—say eight. That is still a reasonable counting game and it will indicate how things will go as the number of molecules increases. Table 3 shows the outcome (without showing all the arrangements in detail) and Figure 4 shows it graphically. The dashed line permits comparison with the six-molecule probability picture taken from Figure 3. We see that the eight-molecule behaviour agrees with that deduced for six molecules behaving randomly in that the *most probable* outcome is that there will be equal pressures in the two bulbs. Other possibilities still exist that correspond to unequal pressures, but the *unequal pressures are less likely* than were those obtained with only six molecules.

Obviously the thing to do is to consider larger and larger numbers of molecules. Since the counting gets more and more tedious, the results can be presented without actually showing details, but emphasizing that no new ideas are in use more complicated than the simple counting displayed in Table 2.

Figure 5 superimposes the probability distribution associated with successively larger numbers of molecules, each behaving randomly according to the simplest laws of probability. First, we note that for *any* number of molecules initially in V_I, the most probable outcome of random behaviour is that half of the molecules, hence equal pressures, will be found in V_{II} and half in V_I. Second, the figure shows that as the number of molecules increases, the probability that there will be a substantial difference in pressure between the two bulbs gets smaller and smaller. If the number of molecules approaches Avogadro's number, as it does at normal pressures, the probability that a measurable pressure difference will be found approaches zero.

Thus, we see that the expansion of a gas into a large volume can be explained as a change toward the condition of maximum probability. That conclusion is precisely the Second Law of Thermodynamics. It has been obtained without more difficult logic than applicable to flipping coins, rolling dice, or playing any game of chance.

This same level of analysis can be carried all the way to the quantitative statement of the Second Law, relating (for constant temperature processes) the probability that governs spontaneous change to the reversible heat. This involves carrying out the change in a simple apparatus that allows extraction of work. For gas expansions, a piston in a cylinder that lifts a weight can be pictured. Both for some gas expansions and for many chemical changes, an electrochemical cell can be used. In either apparatus, it is readily seen that there is a limit to the amount of work that can be extracted from a spontaneous change. The ideal gas expansion shows us that this maximum work is determined by the probability change. Approached in this fashion, the principles are not made more difficult by adding the word 'entropy' to the student's vocabulary at this time.

The student has no preconceptions that 'entropy' will be a formidable concept. If we introduce it properly, the student will be no more intimidated than if he were taught to play Parchisi. If he grasps the idea that change occurs in the direction of increasing probability, he has the essence of the Second Law.

This example is not offered as the *only* way to provide an understanding of what governs tendency for change. It is offered to show that unifying principles that *might* be presented at too high a level of sophistication do not *have to be* so presented. As *teachers, it is our task to search for new ways to display the meaning of these principles within the capabilities of our students.*

(c) The above method is strongly connected with problem solving in the *form of school-level students' research projects*, in which a real problem has to be solved.

Such projects do not always need to be highly sophisticated; materials for them can be found in our surroundings. For example, when discussing the energy problem in school science, we usually mention the efforts to grow plants which produce hydrocarbons, having in mind *Euphorbia*, *Ficus* and similar species in fashion, and usually forgetting such common plants as the good old pine tree. Teachers could encourage students to collect plants in their surroundings and, bearing in mind the students' safety, distil their parts to find out if they contain components which could be used as fuels.

Students should observe the flame of such liquids from plant material to estimate the carbon content from their sootiness. In a more demanding, semi-quantitative approach, they could compare the distillates from different plants for their relative carbon content, as well as the quantity of liquid fuel per weight unit of plant material. They could draw a graph with different plants on one axis and the quantity of liquid

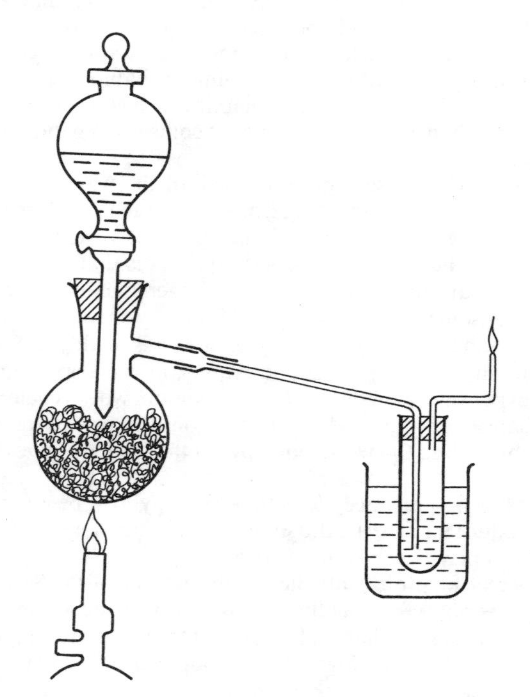

FIG. 6. Cracking.

fuel per kilogram of plant material on the other. Combining their knowledge of biology and chemistry, they could discuss a better use of plant hydrocarbons. More ambitious students could even try to crack such liquid hydrocarbons from plants and to detect the gaseous, liquid and solid phases.

The iron wool used for cleaning cooking utensils is an excellent catalyst for this cracking on condition that the dry flask is almost filled with it, and only after strongly heating the iron wool is the liquid distillate from the plants added with a dropper (Figure 6). By means of this process cracking takes place and in the cooled flask a lighter liquid is obtained, and a gaseous fraction produced which can be collected under water or directly ignited.

Another simple experiment could be the basis of a small students' research project on halogen content in different pesticides. Beilstein's reaction in which a copper rod, which is covered with oxide by being held for a short time in a flame, is then dipped into a pesticide solution or suspension and again put into the flame, is an especially suitable technique. The flame becomes green if the halogen is present.

The results of the experiments should be collected in a table giving information such as the name of the pesticide, its toxicity, effectiveness as a pesticide and the results of experiments on halogen content. The co-operation of young biologists and chemists would be even more desirable in this case.[1]

What about defining the acidity of a plant material extract? Different plants could be collected and divided into roots, stems, leaves and flowers. A water extraction of such materials is simple; it includes cutting and grinding the parts of the plants, extraction by shaking or mixing, filtering and adding indicators to the filtrate. Using different indicators, students could find out which plants or which parts of plants contain more or less acid. Here, too, at least a semi-quantitative approach is desirable.

The production of sugar in the kitchen could be a small technologically oriented students' project. The sugar beet has to be washed, cut into pieces, ground in the mixer and the sugar extracted from the pieces by cooking them with water for about half an hour. Filtration through cloth gives us the raw extract to which lime water is added to precipitate impurities. Carbon dioxide is passed through so that this addition will produce a precipitate of calcium carbonate to remove the excess of calcium hydroxide. Repeated filtration through a coffee filter follows and then the solution is boiled until a thick liquid is obtained, which is then cooled and centrifuged in a liquidizer. The sugar obtained is washed with very little hot water and sieved. If different sorts of sugar

1. *Safety*: pesticides are of course dangerous substances unless handled with care. See *Education in Science*, November 1979, p. 26.

beet are used in a quantitative approach in these processes, the project could be interesting even from a production point of view.

Soap making from fats and lyes and the preparation of aspirin are other possibilities. All such projects can be extended.

Pollution gives us other topics for student projects. What about a series of experiments using distillation of solid waste in an attempt to produce liquid and gas fuels? Old rubber material and kitchen waste usually give good results. Different sorts of rubber could be compared for this purpose, as well as different kitchen wastes. The results could tell us which solid pollutants could be used as secondary raw materials for small-scale energy sources. Students could collect such materials in their surroundings over a certain period of time, and estimate from data collected on such material how much material would be used in a year and how much fuel could be produced from such a quantity of waste. In a related type of project, energy released from the burning of dried cow dung or converting animal waste into methane and other gases can be explored. Highly polluted water could be tested for nitrogen compounds by adding a few cm^3 of such water to approximately the same volume of a solution of sodium or potassium hydroxide, and heating the mixture. The ammonia which is liberated could be tested with moist red litmus paper, attached to the inside of the lid. Such water could also be filtered through different materials, sand and coal being the most common ones. Different sorts of sand could be tested and their adsorptive capacity discussed in proper terms. The results should be collected in tables and presented in diagrams whenever possible.

Collecting and measuring the pH of rain water in urban and rural areas and tracing sources of pollution leads to interesting investigations.

As teachers we should not underestimate children's capacity—and need—for an early development of a quantitative experimental approach and presentation of such results. A brick or tiled wall could help us to present the idea of networks or of a diagram and to bridge the first difficulties in its understanding. Of course we should be careful not to overload students' capacities. But there is also some truth in the approach: 'Aim high, your shot will hit lower in any case.'

3.3 Problem solving

This section is based on a paper published in the *Proceedings of the Sixth International Conference on Chemical Education* [4].

In attempts to achieve more precise aims in chemical education, problem solving deserves our full attention as an efficient educational method. In spite of a number of efforts and publications in this field, it is, however, still little present in teaching practice. This is a consequence of the fact that few teachers are acquainted with its theoretical background and even less with the ways it may be used in everyday teaching.

Three authors of the fundamental ideas on problem solving are referred to in most papers: Gagné [5], who considers problem solving as reaching the cognitive category of the highest order; Ausubel [6], who regards problem solving as a form of discovery learning; and Jackson [7], for his broad definition of problem solving as bridging the gap between a problem state and the solution state. More recently, Herbert Simon, Nobel Laureate in Economics, and research teams at Carnegie-Mellon University and the Massachusetts Institute of Technology have turned their attention to this research area [8]. There are, of course, also a number of other authors, in addition to those listed above, with similar definitions.

Without the intention of diminishing these general ideas present in educational theory on problem solving, we should not forget that this educational method has a much longer tradition. It was in fact the sign of quality for most good schools throughout history and was highly esteemed as far back as the ancient Greek philosophers [9]. Plato certainly had been considered as one of the first authors and also enthusiastic users of the problem-solving method in education.

A comprehensive presentation of problem solving as a structural/process approach in education is given in the book by J. M. Scandura [10], which involves fundamentals of content analysis, cognitive mechanisms and individual differences, as well as instructional applications.

In chemistry, problem solving and problem-solving networks are presented by Ashmore, Frazer and Casey [11]. They define problem solving as the result of the application of knowledge and procedures to a problem situation and propose four stages: (a) definition of the problem; (b) selection of appropriate information; (c) combining the separate pieces of information; and (d) evaluation of the solution. The authors claim that the best chances for success in chemical problem solving rest on a combination of a strong background knowledge in chemistry, a good knowledge of problem-solving strategies and tactics and on confidence. Examples of networks as presentations of interconnections in solving chemical problems are given. The networks are designed as triangular structures for 'closed' problems, i.e. problems with a unique solution at the apex, and as trapezoidal networks for 'open' problems with several possible solutions.

Guided design as a method of teaching and learning originated with Wales and Stager [12–15] should also be mentioned since it is very close to problem solving. An example in the teaching of general chemistry is described by Hoggard [16]. Nine steps are presented for this method in an example of fertilizer selection: state the problem; define the goal of the project; gather information; analyse existing solutions; generate possible solutions; determine constraints; evaluate possible solutions; analyse all solutions; prepare synthesis and recommendations. The author also develops topics on atoms and atomic weights, gas laws and stoichiometry, thermodynamics, colligative properties and solution equilibria.

When numerical problems are considered, the strategies of teaching data manipulation and interpretation might be considered as the prerequisite process for problem solving.

Yeany and Capie [17] develop an operation/interaction model of data processing activities, giving ten stages: collecting; recording; ordering; condensing; preparing; displaying; translating; comparing; expanding; and inferring.

Teaching and learning problem solving in science is also discussed in the papers of Mettes *et al.*, who give first a general strategy [18], which is followed by an example of learning problem solving in a thermodynamics course [19].

In addition to these chains of stages we have also constructed others, e.g. identifying the problem; observing; measuring; classifying; ordering; inferring; predicting; forming hypotheses; searching for patterns; designing and carrying out experiments; interpreting and analysing data; verifying conclusions; defining unanswered questions; and expressing doubts. Some authors enumerate over twenty stages.

In spite of the enthusiasm for designing stages in problem solving, and the richness of its results, this has had hardly any impact on the practice of chemical education. Chemical publications with problem solving in the title are often a collection of more or less traditional stoichiometric, thermodynamic, kinetic, organic, structural and similar well-known items, presented with hardly any strategy of problem solving. The few exceptions with a number of examples of the problem-solving approach, e.g. Peters [20] and Finar [21], only confirm the observation that it is not easy to generate examples of problems to be used in chemical education.

The use of research problems for teaching problem solving

The heuristic approach, i.e. designing a problem by following the historical pathway towards discovery, is certainly an important way, but since students know the answers, or at least the ways to find them, this is better for teaching the construction of problem-solving networks than for the solving of these problems.

Our problem in developing problem solving for learning chemistry was manyfold:

How to formulate problems which will be real but still have only one solution, i.e. be of the 'closed' type and yet not the usual items.

How to select such problems which include relatively simple chemistry, since students cannot struggle at the same time with chemistry knowledge and with the strategy of problem solving.

How to get teachers interested in research, which is an excellent method of developing problem-solving skills, and, last but not least:

How to construct problems for chemistry teaching and learning which will be of help not only at tertiary, but also at secondary level.

We decided therefore to construct problems for teaching problem-solving skills on the basis of real, contemporary research results. For this the following procedure was developed:

A research paper is selected from research journals which are not easily available to students. This is important, otherwise some of them, instead of developing problem-solving skills, prefer a small search in the faculty library.

The selected research paper is then submitted to 'extraction', i.e. research data are taken out and presented to the students as a list of experimental results. This list is adapted to the course.

Students are asked to attempt to solve the problem, as well as to design their problem-solving network.

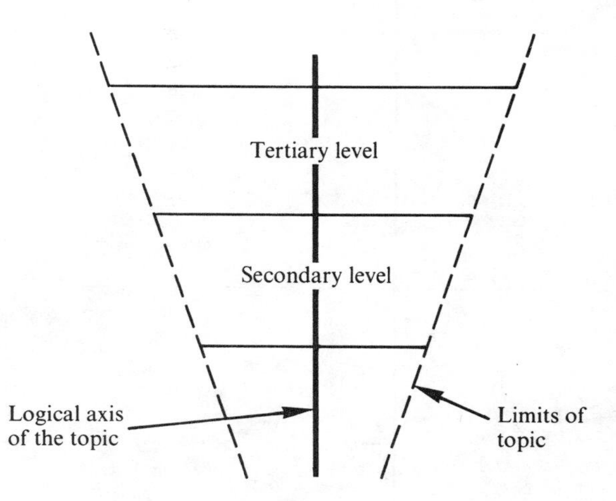

FIG. 7. The 'funnel' approach.

Students' networks are then compared and analysed, to identify the reasons for success and failures. This analysis is the basis of the discussion on the problem, problem-solving networks and problem-solving skills, in which students and teachers co-operate. In some cases, the same problem is submitted to different groups, and results are compared for evaluation.

In well-structured examples, in which students use more or less the same network independently, efforts are made to develop the same sort of problem for the lower level, what we usually call the 'funnel' approach.

The following example, developed in the chemistry of natural products undergraduate course, is given for illustration.

1. Selection of research paper. The paper on 'The Structure of Viscumamide, a New Cyclic Peptide Isolated from *Viscum album* Linn. var. *coloratum* Ohwi' [22] was selected because of its high quality, its relevance to the course, its demanding relatively simple chemistry knowledge and its having a sufficient amount of experimental data.

2. The 'extraction' of research data, adapted to the course, gave the following list:

(a) the unknown compound was isolated from the neutral fraction of the methanol extract of *Viscum album* Linn. var. *coloratum* Ohwi by column chromatography on activated alumina;
(b) mp of the crystalline substance is 622.7–624.2 K;
(c) the substance is optically active, $[\alpha]_D^{25} = -49.1°$ ($c = 0.199$, ethanol);
(d) elemental analyses and the mass spectrum show the composition to be $C_{30}H_{55}N_5O_5$ with a molecular weight of 565.8;
(e) the infra-red spectrum of this substance shows absorption bands at 3,315, 3,050, 1,658 and 1,530 cm^{-1}, and shows no absorption band indicative of the existence of a free amino group or a carboxyl group;

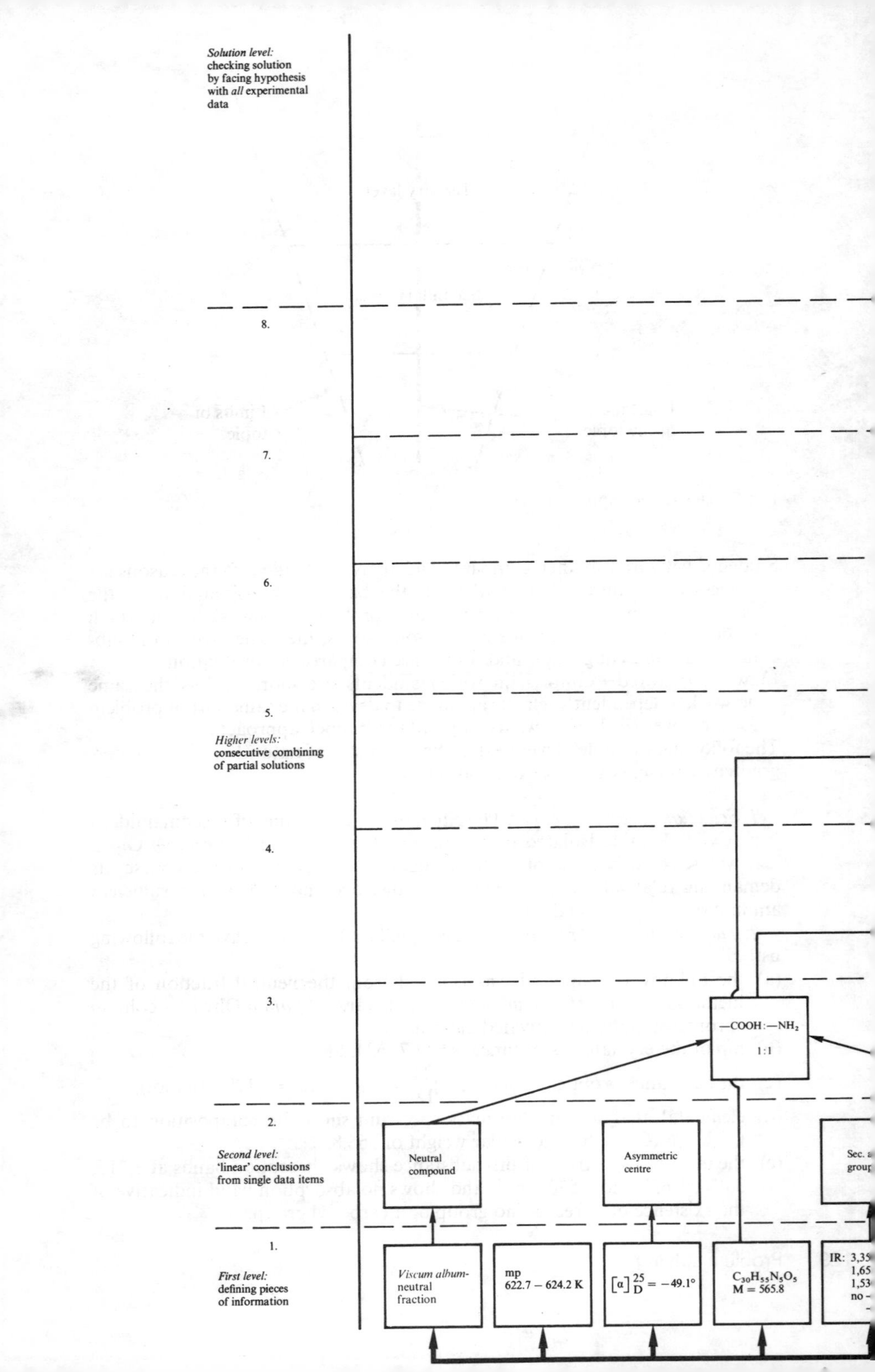
Solution level:
checking solution
by facing hypothesis
with all experimental
data
8.
7.
6.
5.
Higher levels:
consecutive combining
of partial solutions
4.
3.
—COOH:—NH2
1:1
2.
Second level:
'linear' conclusions
from single data items
Neutral
compound
Asymmetric
centre
1.
First level:
defining pieces
of information
Viscum album-
neutral
fraction
mp
622.7 – 624.2 K
[α] 25 D = –49.1°
C30H55N5O5
M = 565.8

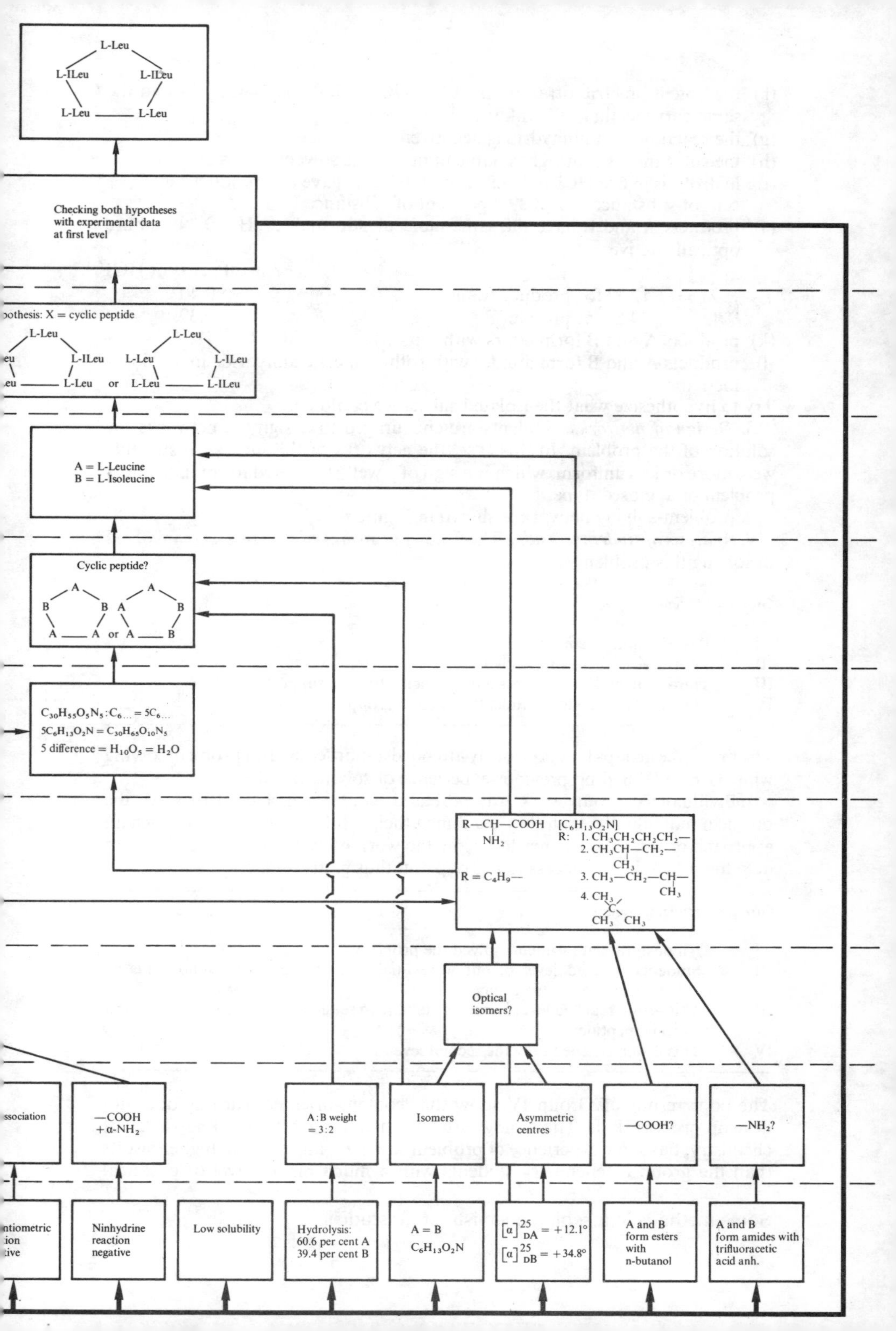

L-Leu
L-ILeu
L-ILeu
L-Leu
L-Leu
Checking both hypotheses with experimental data at first level
pothesis: X = cyclic peptide
L-Leu
L-ILeu
L-Leu
or
L-Leu
L-ILeu
L-Leu
L-ILeu
A = L-Leucine
B = L-Isoleucine
Cyclic peptide?
A
B
B
A
A
or
A
B
A
B
$C_{30}H_{55}O_5N_5 : C_6... = 5C_6...$
$5C_6H_{13}O_2N = C_{30}H_{65}O_{10}N_5$
5 difference = $H_{10}O_5$ = H_2O
R—CH—COOH
NH_2
$[C_6H_{13}O_2N]$
R: 1. $CH_3CH_2CH_2CH_2$—
2. CH_3CH—CH_2—
CH_3
R = C_4H_9—
3. CH_3—CH_2—CH—
CH_3
4. CH_3
C
CH_3 CH_3
Optical isomers?
ssociation
—COOH + α-NH_2
A : B weight = 3 : 2
Isomers
Asymmetric centres
—COOH?
—NH_2?
ntiometric
ion
tive
Ninhydrine reaction negative
Low solubility
Hydrolysis: 60.6 per cent A 39.4 per cent B
A = B
$C_6H_{13}O_2N$
$[\alpha]^{25}_{DA} = +12.1°$
$[\alpha]^{25}_{DB} = +34.8°$
A and B form esters with n-butanol
A and B form amides with trifluoracetic acid anh.

(f) the potentiometric titration in 0.01 M HCl with 0.1 M NaOH showed the same curve as that of blank titration;
(g) the reaction with ninhydrin is negative;
(h) the substance is sparingly soluble in numerous solvents;
(i) hydrolysis in 6 M HCl at 383.2 K for 120 hours gave two products: 60.6 per cent of substance A and 39.4 per cent of substance B;
(j) products A and B have the same molecular formula $C_6H_{13}O_2N$ and are optically active,

	c in 20 per cent HCl
$[\alpha]_D^{27} = +12.1°$ for product A and	0.40
$+34.8°$ for product B	0.39;

(k) products A and B form esters with butan-1-ol saturated with HCl;
(l) products A and B form amides with trifluoroacetic anhydride in dichloromethane.

Try to hypothesize what the isolated substance could be.

3. Design of networks. Students are encouraged to design the networks for solution of the problem. In this case, the networks of the successful students were more or less uniform, which is a sign of a well-structured list of data and a problem of a 'closed' type.

A problem-solving network is shown in Figure 8.

4. Analysis of students' results. The following groups of students were involved in solving this problem:

Group	Students
I	Postgraduate chemistry students
II	Fourth year chemistry students
III	Third year biology students having chemistry as a supporting subject
IV	Fourth year chemistry students—second group

The first three groups had previously attended a short course on problem solving while Group IV had no previous experience of solving problems.

Individualized group work was practised: students first tried to solve the problem individually, then they combined their results in group work. Such an approach enables the teacher to follow the work of every student, and also to help them to achieve success in a group and thus gain confidence.

Group	Students' results
I	All postgraduate students solved the problem.
II	Students reached level 6, but were unable to define both amino acids as L-Leucine and L-Isoleucine.
III	This group reached level 5, but was unable to set up a hypothesis to check for a cyclic peptide.
IV	This group reached only the second level.

The poor results of Group IV show that content-oriented learning does not develop process skills. The biology students had a much poorer knowledge of chemistry, but some experience of problem solving—and reached higher results than the group of chemistry students with a much higher level of chemical

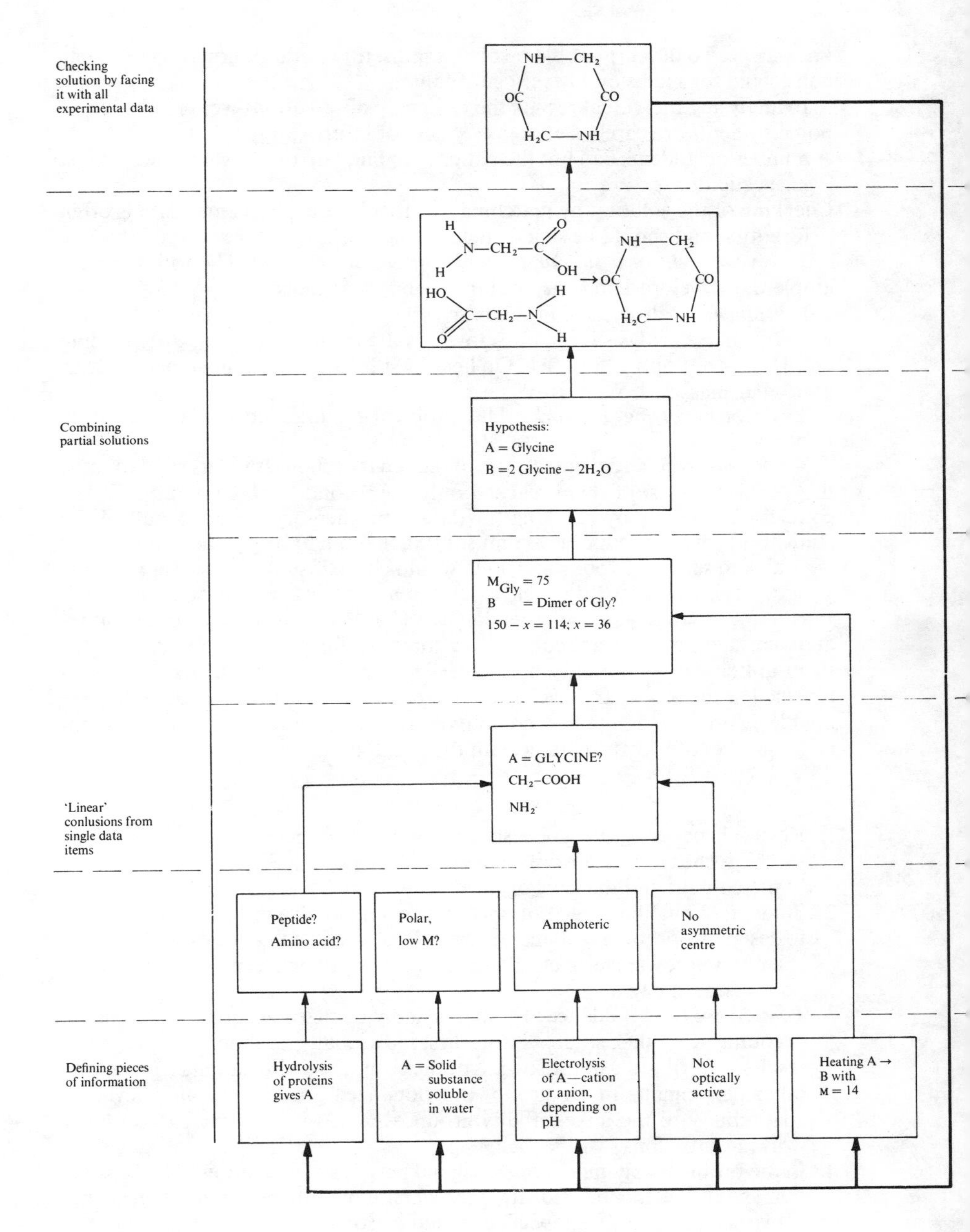
Checking solution by facing it with all experimental data
NH—CH2
OC
CO
H2C—NH
H
N—CH2—C
O
OH
H
→
HO
C—CH2—N
H
O
H
Combining partial solutions
Hypothesis:
A = Glycine
B = 2 Glycine – 2H2O
MGly = 75
B = Dimer of Gly?
150 – x = 114; x = 36
A = GLYCINE?
CH2–COOH
NH2
'Linear' conlusions from single data items
Peptide?
Amino acid?
Polar,
low M?
Amphoteric
No asymmetric centre
Defining pieces of information
Hydrolysis of proteins gives A
A = Solid substance soluble in water
Electrolysis of A—cation or anion, depending on pH
Not optically active
Heating A → B with M = 14

FIG. 9

knowledge. To develop problem-solving skills, this method should therefore be considered for inclusion in every curriculum.

In analysing individual results the following observations were made:

Some students are careless and 'lose' pieces of information;

A number of students had not the patience to build up the network and got lost in speculations;

Checking of the solution by matching it with all the experimental data is often forgotten and should be strictly demanded.

5. *Development of a similar example for secondary level.* The following example was developed and tried out in secondary schools:

1. *Problem.* Hydrolysis of proteins gives a solid compound A which dissolves in water. In electrolysis it migrates towards the cathode or anode—depending on pH. It is not optically active. On heating it forms a solid substance, B, with molecular mass 114. What is A?

2. *Problem-solving network.* The problem-solving network is shown in Figure 9.

3. *Students' results.* Several classes of secondary school students tried to solve this problem. Most of them reached only the second level, coming to 'linear' conclusions, mainly by recalling the data from memory. About a half of the students reached the hypothesis that substance A might be glycine, but very few were able to set up a hypothesis which would check if substance B were a cyclic peptide. The reason for this deficiency lies in learning chemistry by memory. Furthermore, students at the secondary level have never heard of cyclic peptides and they have never been encouraged to construct formulae of potential, but for them unknown, compounds. The experience we had at the tertiary level was repeated at the secondary level: students memorize the chemistry data, but are unable to use it to create new combinations as hypotheses and to check them using sources of literature other than their textbooks.

Conclusions

1. Research papers are a good source for the construction of problems for teaching chemistry via problem solving. They should be selected according to the content and the level of the course.
2. Students should be asked not only to attempt to solve the problem, but also to design the problem-solving network. Poor results might be caused by poor chemistry knowledge, lack of basic problem-solving experience and/or a poor research paper.
3. To increase the confidence of students, individualized group work should be encouraged. In this, students work first individually and then combine their results, contributing to the success of the group, and also gaining from all the other participants in the group. Individualized group work also develops adaptability to the mixed ability group, which is usually the case in most working situations.
4. In the problem-solving process, original networks leading towards the solution should be encouraged. In cases when most attempts lead to the same networks, patterns of knowledge should be sought [23].
5. Positive and especially negative results should be carefully analysed and discussed with students.
6. A speculative approach should not be encouraged. Even if students reach the

solution without a network, they should be encouraged to design it and explain to their colleagues their intuitive 'jumps'.

7. Whenever possible we should try to develop problems of a similar type for different school levels.
8. It is not easy to involve the majority of teachers in research, however desirable it might be. But by encouraging this approach to problem solving, we bring contemporary research closer to the teacher.

3.4 Structuring of chemical knowledge

The main approaches can be met when the structuring of knowledge is considered:

3.4.1 Educational theory.
3.4.2 The computer sciences.
3.4.3 The specific subject, e.g. chemistry.

3.4.1 *Educational theory*

Attempts to structure subject-matter have been present in general educational theory for more than twenty years. The concept is, however, vague and allows many interpretations. A number of authors consider Ausubel [6, 24] to be the originator of this approach for his claim that the organization of knowledge supports learning and the retention of knowledge learned. Some authors reach even further back and declare Bloom to be the first with his *Taxonomy of Educational Objectives* [25] which has formed the basis for Gagné's structuralist theory, considering that there are different types of learning behaviours, which are related. Bruner [26] describes as 'the optimal structure' a set of propositions from which a larger body of knowledge can be generated, and claims that the merit of a structure depends upon its power to simplify information, to generate new propositions and to increase the manipulability of a body of knowledge. Like most authors in the field of educational theory, Bruner claims that the structure must always be related to the status and gifts of the learner, and that the optimal structure of a body of knowledge is not absolute but relative.

Gagné's idea of learning hierarchies has played a major role too [5]. For example the entire curriculum *Science, A Process Approach* (SAPA), developed by the American Association for the Advancement of Science, used this theory to construct learning networks or hierarchies [27].

The concepts expressed in the last sentence are also present in most newer publications written by educational psychologists, who group organized knowledge into formal structures and cognitive maps. Formal structures reflect the structure of a discipline, in which the knowledge is organized into a hierarchical or other type of organizational framework. Cognitive maps, however, are the organizing structures of a particular individual, which facilitate learning by helping the learner to incorporate a large volume of information and/or concepts.

Reigeluth, Merrill and Bunderson [28] recommend the structuring of subject-matter as the basis for deciding how to sequence and synthesize the modules of a subject-matter area, and describe four types of fundamental structures: the learning hierarchy, the procedural hierarchy, the taxonomy and the model.

The learning hierarchy, also called the learning structure, shows the learning prerequisite relations, i.e. what the learner must know to be able to acquire new knowledge. The procedural hierarchy shows procedural relations among subject-matter components, i.e. steps in acquiring new knowledge. The most common kind of content structure is the third one, the taxonomic structure, based on similarities and chains of knowledge. The theoretical structures or models show chains of causal relations among concepts, often on the basis of mathematical representations. They are more demanding, but also more productive than the others.

3.4.2 *The computer sciences*

The extensive possibilities for the use of the computer in chemistry were presented at the International Conference on Chemical Education [29]. Since then, this field has been constantly growing in scope and quality (Chapter 4). Today we still meet the computer in drills and programmed learning, but also at higher levels of its use, e.g. in the construction and use of data bases, in simulations, curriculum planning, problem solving, research design and automation processes.

Most of the uses of the computer at higher levels require a good structure of knowledge. This is of special importance when computer data banks are developed and used, since computerized chemical information systems are needed for: (a) comprehensible, analytically ordered, coherent storage of information; (b) organization of information, enabling correlation of facts; (c) orientation of research design to avoid duplication of work through lack of knowledge; (d) development of learning methods with special support for the memory; (e) support of decision-making by rapid and complete information and its comparison, enabling its optimization.

Two main types of chemical information system, besides the bibliographic data bases, exist today according to Ash and Hyde [30]: chemical structure information and structure/properties relationship information.

There are a number of computerized chemistry data banks which differ in the size of the collection and the degree of sophistication of retrieval. Usually they consist of fragment codes, linear notations, systematic nomenclature and connection tables—with algorithms for their interconversion. Good examples of the classification of molecular structures on the basis of their constituent atoms for the needs of education are given by Luft [31]. The author pays special attention to Chemical Abstracts and the DARC System.

But, according to many authors, e.g. Rowland [32], knowledge and experience in the use of structure-retrieval methods is poor even amongst academic chemists. This is true also for the patent information area, which is, in spite of its key importance for industry, hardly known outside highly specialized circles.

It is obvious that the computer alone cannot help chemists. A new educational approach is needed, using the development of structures of chemical knowledge also as a part of the strategy of handling chemical information.

3.4.3 *The specific subject*

The interest of chemists in structuring chemical knowledge has increased in recent years. Two orientations mainly occur: the educational and the research approach. The first is intended for better learning, the second for a better ordering of existing knowledge to facilitate the production and management of new knowledge.

Among the first who developed the structuring of chemical knowledge for chemical education is M. J. Frazer with his example for teaching Faraday's laws. The next part of this section will be based on a paper he presented recently at an International Symposium [33].

There are always two main reasons for including a topic in the curriculum. The first is with the intention that students should acquire knowledge of the topic (i.e. at the appropriate level students should be able to recall, understand and apply the facts, concepts and principles of the topic). The second is with the intention of helping students to develop appropriate skills and attitudes. To summarize, the first reason is concerned with the *content* of the topic and the second is concerned with the *processes* of science.

Problems of content and process are discussed here with the aid of the specific example of Faraday's laws.

Content

With respect to content, the major problem of curriculum design is how to structure the knowledge of any particular topic. In other words, what is the relationship between the facts, concepts and principles associated with the topic?

There is an analogy between the structure of knowledge and a geographical map. This analogy is developed in Table 4.

Making new discoveries in science can be linked to an explorer arriving in a

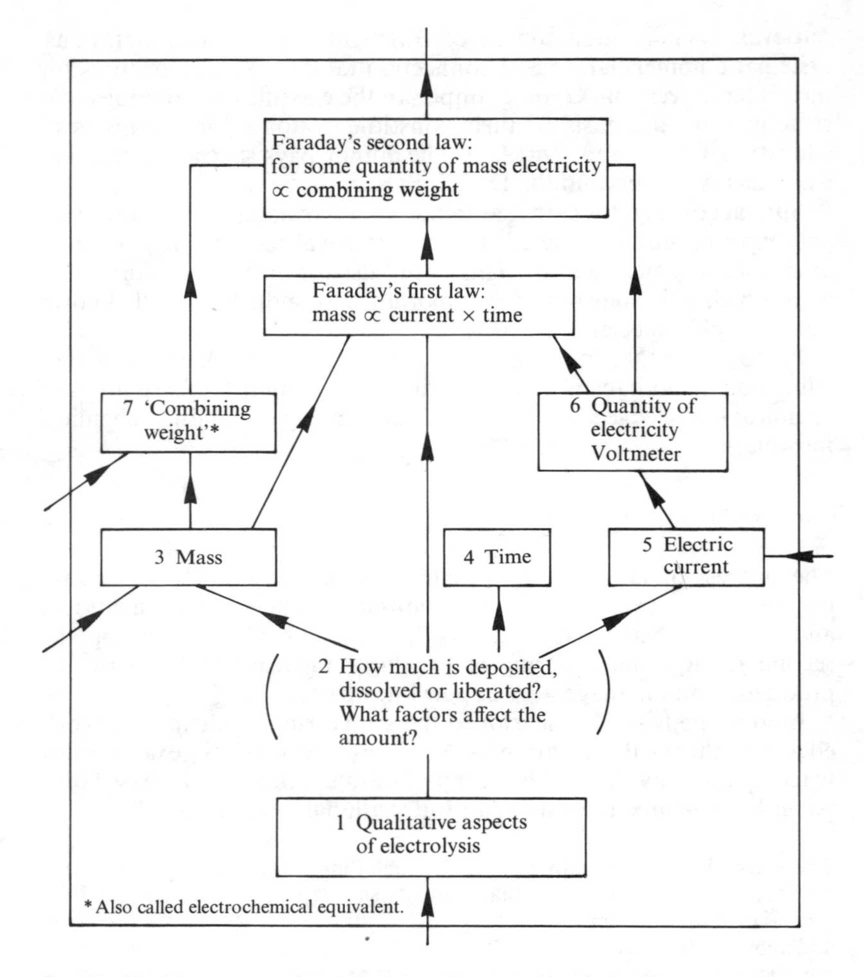

FIG. 10. The history map of Faraday's laws (note the connecting lines to neighbouring knowledge maps.)

new and unmapped land. He stands at a point, he knows what is behind him (his existing knowledge), but he does not know what is in front of him. He is curious to find out, he explores and records what he finds. We can represent Faraday's research on electrolysis in this way and plot the connections between the observations and concepts which led eventually to the two laws of electrolysis. This *history map* is shown in Figure 10.

The starting point for Faraday (1 in Figure 10) was the existing knowledge of the qualitative aspects of electrolysis. He knew that when an electric current was

TABLE 4. Geographical maps and knowledge maps

Geographical maps . . .	Knowledge maps . . .
1. . . . show spatial relationships between places	. . . show the conceptual relationships between facts and theories
2. . . . have different scales	. . . can show different amounts of detail
3. . . . are needed if you don't know your way	. . . can be used by teachers to help students find their way
4. . . . are joined on each side by other maps	. . . have connections with items of the knowledge in maps of other topics
5. . . . are of different types (e.g. road maps, railway maps, etc.)	. . . are of three types (see this paper), viz. history maps, experts' maps, Gagné hierarchies, learners' maps

passed through many salt solutions metals were deposited or dissolved at the electrodes.[1] One of the essential processes of science is curiosity (i.e. the scientist asks: Why? What happens if . . . ? How? etc.) Faraday asked the question (2): How much metal is deposited or dissolved and how much hydrogen is liberated? What factors affect the amount deposited? To even ask such questions required the concept of mass (3).

Faraday started his explorations and the relationships he found are plotted in Figure 10. He found that the longer the time (4) the current was passed, the greater the mass of element liberated. There were in those days some crude methods of measuring electric current (5). It is a tribute to Faraday's experimental skill that he was able to show a relation between mass liberated and current. These two discoveries were combined as shown by the connecting lines on the map which lead to the first law, viz. the amount of element liberated is directly proportional to the amount of electricity (current × time) passed (6). Furthermore, this led to the development of the voltmeter. By measuring the volume of hydrogen formed on electrolysis of dilute sulphuric acid solution he had developed the most accurate method then available for measuring quantity of electricity. Armed with this device he was able to make further exploration. By conducting experiments with a dilute sulphuric acid cell in series with solutions of different metals he was able to show that for constant quantity of electricity passed (i.e. volume of hydrogen liberated) the mass of metal deposited was proportional to the 'combining power' of the metal (*I*).[2] This is Faraday's second law.

This history is well known. Why is it important then for science education to plot out the history in the form of a knowledge map? The answer is that the learner is like the researcher (the explorer). He is in possession of certain knowledge but before him is the unknown. The teacher (who, in this analogy, can be linked to a local guide) knows at least one way forward—the historical way—and so can help the student to discover and learn the knowledge ahead.

1. The word 'electrode' was subsequently invented by Faraday. He also knew that electrolysis of acidified water gave hydrogen and oxygen and that the volumes of these gases could be measured.
2. 'Combining power' or equivalent was later known as electrochemical equivalent.

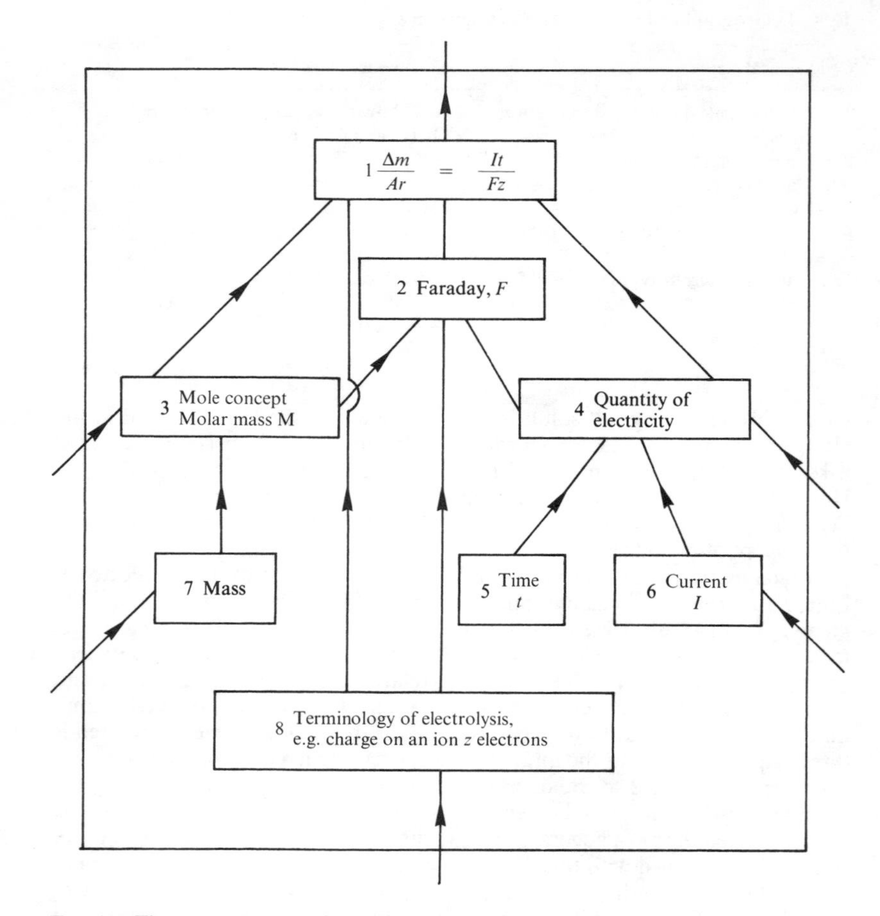

FIG. 11. The experts' map of Faraday's laws. Δm is the mass of element deposited, dissolved or liberated.

Much of the tradition of science education (the curricula, the textbooks, the order of teaching topics, etc.) is based on the history of scientific discoveries.

There is a danger, however. Early explorers' maps were often inaccurate. Their perception of their territory depended very much on where they started. How different our view of certain topics in science might be if by chance the great discoverers had started in different places or taken different routes. This leads us to the second kind of knowledge map. This is the one which describes the experts' view of contemporary knowledge of the topic. Such an *experts' map* for the quantitative laws of electrolysis is shown in Figure 11.

We start with a statement of Faraday's laws expressed in SI symbols (1 in

Figure 11).[1] We, as experts who know the 'territory' well, can start at the top and ask ourselves what has to be understood in order to appreciate (1). Clearly the nearest related concept is the faraday (2), which needs to be understood as a 'mole of electrons'. This leads back to the need to know about the mole concept (3) and the terminology (8) of electrolysis (e.g. the word 'electron'). The units of the faraday are coulomb mol^{-1}, and so in order to understand (2) we need to know also about the idea of quantity of electricity (4). In turn, this leads back to time (5) and current (6). Part of the connection between (1) and the mole concept (3) is the idea of molar mass and so we need also the concept of mass (7) and the ability to convert mass to amount of substance. Finally, in order to use the mole concept in this context we need to understand that ions can have different charges (8).

These seven concepts (2 to 8) could be broken down into more detail—that is, a larger scale map could be drawn for each one. However, at a reasonable level of detail, these seven concepts and their interconnections describe the knowledge map of the quantitative laws of electrolysis. As experts we started at the top and mapped out all that has to be known to reach the statement (1) of Faraday's laws. The idea is that the learner would start at the bottom.

Such knowledge maps of the expert are easy to construct. One approach to their construction is to use the ideas of Gagné. Each point on the map is translated into a specification of student performance (e.g. . . . 'express amount of electricity in coulombs'). Then for each point the question:

What must the student already know how to do in order to learn this new performance?

The answers to this question define the lower subsidiary points on the map. In this way a hierarchy of facts, concepts and principles (in terms of student performance) is built.

There are dangers in this approach too. Such hierarchies are often called 'learning hierarchies', but there is little evidence that students actually learn in this logical, expert's way. Research which is said to validate the hierarchy by means of test items is often circular because the wording of the items assumes the hierarchy.

In other words, the map of knowledge drawn by the expert may satisfy the expert but may not necessarily guide the teacher or help the learner. A complete view from 'above' of knowledge of the topic is not the same as the view of the learner from 'below' as he struggles to find his way in unknown territory.

This leads us to the third kind of map—the *learners' map* (Figure 12). The absence of any numbers or connecting lines in Figure 12 is intended to convey how little we know about how students learn.

One of the most important tasks in science education is to make progress with learners' maps of the structure of knowledge topic by topic. This will not be easy because for any one topic there will be many maps—perhaps as many as there are learners! We must not expect learners' maps to be precise like the history maps and the experts' maps.

1. The symbols are defined in the lower boxes in Figure 11. Δm is the mass of element deposited, dissolved or liberated.

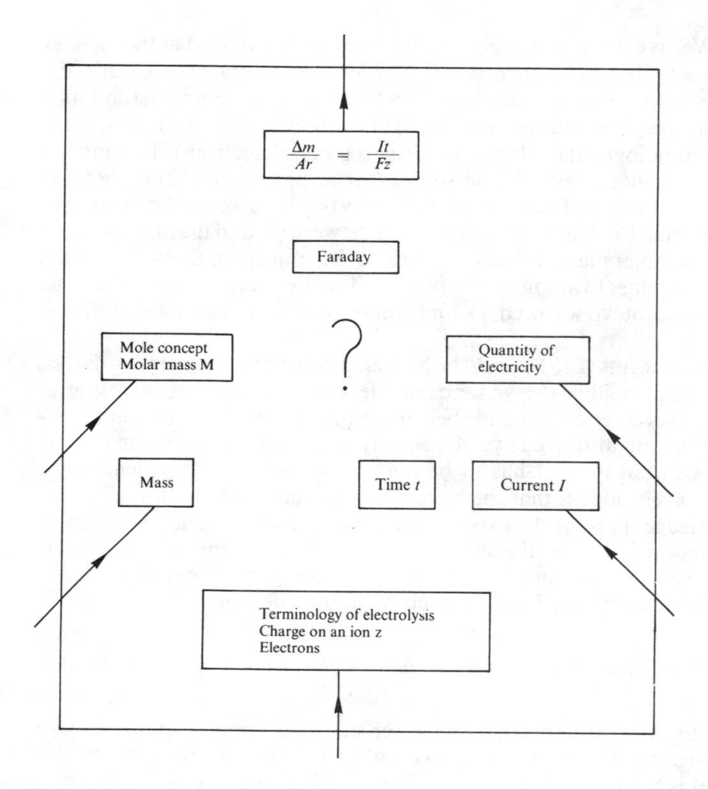

FIG. 12. The learner's map of Faraday's laws. Δm is the mass of element deposited, dissolved or liberated.

In order to produce learners' maps we shall need to use as many methods as possible but the approaches of developmental psychology will be particularly helpful. Clinical methods—observation and interview, as well as test items and computer simulations and analyses—, should be used to explore how students learn and to discover the barriers to learning.

One example will illustrate the difference between an experts' map and a learners' map. This will be expressed in terms of Piaget's stages of cognitive development. In Figure 11, the experts' map, the arrows on the connecting lines all point upwards. This is meant to represent, for example, that an understanding of the mole concept at the Piaget formal operational level is necessary before Faraday's laws can be applied. Although we lack objective evidence, it is the subjective view of many teachers that practice at applying Faraday's laws at the concrete level leads in time to a formal operational level of understanding and application of the mole concept. In other words, in the learners' map, the arrow on the connecting line is reversed. The subjective views of teachers based on their classroom observation of students could be important in constructing learners' maps. There is a great temptation when groups of teachers meet to discuss the structure of knowledge to produce an experts' map. This temptation should be resisted because it is the learners' map which is much more important.

The learners' map is perhaps closer to the history map than the experts' map, but as explained earlier the chance nature of the history of discovery may cause confusion for some topics and be a barrier to learning.

To summarize, in curriculum development we need to structure the knowledge of each topic according to (i) the history of discoveries, (ii) the contemporary experts' view (Gagné's hierarchy) and (iii) the various ways we believe students learn.

Processes

There are many intellectual skills, manual skills and attitudes which we hope might be developed in our students by a study of science. Some of the ways in which learning about Faraday's laws could be used to achieve these process aims are now described.

Curiosity. Science lessons should develop students' curiosity. A good teacher, having presented the qualitative aspects of electrolysis, will leave time for curiosity to grow in students' minds. He may even stimulate this curiosity by asking questions. But an encouraging teacher is almost certain to find someone in the group ready to ask: 'How much metal or hydrogen is liberated?'

Certainly there will be other questions too—all should be encouraged: Does the amount depend on time, current, size of electrode, concentration, pH, temperature, metal, etc?

Experiments can then be planned to try to find answers to these questions.

Planning an experiment. This is where a teacher needs the greatest skill. Without destroying the students' curiosity and enthusiasm they must be guided into what is feasible. Many laboratory hand-books describe students' experiments in electrochemistry, so they will not be repeated here.

Laboratory skills (e.g. careful weighing, measuring volume, reading an ammeter and clock, constructing an electrical circuit, (etc.). All these skills can be developed by simple experiments on Faraday's laws. The justification is that *all* future citizens (not just future scientists) should know how to make measurements such as reading an ammeter. Furthermore, by developing such skills two attitudes can be generated: (a) confidence and pleasure in the ability to work with the hands and to make measurements accurately; and (b) concern for the safety of others in the laboratory.

Interpreting results. Part of planning the experiment will involve a plan of how to record and handle the data. These will be an opportunity, for example, to practise graph-plotting. The volume of hydrogen, which can be conveniently collected in a syringe connected to a cell for the electrolysis of dilute sulphuric acid, can be plotted against time. If the current has been kept constant, this plot is a straight line and there is a clear interpretation. There are many similar experiments which give students the opportunity to present and interpret data they have collected. Possible causes of error can be discussed. Furthermore, all experiments must be reported and so students can develop the important skill of concise and accurate report writing.

Team work. Controlled experiments with different students taking different values of the variable can have a strong motivational effect particularly if every result is plotted on the board in front of the class. Some of the variables to be tested were listed in the section above under 'Curiosity'.

Problem solving. As an example of the applications of Faraday's laws, a problem involving the time required to produce chromium plating on a car bumper to a specified thickness under specified conditions can be given. There are many other examples within this topic which can be used to develop the problem-solving skills of students.

Appreciation of the human factors in scientific discovery. The story of Faraday's life, or even just the story of his discovery of the laws of electrolysis, will give students a feeling of the excitement of science and the realization that it is part of our culture and is a very human activity.

Appreciation of the fun of science. If none of the foregoing excites some students, perhaps they can be motivated by the aesthetic beauty of the 'lead-tree' experiment. Many teachers will know the satisfying and reversible effect of the electrolysis of lead acetate solution between carbon electrodes.

A number of other authors have tried in recent years to develop examples of structuring chemical knowledge. Basolo and Parry [34, 35] give examples for teaching systematic inorganic reactions without having students memorize specific individual reactions. Instead, they should discuss these in terms of general reaction types, i.e. of a system, making extensive use of the Periodic Table of Elements. A reaction can be classified as: combination, decomposition, replacement, metathesis and neutralization. The authors give classification examples for each type.

In organic chemistry attempts to structure reactions have already been in existence for a number of years. Wilson [36] gives an example of classification of electrophilic addition reactions of alkenes (olefins) and alkynes (acetylenes). Patterns in organometallic chemistry with applications in organic syntheses are discussed by Schwartz and Labinger [37], drawing our attention to the fact that, in spite of the complexity of this field, much of the chemistry of transition metal organometallic complexes can be classified according to a small number of general reactions.

Hall [38] goes further and gives a proposal for an *Organic Chemist's Periodic Table*, based on the movement of electrons, and defining molecules as donors and acceptors. According to Hall, the Periodic Table of the Elements should be regarded as a Reactivity Map. When plotting the acceptor ability (abscissa) against donor ability (ordinate), the location of the resulting point gives us a quick idea of the tendency for electron transfer to occur. Similarly we can try to systematize organic chemistry data by constructing an *Organic Chemist's Periodic Table*. Stronger donors should be located further down the table, stronger acceptors further across. The more reactive they are, the more likely the occurrence of transfer. The author himself, however, warns that one master table of this type would be impractically large and recommends the construction of limited sections of the table.

3.5 Pattern recognition

This next section is based on a paper written for the Sixth International Conference in Chemical Education, entitled *Learning Chemistry by Pattern Recognition* [23].

The growth of chemistry in both the quantity and quality of its data and concepts is especially obvious in some of its branches. Organic chemistry, for example, offers a great number of facts on compounds—their structures, physical properties, reaction mechanisms and their utility in society. Memorizing such facts would not be real learning, since facts without their interrelationship are not knowledge. Memory is also a very unreliable companion of ours, functioning only if we use what it has stored.

Searching for systems is therefore a need, especially in sciences rich in data. A dialectic theory of systems, once mainly the domain of philosophers, is increasingly attracting scientists.

The work in the development of this method for chemistry learning, for which the name *pattern recognition* is proposed, is based on the following hypothesis: chemistry has an excellent example of the structure of knowledge—the Periodic System. In fact, it is more than just a structure of knowledge—it is a *recognized pattern of a number of facts about elements and their relationships.* They exist in nature, are objective and valid in all learning situations. A too enthusiastic teacher or learner may, of course, try to change it towards a 'more individualized approach', but the real learning result would be poor in comparison with the one respecting the natural system of matter.

Next to this pattern of elements, similar patterns of compounds exist in matter. The number of variables is here much higher and the number of possible combinations enormous. It is therefore too early to start with the ambition to create a complete 'Periodic System of Compounds', in spite of powerful computers which might encourage this. Even the much simpler Periodic System of Elements was not built in a short time. A number of sub-systems were designed before the whole pattern was recognized.

When building the pathway towards a bigger pattern of compounds, subsystems should be aimed at first. The main idea of learning chemistry via pattern recognition is therefore to encourage the student to select a specific field of chemical compounds or reactions, or properties, to search for their characteristics, to try to construct patterns and to check them (see Table 5). They should not do this with the ambition of creating a Periodic System of Compounds and Reactions, but with the hope of discovering fragments of it or, what will mostly be the case, just to learn in a more organized and independent way and to reach in this learning higher cognitive levels.

TABLE 5. Steps in learning via pattern recognition

Step task	Results observed
1. Collecting data, e.g. on chemical structures, chemical reactions, properties of compounds in selected field	First guided; later independent search for data in different sources: textbooks, manuals, monographic series, etc.
2. Analysing data, defining criteria as constants and variables	Comparison of collected data often shows a number of deviations. Students try to define the accuracy according to the quality of the sources and search for further data for checking. In an analytical approach, they define constants and variables and try to recognize their interrelationships.
3. Defining the hierarchical order of criteria (i.e. from general to specific, C_k has to be more restrictive than C_j); Starting with building a tree structure, which will be completed in the steps following.	Instead of memorizing chemical structures, students develop the ability to analyse structures. Students also recognize that a 'critical amount of data' is needed for the latter. In building the tree structure, they develop the ability to classify and evaluate data according to their general and specific character.
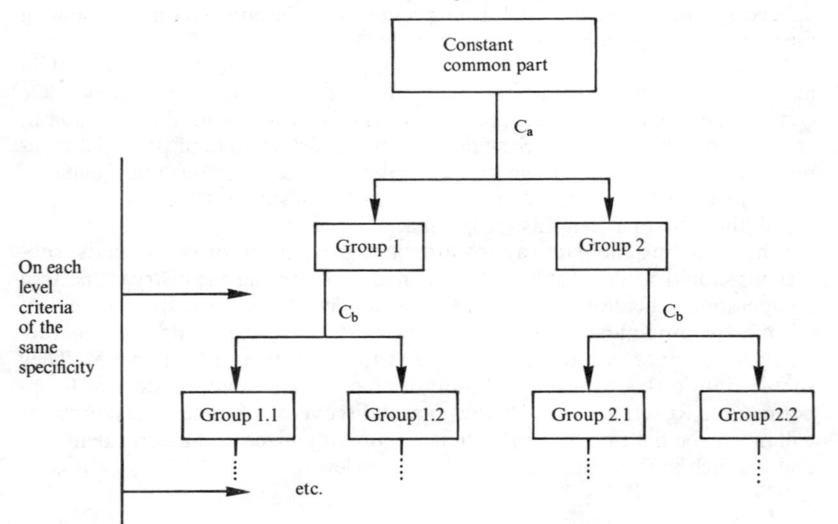 	
4. Designing matrices of possible combinations for each criterion	In an analytical approach, students try to define the pattern for each part of the tree. This supports a synthesis of data, up a hypothesis for prediction. A formalization of knowledge is more organized and the development of abstract thinking supported.

Step Task	Results observed
5. Validation of the pattern designed	Students build the matrices designed into the tree to find out if the hierarchy of criteria was properly defined. This enables them to recognize a clearer picture of data interrelationships and to bring more order into their knowledge. Building a tree and learning about linear and polydimensional structures also assists education in the use of the computer in chemistry.
6. Using the pattern for prediction	Students discuss the 'empty places' in the matrices and the whole pattern. First they search for new sources of chemical information, use research journals and even chemical computer data bases if available. Then they, for example, study the possible eventual compounds or reactions from the view-points of stereochemistry, using models; they discuss pathways of biosynthesis; they may also use their knowledge of chemical thermodynamics and kinetics, etc. to estimate the possibility of the existence of such compounds or re-actions in nature or possibilities for their synthesis. Students involved in research in the field selected also design the way to follow the currently published research results and to build these immediately into their pattern on the way towards its completion. This method also supports the setting up of hypotheses and the design of further research.

The following example has been used at the tertiary level but it will *illustrate* the method:

Example 1—alkaloids (for tertiary level)

When selecting the fields for pattern recognition, two main problems exist: the availability of sufficient data and the definition of the limits. For our first attempts, alkaloids were selected because of our research in this field, in which students are also involved. Since the alkaloids are not a very well structured body of knowledge in most textbooks, the first attempt was to design an overview system. Due to the great number and diversity of alkaloids, this attempt failed.

This experience helped us to recognize that for building a bigger system, patterns of its parts and their interrelationships have to be recognized first. The work was therefore oriented towards specific groups of alkaloids, e.g. Col-

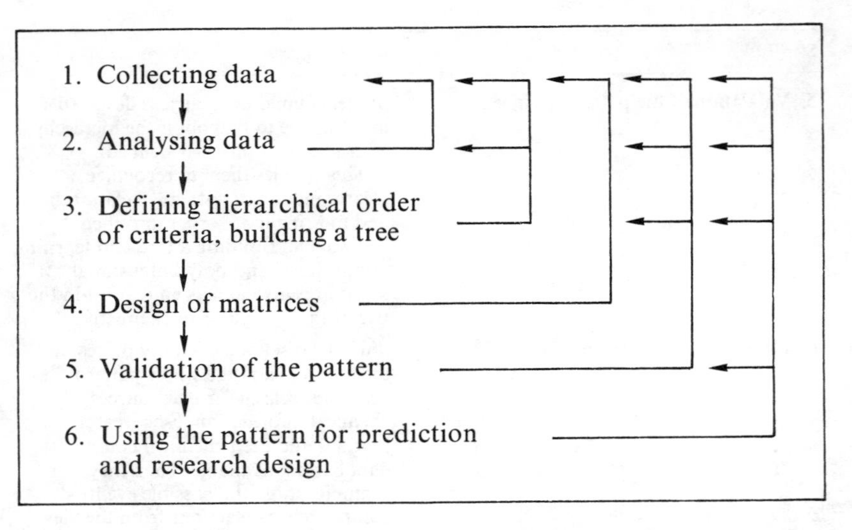

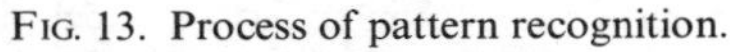
FIG. 13. Process of pattern recognition.

TABLE 6. Pattern recognition exercise using Colchium alkaloids

Steps	Results
1. Collecting data	Students searched for data and found a number of Colchicum alkaloid structures, e.g.

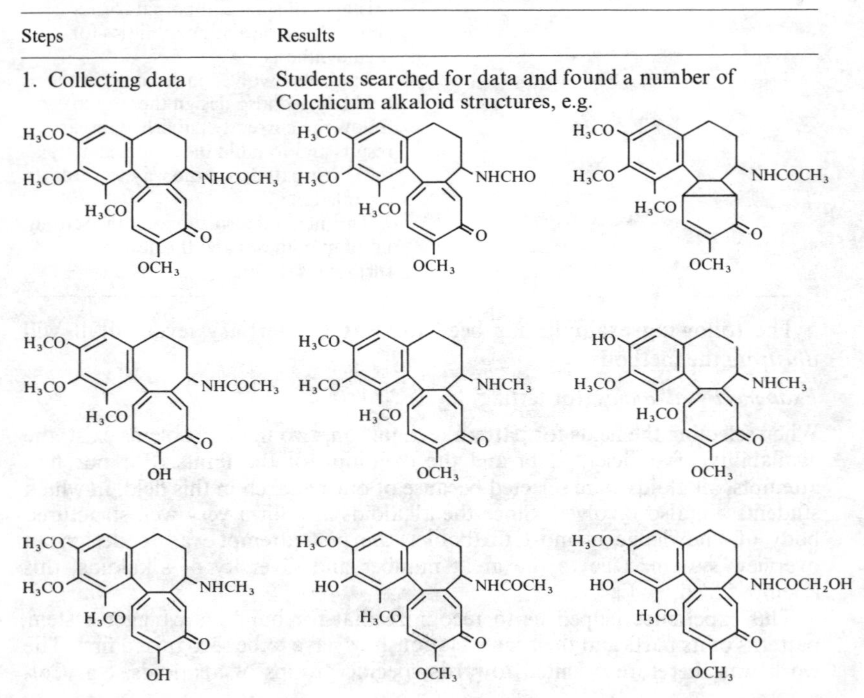

Colchicum alkaloids Steps	Results

2. Analysing data, defining constants and variables

Analysing the above formulae students found that: all these Colchicum alkaloids have the common part (= constant):

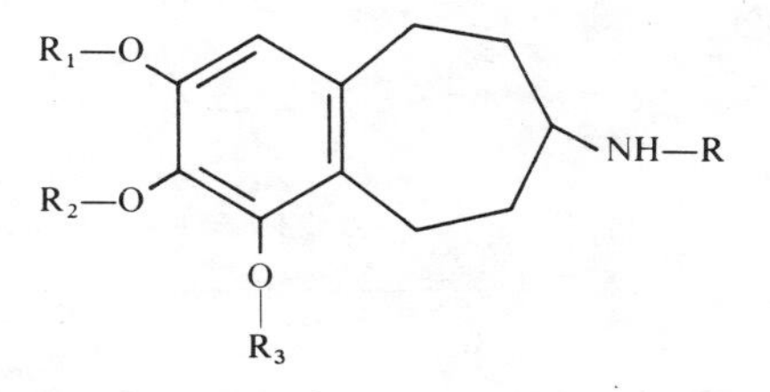

the third, seven-membered ring has two variations:

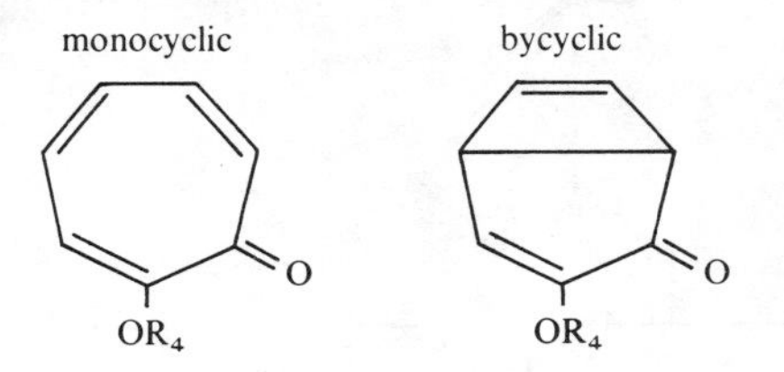

R_1, R_2, R_3 and R_4 might be —H or —CH_3
R might be —CH_3, —CHO, —$COCH_3$ or —$COCH_2OH$

3. Defining hierarchical order of criteria, building a tree

Students estimate which criteria are more general, i.e. corresponding to more alkaloids, and which are more specific for the collected formulae.

They define their hierarchy, e.g. all alkaloids discussed have the ring system:

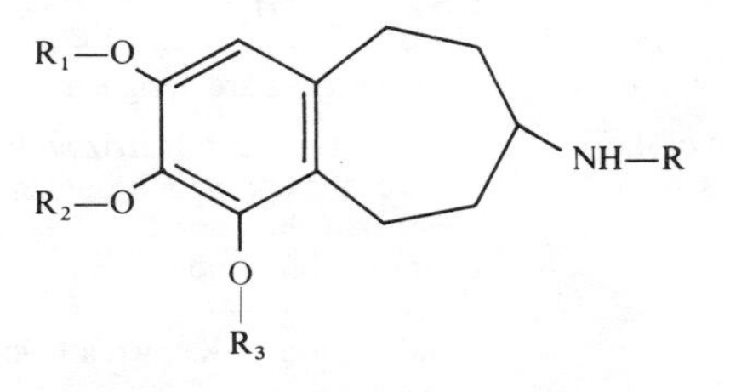

This is a criterion of the highest order, i.e. C_1; the third ring represents C_2; R_1, R_2, R_3 and R_4 enable more combinations, i.e. are more specific than R; the latter is therefore of a higher order, etc.

TABLE 6. (*continued*)

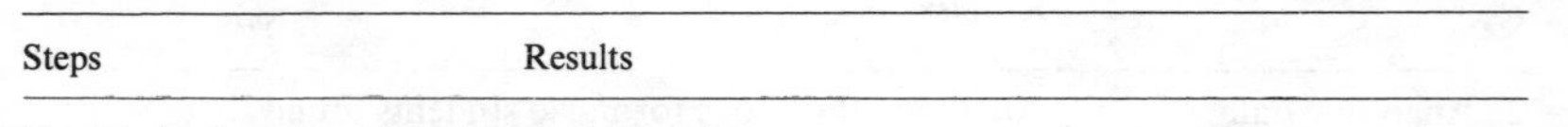

Steps	Results

On this basis, a tree structure is designed, e.g.

Pattern of colchicum alkaloids (first draft)

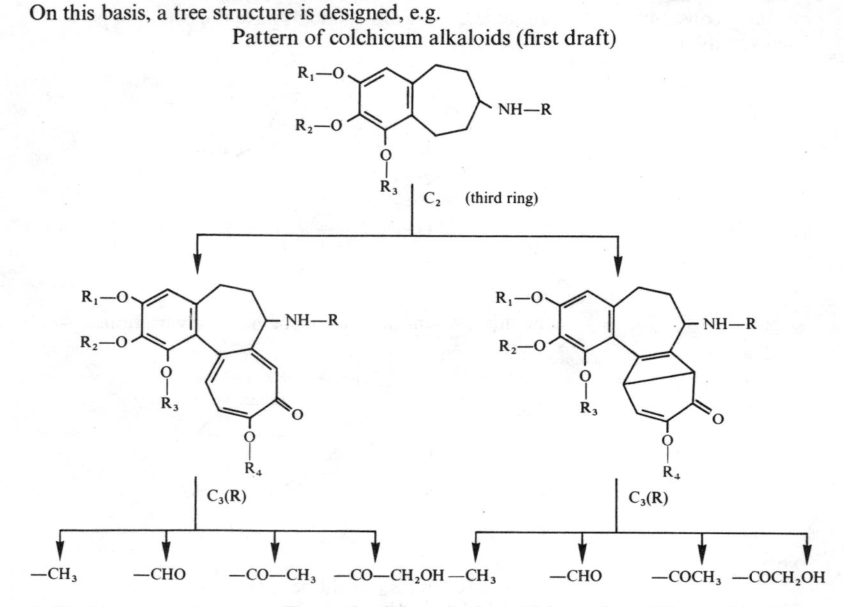

Steps	Results
4. Designing matrices of possible combinations	For each of the criteria, matrices of possible combinations are designed, e.g. for R_1, R_2, R_3 and R_4: for one OH group, i.e. for one radical being H, the following combinations are possible:

R_1	R_2	R_3	R_4	Combination
H	CH_3	CH_3	CH_3	I
CH_3	H	CH_3	CH_3	II
CH_3	CH_3	H	CH_3	III
CH_3	CH_3	CH_3	H	IV

Such matrices are designed for all possible combinations.

Steps	Results
5. Validation of the pattern designed	Students face each matrix with their data collection (e.g. for the left main branch of the tree with two seven-membered rings and R = CH_3) the following data have been found by students:

Combination	Known alkaloid
I	2-Demethyl-demecolcine
II	3-Demethyl-demecolcine
III	?
IV	Demecolcine

Steps	Results
6. Using pattern for prediction	Question marks initiate further work: Does such an alkaloid exist and have we overlooked it during our search for data? The result is a new search, including also research journals and computerized chemistry data bases. If not found, the discussion described in the presentation of this method in step 6 starts. Colchicum alkaloids are interesting because of their strong cytotoxic activity which is of interest in anti-cancer research. Because of their very high toxicity, students also play with the idea that one of these combinations might be less toxic—a good motivation for work.

chicum, Ergot, Valeriana, Veratrum and Vinca. The example of Colchicum alkaloids is given in Table 6 [39].

An extensive knowledge is required for good pattern recognition, and this is time consuming and not always possible in the teaching and learning process. In spite of the fact that with less knowledge these patterns are always full of deficiencies, they have much higher educational value than just memorizing knowledge. An example of this is the reasoning of students when we discussed the pattern of ergot alkaloids:

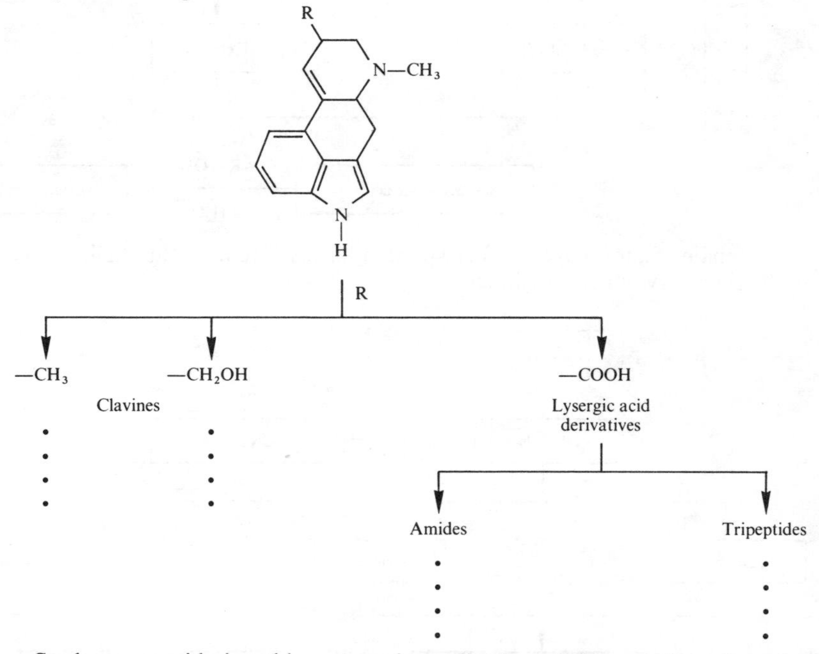

Students, considering this system, immediately put the question: does the biosynthetic pathway run along each vertical branch separately, or are horizontal transformations of clavines into lysergic acid derivatives possible in this system? An important question, which was this time so obvious for the majority of students, would be a rarity in teaching without searching for patterns of knowledge.

Example 2—Organic compounds of oxygen (for secondary level)

Pattern recognition is, of course, more effective if a larger amount of data is available, enabling us to search for common parts and variables. It is therefore usually more adequate for higher levels of education. But there are a number of possibilities of using pattern recognition also in secondary schools. Students, for example, learn organic compounds of oxygen too often only in fragments, in spite of the fact the limited scope of these compounds at the secondary level might be presented as a chain of knowledge:

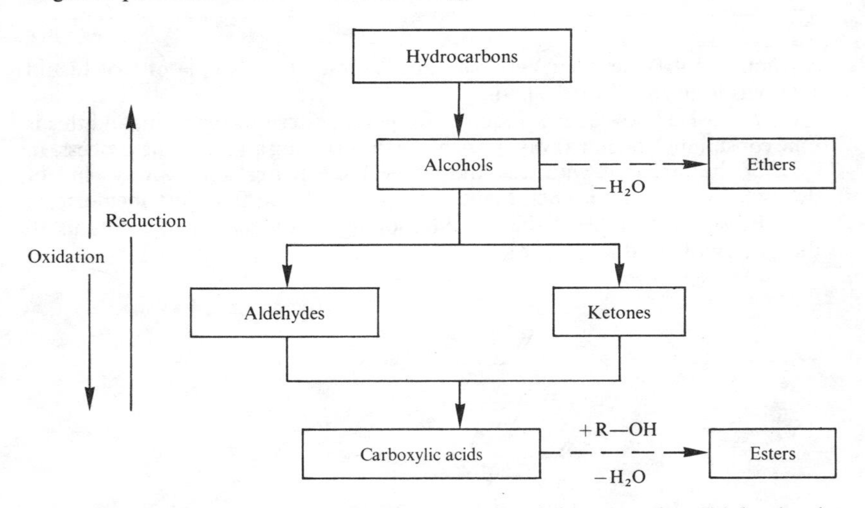

This chain of knowledge is in fact a pattern, since it is repeated at a higher level in the chemistry of carbohydrates:

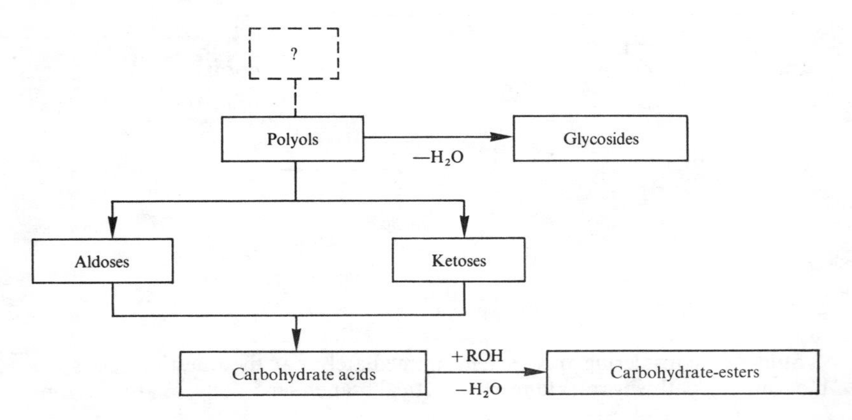

Even some analogy with hydrocarbons is present, if we consider only a fragment of the carbohydrate molecule, i.e. the desoxycarbohydrates.

In this way students: learn more about the carbohydrates and their interrelationships; repeat the previous discussions on oxygen compounds; organize their knowledge and get a feeling for patterns.

Example 3—The chemistry of the OH group

Another example might be the attempt to develop a pattern of compounds with the OH group. Students should write down all compounds with OH group which they can recall. They should classify them according to their knowledge—or even better according to their experimental results—into acidic, neutral and basic, e.g.

acidic	neutral	basic
phenol	methanol	sodium hydroxide

After classifying all compounds already known to the students into these three groups, they should search for new ones in the literature and classify them according to their character, tested if possible by experiment, into one of these three groups. When they have collected a number of compounds for each group, they should start to find what one vertical group has in common to have a similar character. Hypotheses should be set up to explain the compounds listed and new formulae not yet known should be given to students, asking them first to predict the acidic, neutral or basic character—and then always to test the prediction. The latter is of paramount importance since students—and not only them—are inclined to declare hypotheses as scientific truth.

Students usually soon conclude that it is the partner of the OH group which is responsible for the character of the compound. Those with some knowledge of atomic structure and bonds search for the answer in electron attraction by the partner of the OH group.

Example 4—Colour of organic and inorganic compounds

We can also use pattern recognition in the opposite way: instead of a tree branching further and further, students try to link branches of knowledge together to find the crucial answer (e.g. what structure do coloured organic compounds have in common?). They write down a number of formulae which they can recall, then they again search in textbooks for more formulae of coloured compounds. They try to classify them according to their similarities into groups and explain what they have in common. Analysing these formulae bit by bit in a comparative way, they come finally to the conclusion that the accumulation of electrons with higher energy is the only common part, and put forward the hypothesis that these cause the colour. They can test this hypothesis (e.g. by oxidizing the aromatic ring into quinone) using appropriate non-toxic compounds for this reaction. They can discuss why inorganic compounds are coloured and why iron changes colour when heated.

This is the usual way of testing hypotheses. The new approach is in collecting sets of formulae (a similarity with mathematics here) instead of learning some examples, in recognizing crucial parts instead of getting them explained, and of learning that systematic work can help everybody towards success.

Example 5—An example of an advanced project: the properties of a compound via pattern recognition.

The following example can be used at many different levels. At the senior secondary level it can be used as a project. More can be demanded at the university level. The example given is one that is of day-to-day importance—the physiological activity of drugs.

Drugs are mostly classified in three groups:
1. Compounds of natural origin (e.g. alkaloids).
2. Analogues of natural substances (e.g. new penicillins).
3. Purely synthetic substances (e.g. sulphonamides).

This classification according to origin is, of course, of no help when physiological activity is discussed.

The students tried to use as a criterion the pharmacophoric groups.

In this approach, however, the branches were numerous, and criteria were not clear enough. Moreover, the compounds studied showed the students that similar molecules might have different effects and vice versa. They found that in the structure–activity relationship of drugs, several parameters are involved (e.g. electronic state, steric effects, hydrophobic properties, etc.).

The result of pattern recognition in this field was poor, but the 'side-effect' rich: students found quantum chemistry not just a 'demanding and boring abstract subject', but also involved in most interesting research on the influence of molecular shape in biological activity [40], considering the topographical characteristics of the electric field around the molecule to be of paramount importance. In searching for answers, students discovered new questions: What is a molecule? Is it, according to the definition of classical chemistry, a collection of atoms held together by chemical bonds? Or a dynamic system consisting of a number of electrons and nuclei that interact via electromagnetic forces? As a result, the introduction and discussion of electron density maps at the appropriate level of chemistry teaching was their suggestion.

3.5.1 *Conclusion*

There are a number of advantages in the use of pattern recognition in learning chemistry, e.g.:
1. The use of different sources of data, their comparison and critical estimation.
2. Motivation of students, since they search for knowledge independently, and their learning is at the same time productive in the search for more order in data organization.
3. Organizing learning towards reaching higher cognitive levels, especially the level of analysis and synthesis of knowledge and support for problem solving.
4. The development of the ability to organize data, to find and classify parameters, to search for regularities, systems and patterns.

5. Support for the use of the computer in chemistry.
6. Support for prediction, design for checking hypotheses.
7. Orientation towards curriculum design.
8. Design of research.
9. Support for industrial decisions.

The most important feature of this method is a recognition of the development of a selected field, with more questions than answers as the result. Pattern recognition can combine teaching and learning situations with research design and has proved to be of help also in industrial decisions. It therefore belongs among the most efficient methods in education.

But there is also a number of dangers if the method is not used with high doses of common sense. For example, (a) in a too vague limitation of the selected field, students will get lost; if too limited, they might come to the wrong conclusions; (b) without enough data students might end in speculations; and (c) there is a danger of oversimplification, and also of 'model thinking'.

Many of these dangers can be avoided if the process of pattern recognition is seen to be more important than the pattern produced. Students should be encouraged to search for different patterns, the latter should be discussed in a comparative approach, and critical evaluation should be a most important part of the process.

However inexperienced we are, whatever errors we might expect, the way towards recognizing patterns has to be taken, because patterns exist in matter. This also means that every pattern design has to be faced not only with facts and their relationships, but in a dialectic approach, in which the recognition of their changing nature is also essential.

References

1. C. Y. Harben, *Education in Chemistry*, Vol. 11, 1974, p. 2.
2. G. Lind, *European Journal of Science Education*, Vol. 2, 1980, p. 15.
3. G. C. Pimental, in A. Kornhauser (ed.), *Chemical Education in the Coming Decades*, p. 108, Ljubljana, Unesco-DDU Univerzum, 1979.
4. A. Kornhauser, M. Vrtáčnik and G. Djokić, in W. T. Lippincott (ed.), *Proceedings of the Sixth International Conference on Chemical Education*, p. 281, College Park, Md., University of Maryland, 1982.
5. R. M. Gagné, *The Conditions of Learning*, New York, Holt, Rinehart & Winston, 1970.
6. D. P. Ausubel, *Educational Psychology—a Cognitive View*, New York, Holt, Rinehart & Winston, 1968.
7. K. F. Jackson, *The Art of Solving Problems*, London, Heinemann, 1975.
8. J. Warkin, J. McDermott, D. P. Simon and H. A. Simon, *Science*, Vol. 208, 1980, p. 1335.

9. J. R. Partington, *History of Chemistry*, Vol. I, London, Macmillan, 1970.
10. J. M. Scandura, *Problem Solving*, London, Academic Press, 1977.
11. A. D. Ashmore, M. J. Frazer and R. J. Casey, *Journal of Chemical Education*, Vol. 56, 1979, p. 377.
12. C. E. Wales and R. A. Stager, *Engineering Education*, Vol. 62, 1972, p. 456.
13. R. A. Stager and C. E. Wales, *Engineering Education*, Vol. 62, 1972, p. 539.
14. T. S. Tseng and C. E. Wales, *Engineering Education*, Vol. 62, 1972, p. 812.
15. C. E. Wales, *Engineering Education*, Vol. 62, 1972, p. 905.
16. P. E. Hoggard, *Journal of Chemical Education*, Vol. 57, 1980, p. 299.
17. R. H. Yeany and W. Capie, *Science Education*, Vol. 63, 1979, p. 355.
18. C. T. C. W. Mettes, A. Pilot, H. J. Roossink and H. Kramers-Pals, *Journal of Chemical Education*, Vol. 57, 1980, p. 882.
19. C. T. C. W. Mettes, A. Pilot, H. J. Roossink and H. Kramers-Pals, *Journal of Chemical Education*, Vol. 58, 1981, p. 51.
20. E. J. Peters, *Problem Solving for Chemistry*, Philadelphia, Pa., Saunders, 1971.
21. I. L. Finar, *Problems and Their Solution in Organic Chemistry*, London, Longman, 1973.
22. Y. Okumara and A. Sakurai, *Bulletin of the Chemical Society of Japan*, Vol. 46, 1973, p. 2190.
23. A. Kornhauser, in W. T. Lippincott (ed.), *Proceedings of the Sixth International Conference on Chemical Education*, p. 115, College Park, Md., University of Maryland, 1982.
24. D. P. Ausubel, *Journal of Educational Psychology*, Vol. 51, 1960, p. 276.
25. B. S. Bloom, J. T. Hastings and G. F. Madaus, *Handbook on Formative and Summative Evaluation of Student Learning*, New York, McGraw-Hill, 1971.
26. J. S. Bruner, *Toward a Theory of Instruction*, Cambridge, Mass., Harvard University Press, 1967.
27. *Science, A Process Approach (SAPA)*, Washington, D.C., American Association for the Advancement of Science (AAAS).
28. C. M. Reigeluth, M. D. Merrill and C. V. Bunderson, *Instructional Science*, Vol. 7, 1978, p. 107.
29. A. Kornhauser, 'Computer-Assisted Instruction in Chemical Education', in C. N. R. Rao (ed.), *Proceedings of the International Conference on Chemical Education, Madrid, 1975*, p. 107, Oxford, International Union of Pure and Applied Chemistry (IUPAC), 1975.
30. J. E. Ash and E. Hyde, *Chemical Information Systems*, New York, John Wiley, 1974.
31. R. Luft, *L'actualité chimique*, Vol. 33, 1979, p. 37.
32. J. F. B. Rowland, *A Review Study of Information Transfer and Use in Chemistry*, Report to British Library Research and Development, Project S1/G/161, 1977.
33. M. J. Frazer, *Maps of Chemical Knowledge, International Seminar on Chemical Education, University of East Anglia, Norwich (United Kingdom)*, 1979.
34. F. Basolo, *Journal of Chemical Education*, Vol. 57, 1980, p. 761.
35. F. Basolo and R. W. Parry, *Journal of Chemical Education*, Vol. 57, 1980, p. 772.
36. M. A. Wilson, *Journal of Chemical Education*, Vol. 52, 1975, p. 495.

37. J. Schwartz and J. A. Labinger, *Journal of Chemical Education*, Vol. 57, 1980, p. 170.
38. H. K. Hall, Jr, *Journal of Chemical Education*, Vol. 57, 1980, p. 49.
39. D. Glavač, M. Glavač-Ravnik and A. Kornhauser, *Pattern Recognition in Colchicum Alkaloids*, Interim Report, RCPU, University of Ljubljana (Yugoslavia), 1981.
40. D. Hadži, 'The "Shape" of Biologically Active Molecules', in A. Kornhauser (ed.), *Natural Sciences—Social Sciences Interface*, p. 159, Ljubljana, DDU Univerzum, 1981.

4 Practical work and technology in chemical education: three aspects

M. H. Gardner, J. W. Moore and D. J. Waddington

There is often a large gap between our teaching ideals and our teaching practice, and nowhere greater than when we discuss work that is student-centred, be it work in the laboratory or in the classroom. Parallel to this, there have been advances in the technology available to enhance the quality of teaching in the form of hardware (for example, laboratory equipment, projectors, video recorders, calculators, microcomputers) and software (demonstrations and experiments, slides, videotapes, books and wall-charts produced by industrial companies, computer programs).

There is not enough space in the book to explore all the important areas in detail, although they are mentioned throughout the book and interesting references are given. Instead, three areas of both immediate and growing interest in different parts of the world are brought together in this chapter: low-cost equipment, safety in the laboratory and computers in chemical education. The reasons for their choice reflect the interest shown by teachers at recent international and regional conferences on chemical education and from the correspondence between teachers and members of the IUPAC Committee on Teaching of Chemistry.

Laboratory work is an essential component of chemical education. Students can be stuffed with facts and theories but without experiments they cannot experience the reality of chemistry as a science. The development of powers of observation, measurement, prediction, interpretation, design and decision-making are dependent on laboratory and field-work experience, the real world of our discipline. However, schools in most developing countries are ill-equipped for practical work, and even where laboratories have been built the problems of servicing the laboratory (water, power, technical assistance) as well as equipment are becoming more acute. In one area at least, in the provision of equipment, solutions are possible. Yet progress in the development of locally produced, low-cost equipment is of interest to all countries, especially to those in which chemistry teaching at school is least developed.

At the Sixth International Conference on Chemical Education one particular concern of the teachers who attended (and some eighty countries were represented) was laboratory safety. Those teachers who are actively involved in laboratory work inevitably accept more responsibility for the safety of themselves and their students. This has become a particular issue in countries where there is a growing demand to teach chemistry to all students, not to just chemistry specialists, so that classes are larger with perhaps less able students, and in those in which legislation concerning safety has been strengthened (usually aimed at industry but spilling over into schools). Some guidelines for safe practices during demonstrations and student experiments have been selected for presentation here. Attention is also briefly addressed to a related issue, safe disposal of toxic waste.

The use of computers and hand-held calculators provides comparatively new technological resources for chemistry teaching. They are moving rapidly into the secondary schools of many industrialized countries. As relative costs continue to fall, the versatility, capacity and ubiquitousness of this technology continue to increase, and again the reaction by teachers at recent international and national meetings suggests these instruments will be of interest to teachers everywhere. This is a relatively new but rewarding and exciting aspect of chemical education. It is developing so rapidly that it is difficult for any written account to be up to date and encompass the experience of various countries and teaching levels. The article in this chapter reflects the author's experience most closely but provides basic information and direction for us all.

SECTION 4A

LOCALLY PRODUCED AND LOW-COST EQUIPMENT: A SURVEY

4A.1 Introduction

Although much of this section will be concerned with the provision of equipment in developing countries, those of us who are teaching in the more developed parts of the world will learn much from their experiences. For, as the demand for 'science for all' increases so do costs, and successful laboratory courses will depend more and more on the ingenuity and resourcefulness of individual teachers. Moreover, locally produced equipment can serve the needs of the teacher, the student and the curriculum more effectively and is easier to *maintain*, with less reliance on spares which can only be obtained from afar.

'Effective chemistry teaching depends on three factors—teachers, equipment (hardware and software) and chemicals' [1]. However, the supply of equipment in many countries is frequently neglected. For example, at the Unesco meeting of senior officials of the Ministries of

Education of the twenty-five least developed countries, held in 1975, a major problem area for school work was identified as scientific and technical equipment [2]. Similarly, the same problem has been highlighted by the Commonwealth Secretariat, which represents over thirty developing countries among its thirty-six members [3]. Indeed, as stress is laid on laboratory work in school courses, the question of the resources that are needed becomes one in which financial and educational considerations must be discussed side by side. Much apparatus is imported into developing countries. This not only uses up foreign exchange, but the equipment can take a long time to arrive. Even if it arrives in good time, it is possible that the wrong choice has been made as the teacher is unable to know the full range of equipment available. If the apparatus is at all complicated, it needs spares and technical expertise to maintain it.

As we have seen from the many curricula discussed in Chapter 2, there is in all of them the insistence that practical work is an *integral* part of teaching and learning. However, how can we obtain the equipment for the courses? How can the equipment be maintained? How can we reduce the amount of consumable materials and find alternatives to expensive (often imported) chemicals?

Many schools are teaching without significant use of experiments yet following curricula based on learning through experiment. This may actually be worse education than when the curricula was designed for lecturing—'chalk and talk'.

As Hakansson has summarized [4]:

> The high cost of equipment-based science curricula being developed in industrial countries places severe limitations on their use in developing countries. The need for testing and evaluating some of these new ideas and methods, and adapting them to the conditions and economic situation of the developing nations must be recognized. However, by careful selection and development of low-cost equipment, it is believed that most countries will be able to adopt these modern methods of science teaching. If the selection of equipment is based on 'principles' rather than 'precision' of the results, very simple and rather inexpensive equipment can be used.

There are two major ways in which equipment may be produced for schools: by individual teachers for their own schools and at local or national production centres.

4A.2 Equipment made in schools

Among recent courses at the school level are those organized for teachers by RECSAM; collections of apparatus assignments have been published which are devoted principally to the development of primary and junior secondary science and mathematics teaching [5].

Two Unesco books are particularly valuable. In the *Unesco Source-book for Science Teaching*, there are many suggestions for simple teacher-made equipment and a very wide variety of experiments [6]. In the more recent *Unesco Handbook for Science Teachers*, there is a chapter on facilities, equipment and materials and one on resources [7]. The guide-books on *Constructing Inexpensive Science Equipment* [8] produced at the University of Maryland are also mines of information to help the individual teacher make equipment. There is also an excellent pamphlet produced by JETS (Junior Engineers/Technicians/Scientists), based in the School of Engineering, University of Zambia, to help schools themselves produce equipment such as wooden racks and stands for pipettes, burettes and test-tubes, and metal clamps, clamp-holders, retort-stands [9] and similar work has been accomplished by the National Council for Educational Research and Training (NCERT) [10] and the Institute for the Promotion of Teaching of Science and Technology (IPTST) [11]. Details have also been published for a mobile laboratory unit in Lesotho [12] and one in India [13]. There are helpful suggestions in the report on the Asian Symposium on Low-Cost Equipment [14], organized by ICASE [15]. In the report, there are suggestions ranging from the production of a pH meter, simple electrolysis cells for junior secondary schools and alcohol burners, to hand centrifuges and so on.

One example of a manual that contains instructions on how to make simple equipment in schools, and also gives many examples of experiments is *Manual de Química Experimental*, produced in Bolivia [16]. This gives very clear instructions on how to make simple balances, various supports, an alcohol burner and electrochemical equipment; the instructions are written so that teachers with little or no workshop experience are able to follow them. There is also a list of chemicals and where they can be found (in the market, in a pharmacy and so on). The experiments illustrate much of the junior secondary level chemistry (preparation and properties of gases such as oxygen, acids, bases, salts; laws of composition; and course theory).

One of the authors, Hans Schmidt, writes about the seminars he organizes for teachers and about the book he has written [17]:

> Pilot seminars can provide effective short-term answers to the problem of the shortage of teaching materials. The methods and solutions proposed can immediately be put into practice and the materials easily reproduced. These seminars last about two weeks and are divided into two parts.
>
> The first part gives detailed guidance and ideas for the production of multi-purpose teaching kits which can be assembled from low cost materials available even in remoter areas of developing countries. This first part concludes with the compilation of individual transportable kits. The subsequent experimental part is carried out with the items from the kit.

1. A personally produced or compiled collection of teaching materials such as this offers the following *manufacturing advantages*:
   1. The individual items are so constructed, that they can be produced without great technical skills, and with the simplest tools, in a one week (!) seminar.
   2. The collection of teaching materials is compiled and produced predominantly from locally available materials i.e. materials obtained from the environment of teachers and pupils.
   3. The costs are so slight as not to overtake the limited financial resources of local self-help.
2. A collection of teaching materials kept under lock and key in a handy, easily transportable box has *advantages in the teaching situation*:
   1. The items may be used without extensive preparation since the total requirements for experiments are readily available for classes of all levels.
   2. A specialized subject room with energy supply is not necessary; if need be the box lid can be used as a table.
   3. The replacement of lost or used items (e.g. chemicals) or the addition of new materials is possible on the spot and not subject to the irksome official channels of supply.
   4. Because of the minimal monetary value of the materials in contrast with the expense of prestigious imported materials, the risk of theft is slight. In case of loss the teacher is spared frequent restricting liability claims or censure from superiors.
   5. The experiments with the apparatus and materials are simple enough to enable inexperienced teachers to conduct modest experiments after only one week of seminar training.
3. The production of compilation of teaching materials at the level of the teacher consumers has *advantages for the further promotion of experimental science teaching*:
   1. By transferring the procuring of teaching materials to the consumer level one eliminates the administrative regulations which exist in many countries where domestic or imported teaching aids are provided by a centralized distribution agency.
   2. Within about two weeks teachers can produce and compile collections of teaching materials and learn how to use them without any great administrative or financial burden for the school authorities.
   3. One sample kit brought into the schools by the teachers can be copied wholly or partially by pupils without additional burden on the teachers.
   4. The increasing success thus attainable within a very short time serves as a stimulus for further activities in this field.
4. The production of teaching materials by teachers and pupils has *strategic advantages* on a developmental level beyond that related purely to subject usefulness:
   1. Development, planning and application can be carried out in co-operation with local sources of improvisational skills and know-how.
   2. Short-term success can give stability and increased self confidence to local self-help groups.

3. Simple materials produced by the consumers themselves can reach more schools within a much shorter time than administratively organized and controlled deliveries, of use only to development centres.

5. The use of simple teaching materials has *economic advantages*:
   1. The use of such materials can serve as an effective emergency solution in the event of a mass shortage of teaching materials.
   2. Their use does not lead to a sort of technological stagnation but rather serves as a preparatory phase for the later utilization of sophisticated and expensive, imported, or locally manufactured materials.
   3. Previous basic knowledge can promote the proper, efficient use of sophisticated materials and limit the misuse of equipment and chemicals so costly in foreign exchange.

This note from Schmidt summarizes the *advantages* of teacher-made equipment. One particularly important point is that if teachers make their own equipment, they are also able to maintain it. The seminars are an important example of in-service education and training of teachers (Section 6.4) and should, if we are really serious about experimental work in schools, become a cardinal feature in pre-service work (Section 6.2).

Equipment for the senior secondary schools is also being developed, in conjunction with a programme for first year university courses. This has had much less time and effort devoted to it, but since 1975 Unesco has arranged six international meetings, four of which have been in collaboration with the IUPAC Committee on Teaching of Chemistry. The first two meetings were held in the summer of 1975 and in early 1976 in Seoul, Republic of Korea, and in Amman, Jordan. The Seoul workshop was attended by seventeen teachers from nine countries in the South-East Asian region, and eight teachers from the Arab region went to the Amman workshop. Two manuals of experiments for laboratory courses in chemistry were produced [18, 19]. Dominant aims in the planning and execution of the first workshop included the following: practical work was chosen and written up to show how it could interact with classes; to direct attention to areas in the curriculum that are either difficult to teach in the laboratory and/or need to be introduced because of the changing nature of what we teach; to show that chemistry is a unified subject; to produce an account of the workshop that would be really useful for teachers; to simplify experiments, and this is particularly cogent to our discussion.

In considering new equipment, we must first define the purpose for which it is to be used. For example, and this is particularly important, the accuracy of an experiment too often in turn demands sophisticated apparatus. Thus, either the experiment cannot be done at school or, if it can, the student forgets the essential points of the chemistry. Further, apparatus is often used unnecessarily and unscientifically, producing data whose accuracy is incompatible with the other measurements taken during the experiment. Particular examples from the first workshop included a detailed study of the types of calorimeter that are required to get consistent and meaningful results for, say, the enthalpy changes for reactions between ions in solution and metals, between acids and bases, and the strength of intermolecular hydrogen bonding. Thus, the calorimeter constants of a series of different calorimeters were determined by some of the participants for experiments in thermochemistry, showing that an ordinary plastic cup is effective for these experiments at room temperature. In other experiments, measuring cylinders were used instead of pipettes; an accurate overall result is obtained and a potentially dangerous operation is averted.

Another area in which simplified equipment was shown to be very helpful to the understanding of the experiment was electrochemistry, where the introductory work on redox reactions, the effect of concentration on electrode potential, and experiments on transition metal systems were done with a very simple voltmeter, in place of a digital voltmeter, and on a 'test-tube' scale. The preparation and properties of many organic as well as inorganic compounds can be examined with very simple equipment on a test-tube scale.

This does not mean that we should encourage teachers to simplify for the sake of it, and spoil the point of the experiment. There is also a place for the training of students in making accurate measurements. Ideally, in quantitative experiments, one should simplify if the answer obtained is as accurate as that required. Some equipment may appear expensive, but if it is frequently used it can prove to be less expensive per student than existing equipment. The digital voltmeter (or pH meter) is simple to operate and can be used with more effect than, say, five conventional potentiometers. However, if there is no foreign exchange available, how can one obtain such equipment? In collaboration with Unesco, the IUPAC Committee on Teaching of Chemistry has a team based at the University of Delhi developing equipment such as pH meters and conductance bridges which can be built for as little as US $50 each, and give the accuracy of conventional industrially produced equipment. K. V. Sane, the leader of the project, has produced a manual for teachers giving details on how to make this equipment and also giving examples of experiments [20]. The manual, based on in-service seminars, contains a wealth of information such as, for example, the use of carbon electrodes (taken from old batteries) which after treatment can be used in place of platinum.

The experiments considered in the Amman workshop were an excellent complement to those at Seoul; they were generally simpler in terms of chemical concepts and used simpler equipment. As with the first workshop, 'the experiments were chosen with regard to cost, ease of performance, ease of variation, a high probability of providing meaningful results in inexperienced hands, and, above all, teaching and learning potential'. The background information given to the student and teacher was more limited than in Volume 1, for the authors did not wish 'to pre-empt the role of the instructor' on the one hand, or 'to teach the teacher' on the other. Over fifty experiments on stoichiometry, energy, rates of reaction and equilibrium are described all using simple equipment, for example, an acid-base titration using a dropper, titration curves using a Universal indicator, analysis of cations by paper chromatography, kinetics illustrated by the drawing of a burette.

4A.3 Chemicals

It is also important to consider how much and what chemicals should be used. Small-scale work not only is a cheaper way of running a practical course, but it enhances safety and it can increase the manipulative skills of students. For example, a recently published text in Hungary shows how these techniques can be used throughout the school course [21] and similar texts have been published in other countries.

However, as Alonge has written [22]: 'A set of chemicals that would be classed as locally available is yet to be given the attention it deserves.'

Alonge surveyed 180 reports in the literature on such material but found only eleven which showed how such materials were used in teaching chemistry by experiment. The reports were based on such items as geochemical minerals, disused dry cells, metal scraps, vegetable oils, orange peel, root extracts, soap detergent and baking powder. He points out that the chemistry of the experiment is often over-simplified and that the cost–benefit of the materials is rarely shown. Alonge stresses that much more work in this area is needed and that the reports are found in journals to which teachers will not have access. Nevertheless, he himself gives a list of useful references. IPST, in Thailand, has produced a detailed list of chemicals readily available from local markets, reflecting differences in availability in different parts of the country [11]. Other institutions could usefully provide similar lists in their own countries.

4A.4 Equipment made in production centres

But to be realistic, many teachers, faced with the day-to-day difficulties of teaching, find the burdens of making and servicing equipment for experimental work daunting. Others are unconvinced that it is all worth the trouble; there are many cultural as well as economic hurdles to surmount. Many have had no in-service or pre-service training. Many have not even been given designs, instructions or the simplest tools. Many have no access to 'petty-cash' to buy even small amounts of materials from the local markets—bureaucracy reigns supreme. Finally, teachers really should not be expected to be sole providers of equipment but should be expected—and trained—to supplement equipment manufactured and supplied to the school.

We must consider whether teachers are really encouraged to make and maintain their own equipment. Nevertheless, we must harness the work of these dedicated and talented teachers; it is too valuable to waste. From their efforts we can develop local production units (while at the same time remembering the suggestions made by Hans Schmidt and utilizing them).

A brief summary of experiences in Bangladesh, Fiji, Hong Kong, India, Indonesia, Japan, the Republic of Korea, Pakistan, Philippines, Singapore and the Socialist Republic of Viet Nam has been published [23].

What now are the requirements and prerequisites for providing science equipment on a national scale? Hakansson suggests [4]:

> First of all, there must be a genuine interest from the side of the educational authorities to raise the quality of science education in general. A further basic criterion is the establishment of standard lists of teaching aids and equipment which hopefully should be available in the schools. These lists should be developed and prepared by the curriculum authorities and carefully evaluated by science specialists, senior science teachers, equipment experts and educational financing authorities.
>
> When the requirements have been established there are mainly three courses of supply, namely import, organized local production and equipment making at the school level by science teachers and students, or school technicians. I have deliberately excluded donations and foreign aid in the form of ready-made equipment as these means of supply cannot, in any way, cover the needs of the schools except for very small countries where the criteria are altogether different.
>
> As mentioned earlier, large scale import of basic science teaching equipment is most often not possible due to shortage of foreign exchange. Import must therefore be limited to specialized items which cannot be obtained or manufactured locally. Items such as chemicals, electrical and electronic components, and sophisticated instruments such as microscopes and electrical meters or components for such instruments must probably be imported for some time to come. Preferably, imports should be limited to components and raw materials which later can be processed locally to finished products. The provision of imported foreign equipment to teacher training institutions has also proved to have some adverse effects, if simple low-cost equipment is not also available. Student teachers are then trained to conduct lessons in well-equipped laboratories and with imported equipment, very often rather sophisticated. When the same teachers come to conduct the same lessons in schools, they are often frustrated in their efforts because they find that even the most elementary pieces of teaching aids are not available.

There are some very interesting large-scale projects in developing countries where locally based equipment is produced, for primary and lower secondary school courses. Warren and Lowe review some of the current development in some countries in *The Production of School Science Equipment* [24].

Four production centres that spring to mind immediately are NCERT in New Delhi, India; IPST, based in Bangkok, Thailand; the Science Education Production Unit (SEPU) in Kenya; the National Educational Equipment Centre (NEEC) in Pakistan [25]. The concept of centres is not new. For example, one was set up in Chile in 1964 as a prototype [26] while, on a much bigger scale, the Science Workshop of NCERT was able to report that they were making batches of 1,500 lots for primary and middle schools and that contracts were being made by UNICEF for 50,000 kits [10].

The driving force behind SEPU was the academic staff of the Kenya Science Teachers' College (KSTC), and although SEPU is now officially a separate organization, the workshops and sales department are in the KSTC compound. SEPU produces kits in biology, chemistry and physics for secondary schools and these are designed to enable children to do all the practical work needed for the East African Certificate of Education. The materials used are all local, and with the kits come manuals for teachers and pupils; teaching aids (slides, photographs) are also produced. The chemistry kit uses a peg-board and terry clips as the stand; glass tubing is replaced by plastic. The emphasis is on pupil participation with small-scale work; the kits are not suitable for demonstrations. The kit has been described in detail in an article in *School Science Review* [27].

IPST has produced materials which include more sophisticated equipment, such as electrical meters, than in many of the other centres. As IPST is 'mainly concerned with innovations in education, it produces the equipment for its own centres, thirty-six of which are established in Teacher Training Colleges, and for trial schools' [28].

A Directory of Educational Equipment Production Centres has been published, giving outline details of centres in over forty countries [28]. There are also useful sections on the education and training of personnel for the workshops; curriculum developers, designers, draughtsmen, technicians, purchasing and supply, storekeeping and distributors. Several very helpful articles on setting up centres are included in the *Proceedings of the Asian Symposium on Low-Cost Equipment* [14].

To summarize, the following appear, from past experience, to be the most essential elements when setting up activities leading to the production of equipment and teaching aids:

1. The centre must have expertise in design (professional engineers in equipment design can make use of modern production methods and materials) and in management and distribution. The training of management staff must be given high priority as past experiences have shown that some potentially promising projects have failed due

to lack of proper management and not because of technical difficulties, which is often the belief [4]. The training of technicians too is of vital importance for the development and production of equipment, and its subsequent maintenance was seen as one of the major concerns at two important conferences held in Adelaide [29] and Perth [30]. The Committee on Teaching of Science of the International Council of Scientific Unions (ICSU), in conjunction with Unesco, is endeavouring to find out the extent of the shortages and to find ways of alleviating them [31].

2. The production centre must be cost-conscious; it must work to realistic budgeting. Marketing must be done properly.
3. The effective way is by establishing a production centre in which the minimum amount of equipment is used for making the apparatus.
4. A distribution infrastructure must be built up between the production centre and the educational establishments.
5. It may be necessary and indeed desirable that for small countries (where a regional centre is out of the question), there should be a production centre to cater for both university and school courses.
6. The locally produced equipment must be designed in conjunction with teachers and curriculum designers. (It is no good, for example, producing small-scale equipment if the textbooks are promulgating use of larger scale experiments.) Teachers must be encouraged to consider what is required and to see that their ideas are developed.
7. The apparatus made must be researched beforehand, trials organized, effective spares made, back-up material written. The apparatus must be of good quality.
8. Facilities must be available for in-service and pre-service training of teachers.
9. It must not be forgotten that in many countries these problems are not unique to schools and that universities suffer from them too.

Finally, the flowchart produced by Hakansson [4] shows the different routes by which locally produced—and hopefully low-cost—equipment can reach schools.

For example, the country statements on the role of laboratory teaching in university courses [30] given in the Unesco International Chemistry Congress, held in 1978, mention these points time and again. Indeed, it is significant that in these country statements, the ones from Argentina, Egypt, India, Indonesia, Iraq, Jordan, Kenya, Saudi Arabia and Sri Lanka among others, commented that the lack of experimental work in schools hindered progress at university [31]. Teachers in other countries feel the same, to judge from comments in discussions with teachers at the ALECSO seminar [32] and the Unesco meetings [33–35].

SECTION 4B

SAFETY IN THE LABORATORY

4B.1 Introduction

As discussed in the introduction to this chapter, one of the most significant aspects of the Sixth International Conference on Chemical Education was the concern of teachers about safety matters in the laboratory [36]. This is not only because of new safety legislation that has been introduced into many countries [37] but also because teachers are being encouraged, through new curricula, to base their lessons on students' individual laboratory experiments.

This section therefore outlines some of the areas which the chemistry teachers must consider—procedures, rules, advice and education.

The section is based on articles published previously, primarily in the *Unesco Handbook for Science Teachers* [7]. Even so, as with so much in chemical education, what will be appropriate is one country, or even in one school, will not be appropriate in another. It is therefore very important to translate the ideas in this section into action within the constraints of the teachers' own workrooms.

As science progresses, hazards progress with it. Within the home and community the child learns safety by being told about the hazards rather than by experiencing them at first hand. Bearing in mind the developmental stages of the late-primary/early-secondary school pupil, it is equally as necessary to focus attention directly on the hazards of learning about science. Such a positive focus of attention on safety is not only a precaution against accidents in the school laboratory but also a stage in encouraging an attitude of mind which will be reflected throughout the individual's life. Until the individual has reached the appropriate level of development of reasoning and awareness, *safety has to be taught; it cannot be learned.* Fortunately, the majority of teachers are aware of their responsibilities to inform their students of the need for caution in practical chemistry; for example, the explosive properties of nitric acid and ethanol if they are poured into the same container; or the danger from hydrochloric acid, formed from hydrogen chloride gas, as a result of disposing of sodium chloride (common salt) and used sulphuric acid in the same waste bin.

Such is the responsibility placed on the teacher. Given the number of pupils learning science, it is a tribute to teachers that very few serious accidents occur, but a continuous awareness is necessary to avoid that combination of:

$$\text{Hazardous situation} + \text{human error} = \text{accident}$$

With the overall theme of 'Good Housekeeping', the following approach may help the new (and not so new) teacher to maintain an accident-free environment:

1. *Timetable* a period early in each term to discuss aspects of safety.
2. *Examine* your laboratory and classroom for all the obvious hazards and implement methods to reduce them.

3. *Examine* your teaching technique and identify hazards relating to subject-matter.
4. *Prepare* and apply rules to the laboratory which each student can understand.
5. *Obtain* appropriate fire-fighting, safety and first-aid equipment.

Finally, in imparting knowledge about safety, over-emphasis is undesirable. Too many restrictive rules can bring about an accident-prone area simply because nobody will take the trouble to read the rules, or because they are too restrictive. Maximum effect will be obtained: by a clear identification of the hazards as new materials are introduced; by having a clean and tidy working environment; and by setting a clear example to the students by your own techniques of working.

4B.2 Potential sources of accidents

The following indicate some areas for examination by the teacher [7]:

Windows and doors in good repair: Maximum ventilation may be necessary in the event of excessive fumes; minimum ventilation in the event of fire.
Benches, seats and floors: All in good repair to avoid spillage and falls.
Gas taps and pipes: All operative and without leaks.
Water taps and sinks: Leaks cause slippery floors and rotting woodwork.
Electrical sockets and wiring: Broken sockets should be replaced immediately or sealed off so that they cannot be touched.
Storage cupboards and shelves: These should all provide safe storage for apparatus and chemicals. Poisons should be stored in a locked cupboard. Flammable materials, including liquids, should be stored in a fireproof, well-ventilated, lockable area away from the main storage area and preferably outside the building.
Apparatus and equipment: All should be in good repair, broken or cracked glassware should be repaired or discarded.
Safety equipment: All applicances should be regularly checked. Make sure that your first-aid box is fully stocked.
Some questions to be answered (this list is by no means exhaustive):

1. Non-luminous flames are not obvious in bright sunlight—can they set fire to anything if left unattended?
2. Are gas cylinders stored in direct sunlight (or fitted unprotected to an outside wall)?
3. Are vapours from flammable liquids likely to be ignited by a heat source remote from where they are being used (remember they are mostly heavier than air)?
4. Are flammable liquid containers stored in trays large enough to contain the liquid in the event of breakage?
5. Are safety goggles, safety screens and gloves readily available, and used?
6. Are all electrical equipment and fittings safely wired and, where appropriate, earthed?
7. Are any pupils suffering from physical (or other) disabilities?
8. Are all chemicals correctly labelled? (see Fig. 1).

9. Do you know the hazards and safety precautions necessary for the materials you are using?
10. Have you tried out all the investigations which you are asking your pupils to perform?
11. Has provision been made to communicate to the pupils the particular hazards involved in their investigations? (For example, do they appreciate the dangers of mouth pipetting?)

4B.3 Safety rules

In promoting safety in the laboratory, students should be encouraged to think of this as a *partnership* responsibility. Since experiments are done by individuals or in small groups, the teachers cannot know what is happening in all parts of the room. It is therefore very important for students to be given guidelines.

The following 'dos and don'ts' were some of the suggestions during a course for teachers [7]. Teachers can *select* appropriate instructions from this list for *their own* laboratory rules. Too many rules is almost as great a fault as too few; it is important to be selective. It is also important to remember that the safety of the students—and the teachers—is more important than safety of the equipment.

1. Follow *all* directions exactly as given.
2. Consider the safety of your fellow students. The laboratory is a place for serious work.
3. Perform *only* those experiments assigned by the teacher.
4. Any accident or injury must be reported to the teacher.
5. Do not work in the laboratory and work-rooms without supervision.
6. No equipment will be used until proper instructions are received from the teacher, and you have proven your proficiency.
7. No chemical may be used until the teacher explains use and precautions to be taken.
8. Never touch materials or apparatus on the demonstration table.
9. Learn at once the location and operation of the fire extinguishers, and other first-aid materials.
10. No laboratory equipment may be taken from the classroom unless so directed by the teacher.
11. Always use caution when transferring or pouring reagents from their containers. If something is spilt, the teacher should be notified so that proper cleaning-up procedures can be followed.
12. Study each experiment before coming to class. This procedure not only saves time but will prevent unnecessary mistakes and accidents.
13. Immediately wash with water any part of the skin that has been in contact with any chemical.
14. Chemicals in the eye should be washed immediately with running water.
15. In laboratory work involving the heating of test-tubes, never look into the

test-tube while heating it, or point the mouth of the tube in the direction of any other student during the process.

16. When heating liquids in test-tubes, heat from the upper portions of the test-tube downwards. Otherwise vapour, meeting the head of liquid above it, may cause the contents of the tube to be blown out.
17. Test for odour of gases by wafting the hand over the sample and sniffing cautiously.
18. Always work in a well-ventilated room. Most fumes are poisonous and could cause serious respiratory ailments.
19. An apron and safety glasses must be worn, when so directed, while working in the laboratory.
20. When using a volatile liquid such as ethanol, which is flammable, care must be taken that no flame is used.
21. Large storage bottles of dangerous chemicals such as acids and alkalis are to be handled only by the teacher.
22. Never add water to concentrated acid. If it is necessary to prepare diluted acid, the concentrated acid should always be added in small quantities to the water, stirring all the while.
23. Handle all corrosive substances with the greatest of care. Special precautions should be taken with concentrated sulphuric acid, nitric acid, glacial acetic acid and concentrated solutions of caustic alkalis and other corrosive chemicals such as phenol, bromine, iodine, etc.
24. Never pour reagents back into bottles, or exchange stoppers of bottles, or lay a stopper on the table. Contamination could result in dangerous reactions.
25. When bending glass, allow to cool on a heat-resistant mat or wire screen before handling further.
26. Never force glass tubing, thermometers or any breakable materials in or out of rubber stoppers and tubing. Glycerine is recommended rather than water as a lubricant.
27. Specimens, solid waste, broken glass and other laboratory waste will be placed in their specific containers. The wastepaper baskets are for papers only.
28. Laboratory work areas and equipment are to be cleaned and dried before the end of the class. All missing or broken items must be reported to the teacher before dismissal.
29. Reagents are not to be taken from the laboratory stock. Take containers to the stock-room for materials. All specimens are to remain in the dissecting pans.
30. Do not taste, eat or drink any laboratory materials unless so instructed. Regard all reagents as poisonous.
31. Do not grasp any electrical device which has just been used. Most electrical devices are hot after use and serious burns may result if caution is not exercised before the hot object is grasped.
32. Appliances which depend upon high voltage for operation are to be plugged in *only* with the consent of the teacher.
33. Notify the teacher about all broken glassware. Prevent other students from moving in the area and, above all, do not attempt to clean up the broken glass until given instructions and materials, if chemicals are involved.

$H_2O_2(aq)$	HYDROGEN PEROXIDE SOLUTION	M,34

WARNING	EMERGENCY ACTION
Corrosive—causes burns. Avoid contact with skin, eyes and clothing. Powerful oxidising agent—may cause fire. Store in a cool place. WEAR SAFETY GLASSES AND GLOVES.	Eyes —Irrigate with water. Lungs —Remove from area and keep at rest. Mouth —Wash out with water. Skin —Wash with water. Spillage—Drench with water.

ASSOCIATION FOR SCIENCE EDUCATION

HCl(aq)	HYDROCHLORIC ACID	M,36.5

WARNING	EMERGENCY ACTION
Harmful if taken internally. Corrosive—causes burns. Irritant vapour. Avoid contact with skin, eyes and clothing. Avoid breathing vapour.	Eyes —Irrigate with water and get medical attention. Lungs —Remove from area and keep at rest. Mouth —Wash out with water. Give water to drink followed by milk of magnesia. Skin —Drench with water. Spillage—Drench with water.

ASSOCIATION FOR SCIENCE EDUCATION

Hg	MERCURY METAL	A,201

WARNING	EMERGENCY ACTION
Harmful vapour and liquid. Avoid breathing vapour. Avoid contact with skin, eyes and clothing. Keep in well ventilated space.	Eyes —Irrigate with water. Lungs —Remove from area, keep at rest and get medical attention. Mouth —Wash out with water and get medical attention immediately. Skin —Drench with water. Spillage—Cover with zinc dust and sweep up.

ASSOCIATION FOR SCIENCE EDUCATION

$NH_3(aq)$	AMMONIA (AQUEOUS SOLUTION)	M,17

WARNING	EMERGENCY ACTION
Harmful if taken internally. Irritant vapour. Avoid contact with skin, eyes and clothing. Store in a cool place. WEAR SAFETY GLASSES.	Eyes —Irrigate with water and get medical attention. Lungs —Remove from area and keep at rest. Mouth —Wash out with water and give water to drink. Skin —Drench with water. Spillage—Drench with water.

ASSOCIATION FOR SCIENCE EDUCATION

FIG. 1. Example of labels prepared by the Association for Science Education in the United Kingdom. Ninety-nine different labels have been prepared for the most commonly used chemicals in the laboratory.

There is a most attractive eight-page booklet published by a group of teachers called *Safety in the Lab*, giving a series of rules for students [38].

4B.4 Chemistry experiments: advice on safety

The following section is based on an article in *Education in Science* following a report by a Working Party of the Association of Science Education (ASE) [39]. Since then, the article was republished, with some corrections supplied, in *Chem 13 News* (Vol. 116, p. 3, 1980) and in the *Sourcebook for Chemistry Teachers*, prepared for the Sixth International Conference on Chemical Education [40].

It is a helpful article and we reprint it with kind permission of the Association of Science Education in a slightly abridged form:

The advice presented here has been prepared after extensive consultation with science advisers, practising teachers and other interested persons. The experiments listed are largely those that are known to the working party to be restricted by one or more local education authorities and in many cases it also feels that the experiments are not appropriate for use in schools at all levels. Again, it is hoped that the consensus view presented here will be regarded by responsible bodies as a source of informed opinion on the desirability or otherwise of the experiments, leading to commonly agreed practice and fewer cases of over-reaction to possible hazards.

One function that it is hoped the list will serve is to point out the dangers that are not widely known. Experiments such as the reaction of zinc and sulphur appear in many texts and many teachers have performed the experiment several times without mishap. Nonetheless, cases of quite serious explosions have occurred with this mixture and although the reasons for this are not fully understood, it does seem necessary to suggest that only teachers themselves should perform the experiments with full precautions as set out here.

It is difficult to allow for the widely varying circumstances in which the experiments mentioned here may be carried out and the list necessarily assumes usual school conditions. Cases are bound to occur where a teacher will exercise his professional judgement and not adhere to these recommendations. Thus, an experiment listed here as suitable for teacher demonstration in a fume cupboard might be done by a responsible sixth form group in a school with adequate fume cupboard provision for classwork.

Similarly, a teacher may know of an alternative, safer method to the standard method of performing an experiment. Where such cases are known to the working party, these have been mentioned in the comments column.

Finally, the absence of any experiment from this list should not be taken to imply anything.

The categories of restriction suggested for each experiment are shown by letters as outlined below.

N Unsuitable. The experiment is considered unsafe for use in schools.
T For teacher demonstration only. Teachers should be thoroughly familiar

with the technique to be used. It is assumed that these experiments will have been rehearsed before being done in front of a class for the first time.
S Considered suitable for supervised senior students. Some of these could perhaps be entrusted to *responsible* younger students (16 year olds).
O Considered safe as a class experiment for 13–16 year olds. It is essential here for the teacher to exercise his or her discretion as to the responsibility of a particular class. Any experiment listed here may present dangers to irresponsible students.
F The use of a fume cupboard is recommended. Again, teachers may have to use their discretion and allow experiments classified in this way to be carried out in a very well-ventilated room with small quantities of materials.

EXPERIMENT	*RESTRICTION*
Ammonia, oxidation using oxygen in an enclosed apparatus	**N** Use air (**T**). Oxygen may be used in an open vessel (**T**).
Ammonium dichromate(VI), heat ('Volcano experiment')	**T**, **F** A fume cupboard is needed to avoid possible inhalation of chromate(VI) dust.
Ammonium dichromate(VI), heat with aluminium or magnesium powder	**N**
Ammonium nitrate, heat	**T** Heating a mixture of ammonium chloride and sodium nitrate is considered safer. Use safety screens.
Ammonium nitrate, prepare and heat	**T** in solution only, concentration less than molar.
Aryl and acyl halides, reactions	**S**, **F**
Cadmium iodide, electrolysis of molten	**N** Lead bromide preferable (see below).
Carbon monoxide, reductions with	**T**, **F** Use safety screens.
Carbonyl chloride, preparation	**N**
Chlorine, reaction with ammonia	**N**
Chlorine, reaction of a mixture with hydrogen	**N** This refers to the gas syringe and similar experiments. It is possible to demonstrate the reaction in, for instance, a plastic bag. Burning hydrogen at a jet in chlorine is safe for teacher demonstration.
Chlorine, reaction with ethyne	**N** The reaction where the gases are generated simultaneously by adding dilute hydrochloric acid to a mixture of bleaching powder and calcium dicarbide is acceptable as a teacher demonstration (**T**, **F** Use safety screens).
Chlorine, preparation	**S**, **F** See 'Potassium manganate(VII), reaction with concentrated hydrochloric acid'.

Chlorine, reaction with metals	**S, F**	
Chlorine oxides, preparation	**N**	
Crude oil, distillation	**O, F**	
Cyanogen, preparation	**N**	
Ethene or ethyne, explosion of a mixture with oxygen	**N**	
Ethene or ethyne, igniting in a gas jar or test-tube	**T**	
Explosives (i.e. mixtures of chlorates, manganates(VII) or nitrates with combustible substances	**N**	
Hydrogen, large-scale generation and collection	**N**	
Hydrogen, generation and testing for on a test-tube scale	**O**	
Hydrogen, burning in air	**T**	
Hydrogen, burning in chlorine	**T**	Use safety screens
Hydrogen, explosion with air	**T**	
Hydrogen, explosion with oxygen	**T**	
Hydrogen, reductions using:	**T**	Use safety screens for the normal scale experiment. Reduction of metal oxides may be performed on a test-tube scale using, for example, a mixture of zinc powder and calcium hydroxide in the same tube as the oxide to generate the gas. Such experiments may be classified as **O**.
Hydrogen cyanide, preparation	**N**	
Hydrogen sulphide, preparation	**T, F**	
Hydrogen sulphide, use of gas	**S, F**	
Hydrogen sulphide, use of aqueous solution	**O**	
Iodine, heating in air	**T** or **S, F**	
Lead bromide, electrolysis	**T** or **O, F**, in which case the fume cupboard is essential.	
Lead(II) carbonate, heating	**O**	
Lead(II) nitrate, heating	**O, F**	
Lead oxides, heating	**O**	
Lithium, heating	**T**	Use safety screens.
Mercury, heating	**T, F**	The fume cupboard is essential and must be left on for the duration of the experiment (i.e. while the mercury is above room temperature).
Mercury(II) oxide, heating	**T, F**	The fume cupboard is essential.

Natural gas, enrichment for reductions	**T** if ethanol tetramer (metaldehyde or 'meta fuel') is used as the enriching agent, in a test-tube scale experiment, it may be classified as **O**. The large-scale experiment has proved dangerous, probably because of the extra dead volume introduced into the apparatus.
Nitrations, organic	**S, F** In some cases when only a mild nitrating agent, such as dilute nitric acid, is necessary, a fume cupboard is not needed; e.g. nitration of phenols.
N-nitrosamines, preparation from amines	**N** See previous article on carcinogens (*Education in Science*, September 1979).
Oxygen mixture, use of:	**T** See 'Potassium chlorate(V) and manganese(IV) oxide, heating mixture'.
Phosphine, preparation	**T, F**
Phosphorus halides, reaction with water	**S, F**
Phosphorus, red, burning	**O, F**
Phosphorus, white, burning	**T, F**
Plastics: heating polyurethanes and polystyrene	**O, F**
Plastics: heating PVC	**T, F** The fume cupboard is essential.
Plastics: polymerization and depolymerization of acrylics	**S, F** The fume cupboard is essential.
Plastics: polymerization of phenylethene	**S, F**
Plastics: preparation of nylon 'rope'	**O** Note that if a solution of dioyl chloride in tetrachloromethane is used this must be classified **T, F**. 1,1,1-trichloroethane may be used as solvent if the solution is freshly prepared.
Potassium, reaction with water	**T** Use safety screens.
Potassium chlorate(V) and manganese(IV) oxide, heating mixture	**T** Many safer alternatives for oxygen preparation. Use demonstration as illustration of catalysis only. Use safety screens.
Potassium manganate(VII), heating	**O** Eye protection essential. Heat in small test-tubes fitted with a loose ceramic wool plug to prevent spitting.
Potassium manganate(VII), reaction with concentrated hydrochloric acid	**S, F** Cover the manganate(VII) with water first. This experiment is highly dangerous if sulphuric acid is used by mistake instead of hydrochloric. It is safer to use bleaching powder or sodium chlorate(I) and dilute hydrochloric or sulphuric acid.
Rocket fuels, preparation	**N**
Silicon(IV) oxide, reduction with magnesium or aluminium	**S** The reactants must be dry. Use safety screens.
Sodium, reaction with water	**S** Use safety screens.

Sodium hydroxide (molten), electrolysis	**T, F**
Sodium peroxide, preparation of oxygen from	**T** Use safety screens.
Sulphur and zinc, reaction	**T** Do not confine the mixture in any way, i.e. heat the mixture on a ceramic centred gauze or mineral fibre paper. Use safety screens.
Sulphuric acid, concentrated, reactions	**O** With close supervision, otherwise **S**. Use a fume cupboard if corrosive or toxic gases are likely to be evolved.
Thermite, reaction	**T** Use safety screens (or perform outdoors). Fe_2O_3, Mn_3O_4, Cr_2O_3 are safe oxides to use. Do not use CuO, MnO_2 or CrO_3.
Zinc, burning	**O, F**

The nomenclature used here is that recommended by the Association of Science Education [41]. For example, some teachers may be more familiar with the term permanganate rather than manganate(VII), hypochlorite rather than chlorate(I), ethylene rather than ethene, acetylene rather than ethyne, carbon tetrachloride rather than tetrachloromethane.

There is also a most useful article on restricted uses of chemicals in *Education in Science*, which suggests those chemicals which should be handled only by teachers, those which may also be handled by senior students and those which may also be handled by younger pupils [42].

4B.5 First aid and safety equipment

First aid

All teachers should have a knowledge of simple first aid. It is also important that they should recognize when medical advice is needed. A first-aid kit should be available wherever practical work is being carried out and this kit should include: a few bandages of assorted sizes; a triangular bandage for use as a sling; adhesive plasters; scissors; sterilized cotton-wool packs; bottle of mild antiseptic; antiseptic ointment; safety-pins; eye-bath; eye-bathing solution.

Special antidote mixtures may be necessary if particularly dangerous chemicals are being used. An accident report book and pencil should also be included for reporting the accident details and treatment given.

For fuller details, contact your local branch of the Red Cross or Red Crescent, or fire department or ambulance service—any agency applicable to your country.

Safety equipment

The following is a guide to preferred safety equipment: fire extinguishers; sand buckets; fire blanket; *labelled* waste containers; eyeglasses; rubber gloves; safety screens; laboratory aprons or coats.

4B.6 Lessons in safety

Figures 2 and 3 were published by the Royal Society for the Prevention of Accidents [43] (reproduced with permission). They were produced for the Science Teacher Education Project (J. Haysom and C. Sutton, whom we thank) [44]. They are drawings illustrating two laboratory classes. The students or groups of students could be asked to identify the hazards and make lists of them. A discussion of reasons and responsibilities could then follow, with the teacher guiding the discussion towards an emphasis on the students' relationship with other members of the class, and on how the safety of all depends upon each individual.

Alternatively, the pictures could be used for a class competition followed by a discussion, or the pictures could be used as part of a display. Question to the reader: Can you identify at least eighteen hazards in Figure 2 and twenty in Figure 3?

These pictures will not be appropriate in many countries. They are *examples* and teachers should produce, perhaps with the aid of teachers who can draw (!), pictures that illustrate their own laboratories, workrooms and classrooms.

Some of the main hazards in Figure 2

1. Water on polished floor.
2. High apparatus standing near the edge of the bench—a beaker on a tripod stand with a long pipette sticking out of the beaker.
3. A student inadvertently putting his hand on a hot metal tripod.
4. A heated test-tube pointed in such a direction that boiling liquid could be ejected on to students.

4a. A test-tube being heated with an inappropriately large amount of liquid in it.

5. Bottles of flammable or corrosive materials (e.g. ether or corrosive hydrochloric acid) stored on the floor where they can be bumped into.
6. A retort stand assembled in such a way that when used it will tip over.
7. Rubber tubing trailing among bottles and other apparatus which might thereby be knocked over or swept off the bench. Cluttered benches can be a source of a variety of accidents. Bottles not in use should be removed from the working surface.
8. A metal screwdriver being used to explore an electric mains socket.

FIG. 2. Laboratory safety: spot the hazards.

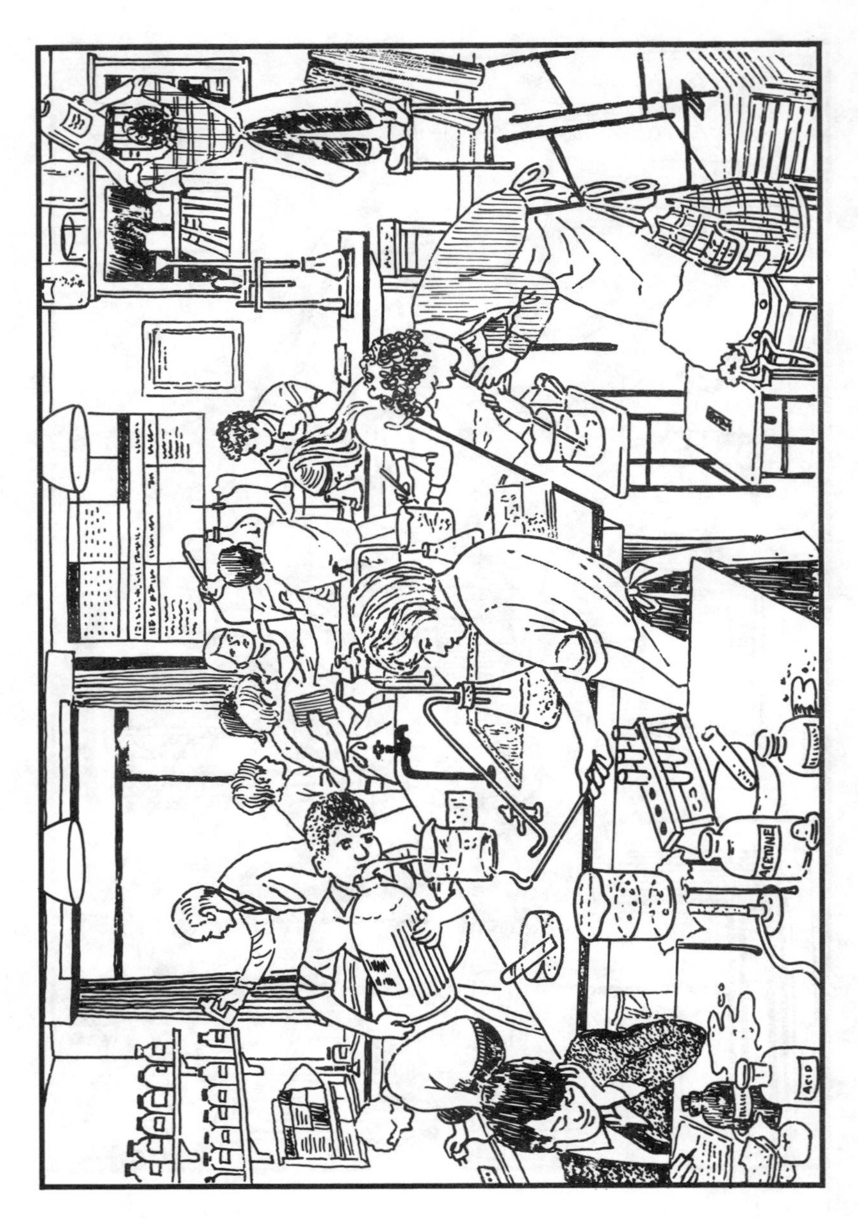

FIG. 3. Laboratory safety: spot the hazards.

9. Heavy metal weights or other heavy objects being supported on thin string or wire (in the experiment with pulleys).
10. Long hair dangling near Bunsen flames: loose clothing.
11. Chemicals kept in bottles which formerly contained food and which still have labels on them.
12. Liquids being poured above eye-level—in this case into a burette.
13. Exit doors blocked.
14. Students too near to a demonstration while watching it.
15. Student flicking paper.
16. A student carrying loads that obscure his view of where he is walking.
17. Safety notices on the power unit too high to be read easily by students.

Some of the main hazards in Figure 3

1. Overcrowding likely to cause students to knock apparatus off the bench.
2. Spilled acid on a bench in a position such that clothing might mop it up.
3. Student pouring a solution from a very large (Winchester) bottle.
4. Gas cylinders stored in positions where they could fall or be knocked over: they should be chained in position.
5. A student standing on a stool to get at heavy apparatus, reaching for it from high up.
6. A jar over the edge of a cupboard could fall off.
7. A student trying to light, with a taper, gas coming from a gas-generation apparatus which includes a thistle funnel. (If hydrogen is being prepared this way the apparatus will contain an explosive gas–air mixture.)
8. A test-tube stand on the edge of a bench.
9. Food in among poisonous chemicals.
10. A flammable solvent (acetone) being heated with a naked flame.
11. Mercury spilled on a bench.
12. Liquid being pipetted by mouth from a vessel in which the pipette tip is only just below the surface.
13. A chart on the wall curling off for lack of drawing pins, capable of being ignited by the nearby burner.
14. Apparatus in such a position that it would be knocked over by opening cupboard doors.
15. Unsupported apparatus.
16. Combined hazards, e.g. stool falls over, hence knocking over other stools in a domino effect.

17. A student with wet hands putting a plug in an electric mains socket—notice also that the socket is upside down.
18. Stools and bags blocking walking space.
19. Long hair, not tied, could be a fire hazard.
20. Loose clothing.

4B.7 Bibliography

General article

1. RENFREW, M. M.; VOLKENING, L. Safety in the Laboratory. In: W. T. Lippincott (ed.), *Sourcebook for Chemistry Teachers*, p. 39. Washington, D.C., Division of Chemical Education, American Chemical Society, 1981. This article contains several helpful articles on school safety published elsewhere.

Publications of science teachers associations

2. *Safety in the Secondary Science Classroom*. Published by the National Science Teachers Association, 1742 Connecticut Avenue, N.W., Washington, D.C. 20009 (United States).
3. *Conditions for Good Science Teaching in Secondary Schools*. Published by the National Science Teachers Association, 1742 Connecticut Avenue, N.W., Washington, D.C. 20009 (United States).
4. *Safety in the Lab*. Published by the Association for Science Education, College Lane, Hatfield, Hertfordshire (United Kingdom).
5. *Safety Check Lists*. Published by the Association for Science Education; for address, see (4) above. Reprinted in *Education in Science*, Vol. 75, 1977, p. 18. A sample two-page guide for teachers.
6. Labelling of Chemicals. *Education in Science*, Vol. 82, 1979, p. 25.
7. Management and Organization of Chemical Storage. *Education in Science*, Vol. 89, 1980, p. 16.
8. Storage and Handling of Flammable Liquids. *Education in Science*, Vol. 89, 1980, p. 20.

Books written especially for schoolteachers

9. ARCHENHOLD, W. F.; JENKINS, E. W.; WOOD-ROBINSON, C. *School Science Laboratories: A Handbook of Design, Management and Organization*. London, John Murray, 1978. The book is divided into nine chapters. Chapter 3 covers safety in school science teaching and gives one of the most helpful views of the hazards which are present in the school laboratory; for example, eye protection, hazards of glassware, use of gas cylinders, chemical hazard warning labels, electrical hazards.
10. JENKINS, E. W.; EVERETT, K. *A Safety Handbook for Science Teachers*. 3rd ed. London, John Murray, 1980. A simple treatment specially written for science teachers. The chapters on safety in laboratories and workshops, fire, first aid are particularly useful. So too is a list of all the major hazards that have been published in the *School Science Review* for 1919–79 and in *Education in Science* since its first publication.

Comprehensive reference books

11. Bretherick, L. (ed.). *Hazards in the Chemical Laboratory*. 3rd ed., 1981. Published by the The Royal Society of Chemistry, Burlington House, Piccadilly, London W1V OBN (United Kingdom). This is an authoritative and detailed compendium of hazardous chemicals, with methods of treatment. There are introductory chapters on safety planning and management, fire protection, chemical hazards and toxicology, health care and first aid.
12. Bretherick, L. *Handbook of Reactive Chemical Hazards*. Sevenoaks (United Kingdom), Butterworths, 1975. This book gives detailed information on the hazards of handling over 4,000 chemicals. There are data on possible violent interaction between two or more compounds.
13. Sax, N. I. *Dangerous Properties of Industrial Materials*. 5th ed. New York, Van Nostrand Reinhold Co., 1979. This lists nearly 15,000 industrial laboratory materials, in terms of their safe handling. There are introductory chapters on topics such as air pollution, control requirements, industrial fire protection, legislation, labelling and identification of hazardous materials. Although very much for the industrial laboratory, the book is a mine of information for those designing experiments for new curricula.

SECTION 4C

COMPUTERS AND CALCULATIONS IN CHEMISTRY[1]

4C.1 Introduction

Computers can play a dual role in the chemistry curriculum. Computer programming, numerical methods, digital electronics and laboratory automation, microprocessors, and computer-based information processing, storage and retrieval are tools that a great many chemists use in their everyday work. These subjects ought to be a part of chemists' basic training, along with other important tools of the trade. On the other hand it is possible to teach with a computer, using the machine as part of the delivery system for the subject-matter of a course. These two aspects of computer use are not separable in practice. A student who is taught with a computer learns a good deal about computers, and to learn about computers a student should be taught with a computer. The main thrust of this chapter is to provide basic information, and sources of further information, that will be useful for teachers who plan to use a computer as an aid to chemistry teaching.

1. Much of this article was written early in 1981; consequently some information on prices and availability of microcomputers will be out of date. Also, since that time many new models and programs have become available.

4C.2 Computer-aided instruction

All aspects of teaching with a computer can be classified under the heading computer-aided instruction. The acronym CAIDI will be used to denote this all-inclusive category. CAI has a narrower meaning, including only those applications in which there is an interactive tutorial involving student and computer. To give an overview of the many ways in which chemistry teachers are using CAIDI, the next few paragraphs consist of vignettes drawn from personal experience that illustrate sub-categories of CAIDI. Most of these are amalgamations of applications developed by several programmers, and thus are not obtainable as specific programs for specific computers, though very similar programs do exist, sometimes at several locations. The intention here is to illustrate the broad scope of CAIDI by giving examples that illustrate its sub-classifications, not to provide information about specific programs.

4C.2.1 *Dialogues*

When a student interacts with a computer in a direct conversational dialogue, and the computer is programmed to present course material, ask questions that require active responses, evaluate answers, and vary its presentation on the basis of student responses, then we have true computer-assisted instruction (CAI). An example of tutorial CAI is the following dialogue between a student (S) and a computer (C).

C: The names of many compounds are simply derived from the names of the groups of atoms that must be assembled to form a molecule. For example, what is the name of NaCl?
S: Sodium chloride.
C: The CH_3 group is called methyl. What is the name of CH_3Cl?
S: Methyl chloride.
C: What is the name of CH_3CH_2Cl?
S: Ethyl chloride.

In this type of dialogue the computer can be programmed to provide help should the student not be able to devise an expected response. The computer can also be programmed to recognize alternative right answers, such as chloromethane, to correct common errors, such as writing methylchloride, and to adjust the pace of the dialogue—fast for students who give few incorrect responses, slower for others. The rate at which material is presented, the quantity of drill and review required, and the time of day or week at which instruction is made available can all be adjusted to each student's needs.

4C.2.2 *Graphics*

CAI, or CAIDI in general, can be significantly enhanced if graphic images, as opposed to verbal descriptions, are presented to the student

(Figure 4). The computer can present to the student an overall scheme for separation of a neutral compound from an acid. Moreover, individual parts of such an overall display can be presented separately, discussed, and built up into the complete picture shown here. At each step of the way student responses can be required, so that the student becomes an active participant in the learning process.

4C.2.3 *Simulation*

Simulation is another important instructional tool. It can be applied successfully in CAI lessons or to other aspects of CAIDI, especially if it is combined with graphic capability. For example, the display shown in Figure 4 is part of a lesson in which students carry out a simulated pre-laboratory exercise on organic separations. The use of the separatory funnel is presented in detail, with illustrations and animations on the screen. The student must tell the computer what operations to perform in which order, and the computer illustrates the simulated results. If an operation is forgotten the computer can remind the student to do it, or simply illustrate the consequences. In one such situation a student who heats a still too rapidly produces a simulated explosion that graphically illustrates the hazards of not paying attention to what one is doing in the laboratory.

4C.2.4 *Interactive tutorials*

Interactive, tutorial CAI is discussed in some detail later in this chapter. It requires considerable expertise on the part of a programmer/teacher, a very sophisticated CAI author language that helps an author to program the computer to recognize and evaluate student responses and expensive terminals if high-quality graphics are to be displayed. Consequently, tutorial CAI is not available to the majority of chemistry teachers, though the revolutionary decrease in costs of microprocessors and the availability of self-contained 'personal' computers may soon belie this statement. Nevertheless, much can be done with non-tutorial computer applications (NTCA) that require less sophisticated student/computer interactions.

4C2.5 *Non-interactive computer applications*

Examples of NTCA abound. Data reduction programs that perform calculations can save students and instructors much time and effort. Such programs also serve to introduce students to computers, providing a good indication of what computers can do as well as what adaptations people must make to use computers effectively. Other non-tutorial programs can perform simple drill-and-practice that help teach students

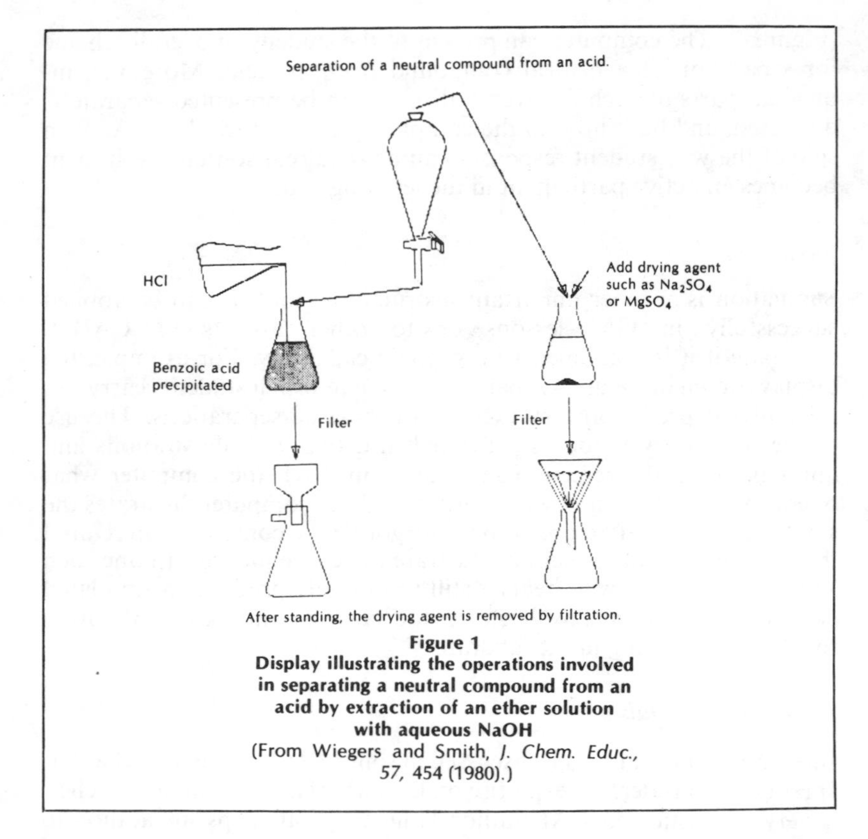

Figure 1
Display illustrating the operations involved in separating a neutral compound from an acid by extraction of an ether solution with aqueous NaOH
(From Wiegers and Smith, *J. Chem. Educ.,* *57,* 454 (1980).)

FIG. 4. Display illustrating the operations involved in separating a neutral compound from an acid by extraction of an ether solution with aqueous NaOH. (From Wiegers and Smith, *J. Chem. Educ.*, Vol. 57, 1980 p. 454.)

the names and symbols of the chemical elements, their positions in the periodic table, nomenclature of inorganic ions and compounds, organic nomenclature and many other rote topics. Such drill is certainly boring for instructors as well as students, and so it is often left undone. The computer cannot become bored, and if programmed to introduce an element of competition the computer can make boring drill into a game that will hold students' interest long enough for them to memorize the desired information. An example of such a game, written for the Commodore PET, is shown in Figure 5. Each of two players is asked to give the name, or the symbol, for an element that has been identified by a flashing rectangle at a particular position in the periodic table. Each student's score is based on the time he or she takes to identify the element, with penalty points added for wrong responses. On the second

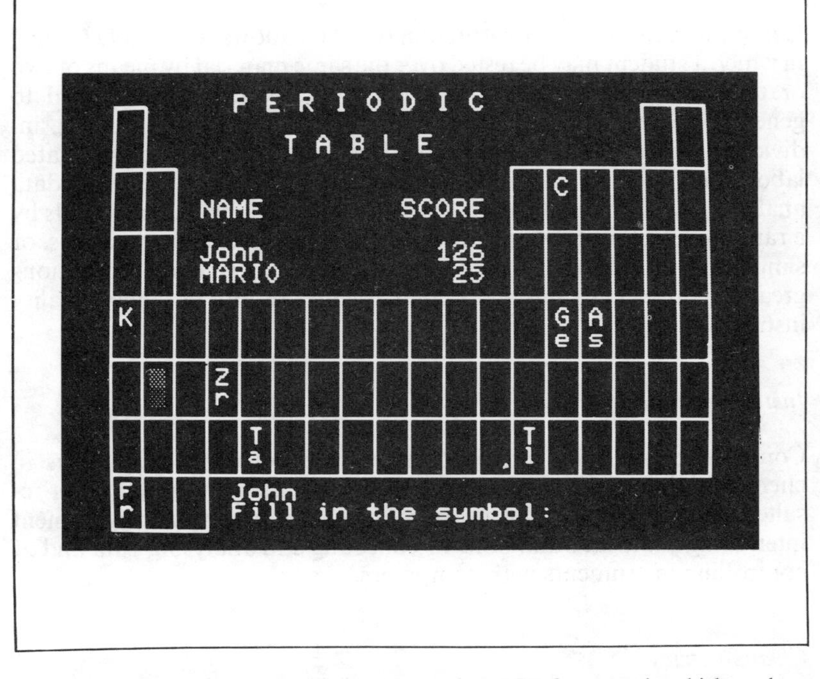

FIG. 5. Example of screen display on Commodore PET for game in which students identify symbols or names of elements based on location in the periodic table. Scores are based on the time taken to identify an element with penalty points added for incorrect answers.

incorrect try the computer provides the name or symbol, but no other tutorial information is available.

4C.2.6 *Computer-managed instruction*

Computers can also play an important role in the management of chemistry courses, in testing students and in providing homework or other student exercises. Computer-managed instruction (CMI) usually involves machine grading of exams, quizzes and/or laboratory results, with all student records being stored in computer data files. Instructors, and in many cases students, can then quickly obtain a summary of progress in the course, remedial work that should be carried out in order to improve performance on tests and so on. Computers can also be used to store and even to generate test questions, to assemble questions from selected categories into tests, and to print large numbers of unique but equivalent tests.

4C.2.7 *Computer-assisted test construction*

Computer-assisted test construction (CATC) allows repeatable testing, in which a student may be tested over the same material by means of two or three different test forms. CATC programs can also be used to generate homework assignments that are unique to each student, individualized pre-lab assignments or tests, or unique sets of simulated laboratory data so that each student must do his or her own data analysis. The computer's ability to generate individualized materials by a random selection of questions, filling blanks on skeleton questions, or simulation of data from randomly selected experimental conditions greatly reduces cheating and copying which results in higher quality instruction.

4C.2.8 *Instrument interfacing*

Computers, especially microprocessors, are becoming ubiquitous in chemical laboratories and in chemical instrumentation. This has resulted in the development of a variety of courses on computer/instrument interfacing—the nuts and bolts of collecting and analysing data and of controlling instruments with computers.

4C.2.9 *Chemometrics*

There is also considerable interest currently in chemometrics, the application of mathematical methods and computers to chemical problems. Both of these topics involve much teaching about computers, but they are invariably taught with computers, often those based on microprocessors.

4C.2.10 *Computer literacy*

An appreciation of the importance of these fields and how they apply to chemistry requires a minimal computer literacy on the part of all chemists. The same minimal literacy requirement applies to persons who want to use CAI, CATC, CMI, computer simulations, computer graphics, NTCA or other computer techniques in teaching chemistry.

In the next section, definitions and facts about computers are summarized. Subsequent sections discuss CAI, graphics, microcomputers, pocket calculators and other sub-specialities in sufficient detail to provide background for teachers who plan to become actively involved in any of these areas. More details can be obtained from the references at the end of this chapter.

4C.3 Computer fundamentals

To solve problems and communicate with the outside world a computer needs the five types of components shown schematically in Figure 6. The input and output units allow programs and data to enter the machine and results to exit. The memory stores programs, data and intermediate results. Calculations are performed in the arithmetic-logic unit (ALU) and the overall operation of the computer is overseen by the control unit. The physical devices that perform the input, output, storage, arithmetic-logic and control functions are called the hardware, while the collection of programs available to the computer is referred to as the software.

4C.3.1 *The control component*

The control unit has a fundamental repertoire of instructions that it can carry out. Each instruction initiates some operation in the ALU or causes the transfer of a number to or from the memory, output or input components. The control unit and the ALU usually occupy the same cabinet or circuit board, and together they are referred to as the *central processing unit* (CPU) of the computer. A microprocessor is a CPU in which all circuits are integrated on to a silicon chip on the order of 5 mm square. This permits a complete *microcomputer* to be built on a circuit board no larger than this page and results in costs as low as US $100 for a device having rudimentary input/output and a small memory. *Minicomputers* are somewhat larger, more versatile and can calculate more quickly. Of course they cost more, too. *Mainframe computers* of the type found in large businesses, universities and research laboratories are the most complicated and expensive of all.

4C.3.2 *Storage or memory component*

A register is an electronic circuit or device that can store some value, usually a binary number. The main storage or memory of a computer may be thought of as a large collection of registers, usually of equal size, each of which is referred to as a *word storage cell*, or *memory location* and is identified by a unique binary number called an *address*.

The time required for the CPU to specify the address or location of a memory register and retrieve the contents is called the access time. The access time for one memory location is essentially the same as for any other, and on a state-of-the-art memory it is usually in the range from 10 to 100 ns.

The number of words and the word size (the number of bits that can be stored in a given word and transferred in a single memory access) vary from one computer to another. Both must be specified if you want to know how much memory is available or needed. For example, a

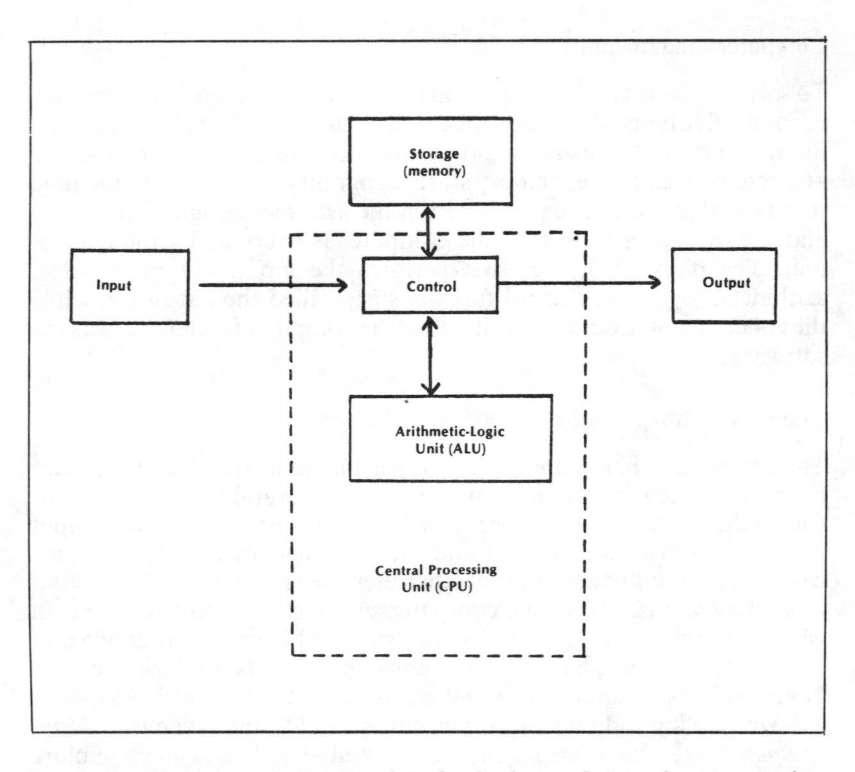

FIG. 6. Simplified schematic diagram of the five fundamental types of components in a computer system. (From Moore and Collins, *J. Chem. Educ.*, Vol. 56, 1979, p. 140.)

program that requires 8K 36-bit words might not run on a computer whose available main storage consists of 16K 16-bit words. The program takes up $8{,}192 \times 36 = 294{,}912$ bits of storage while the computer only has $16{,}384 \times 16 = 262{,}144$ bits available. (Although the K in 8K is derived from the metric prefix kilo, it is generally convenient for the number of words in a memory to be an exact power of two. Hence K refers to 1,024, the power of two that is closest to 1,000). In order to run the program in question, the user of the 16K machine might have to modify the program (or the way the program stores its data) so that fewer memory words would be required.

The word size of a memory affects the precision with which numbers can be stored. A 16-bit word, for example, is capable of representing any integer between 0 and $2^{16} - 1 = 65{,}535$, or, if one bit is used to represent + or −, any integer between $-2^{15} + 1$ and $2^{15} - 1$.

The absolute error in such an integer or *fixed point* representation is always $\pm$ 0.5, because the decimal point (really binary point) is assumed to be fixed in a position just to the right of the right-most digit (bit). Relative errors, of course, range upward from one part in 65,535, depending on what number is represented. Constant relative error can be attained if a *floating point* representation similar to scientific notation is employed. Some of the bits in each word are set aside to store significant figures (a mantissa). The remaining bits represent an exponent (characteristic) of some base (not necessarily ten). Precision is important because the computer often does a great many arithmetic operations in situations that look simple to the programmer. Logarithms, exponentials and trigonometric functions, for example, are usually calculated as sums of infinite series, and rounding errors can accumulate rapidly. To obtain adequate precision, machines using 8-bit words use two or more words to store a single number. This results in slower calculating speeds.

4C.3.3 *Input and output components*

Input and output components provide the computer with a means of communicating with programmers, other users and other computers. From the user's standpoint two main categories of input/output (I/O) may be distinguished.

Batch processing, in which a program and all the required data are provided to the computer in one batch (most commonly in the form of a deck of punched cards) was by far the most common until about 1970. Each batch *job*, that is, each individual task set for the computer, is completed before the next one begins, and when a job is finished the results are output, usually on a line printer that is capable of producing on the order of 1,000 120-character lines per minute.

Since 1965 timesharing has become more and more important. In timesharing the CPU allots a fixed period, such as 50 ms, to each job. During that time the CPU can carry out tens of thousands of instructions. If the job is completed within 50 ms, fine. If not, the current status of the job is stored for completion later, and the CPU goes on to another job.

On a large computer timesharing permits *interactive* I/O in which a user can key in a program or data from a teletypewriter (TTY) or a cathode ray tube (CRT) terminal and receive what is perceived as immediate feedback of results and/or prompting for additional input. Timesharing allows several users to interact with the computer because the responses of the terminal (usually from 10 to 120 characters per second) and of the person using the computer (which varies greatly depending on thinking and typing speed) are orders of magnitude slower than that of the CPU. A second is not very long for a human to wait, but it is long enough for the computer to touch base with many other jobs before it returns to yours. The immediate response and prompting ability of an interactive computer system make it ideal for instructional purposes. Microcomputers often have only a single user and so a microcomputer can provide the advantages of interactive computing without the necessity of timesharing.

A wide variety of I/O devices other than card readers, printers, TTYs and CRTs is available. Detailed, high-resolution graphs and drawings (of molecular structures, for example) can be produced on digital plotters and on CRT graphics terminals, and graphical data may be read into a computer by means of digitizer tablets. Student responses to test questions can be entered on optical scanning forms and read directly into the computer. Input and output can also be done via paper tape readers and punches, magnetic tape units and magnetic discs or drums. Each of these devices has characteristics that suit it to certain applications but not to others. For example, if a student is to be given lengthy homework problems by a computer, a hard-copy device like a line printer, a TTY or a plotter would probably be more suitable than a CRT terminal. On the other hand, a program that was designed to give advanced students an intuitive grasp of quantization by plotting various wave functions of a particle in a finite potential well would probably employ a CRT graphics terminal to provide immediate feedback based on parameters input by the student.

Another criterion for judging a particular mode of I/O is whether, like punched cards, punched paper tape, magnetic tape and magnetic discs or drums, it is *machine-readable*. If so, it can be used for storage of data on a single computer system or for transfer of information from one computer to another. Magnetic discs, for example, commonly serve as an adjunct memory. Like the main storage, they are *random-access* devices—any item of information stored on a disc has essentially the same access time as any other. This makes a disc very useful in a timesharing system, because a user's program and data can be stored on disc, accessed in a few hundredths of a second, and rapidly copied into main storage when the program is to be executed. Storing programs and data on disc *files* of this sort is far more convenient than storing them on punched cards, punched paper tape or even magnetic tape, because the programs are available at a second's notice. If a computer has limited main storage, programs or data that would otherwise exceed the capacity of core memory can be accommodated by transferring data (or even parts of the program) to and from a disc during execution. Since microcomputers often have very little memory available, paging programs or data from disc to memory and back to disc is often an important feature of microcomputer programs.

4C.3.4 *Storage media*

The traditional medium for storing large quantities of data and for transferring that data from one mainframe computer to another is magnetic tape. Typically a reel contains 2,400 feet of half-inch-wide mylar tape that is coated with magnetizable material. Such a reel can store approximately 10 million English-language words.

Bits are recorded in seven or nine tracks that run parallel to one another along the length of the tape. The recording density is normally 800 or 1,600 bits per inch (Bpi) per track. Generally one *character* (a letter, decimal numeral, or special character like /, =, +, etc.) is stored in each row of seven or nine bits perpendicular to the length of the tape. Groups of characters, each of which corresponds to a single line of printed output and is read separated as input, are referred to as *records*. Eighty-character records, corresponding to the eighty columns of a punched card, are common, but other record lengths may be used.

Because a finite time and length of tape (on the order of 10 ms and 0.5 in) are required to start or stop a reel of tape, and because data cannot be read or written except when the tape is moving, every start or stop wastes recording space, consequently records are usually consolidated into blocks, each of which must be read or written in its entirety. Stops are only permitted between blocks. If a block is located 1,200 feet along a 2,400-foot reel of tape, all preceding blocks must be read before the desired data can be accessed. Typical tape speeds are 50 to 100 in per second, and so the average access time for a block may be several minutes, thousands of times slower than for a disc. Because the access time varies by several orders of magnitude from one end of a tape to another, magnetic tape is referred to as a *sequential* rather than a random access I/O device.

When transferring magnetic tape from one computer to another it is necessary to specify the recording density, the block and record sizes, and the number of tracks. Thus one might indicate that a tape was 800 Bpi with ten eighty-character records per block and nine tracks. It is also essential to specify what interchange code has been used for the character on the tape and the parity convention.

The standard code in the United States for information interchange (ASCII) uses seven bits to represent 128 characters; IBM computers use the eight-bit extended binary-coded-decimal interchange code (EBCDIC). Both codes reserve one track of the tape for an error-detecting facility known as a parity bit. Before each coded character is written on the tape the number of ones is counted and the parity bit selected so that every character has an odd number of ones (odd parity) including the parity bit. When the tape is read later, any character that is found to have an even number of one bits must be in error. (A similar, even-parity system generates an even number of ones and notes an error if an odd number is found). A good way to produce a lot of parity errors is to bring a magnetic tape near a strong magnetic field, to pass it through an airport metal detector or customs X-ray device, or to specify odd-parity checking when the tape was recorded with even parity.

4C.3.5 *Remote terminals*

In a timesharing environment most communication with the computer occurs via remote terminals. Each key on such a terminal produces a sequence of seven bits plus a parity bit that corresponds to the ASCII code for the letter, numeral or symbol on the key. (IBM terminals use the eight-bit EBCDIC code plus a parity bit and are therefore not compatible with ASCII installations unless provision is made for translating from one code to the other.)

The DC pulses used to code these bits cannot be transmitted over distances greater than about one mile, but they can be transformed by a *modem* (*mod*ulator/*dem*odulator) or *data set* into AC signals suitable for telephone-line transmission. Although a modem that meets telephone-company standards can be connected directly to the telephone line, an *acoustic coupler* is often employed to produce audible tones that are sent and received via a standard telephone handset. The prospective purchaser of a data terminal should bear in mind that the hundred dollar or more cost of a modem and acoustic coupler is not included in the price of most terminals. This is because such terminals can be *hard wired* (connected directly to the computer), making the modem and acoustic coupler optional extras.

Terminals vary considerably in how fast they can transmit and receive data. Typical transmission rates range from 110 to 19200 baud. In computer applications one baud corresponds to one bit per second. Hard-wired CRT terminals can operate at rates as high as 19200 baud, filling a screen with characters in the blink of an eye and permitting complex, animated graphics displays.

CRT terminals also differ in the amount of data that can be displayed on the screen at once and in the extent of cursor control. The cursor is a mark on the screen that indicates where the next character will go. Convenient cursor control can often be used to produce limited graphics displays on a non-graphics CRT terminal. For program writing and editing it is convenient to have sixteen or twenty-four eighty-character lines on the screen at once.

The video signal that drives a CRT terminal's screen display can also drive additional video monitors. Provided the terminal has a video output jack, you can display the results from a computer program in any classroom equipped with a telephone and an appropriate number of monitors. This is a very effective technique for teaching students how to use a timesharing computer system, and for showing them what programs are available for microcomputers the situation is even better. Most of them are designed to be connected to video monitors or have low-cost adapters for this purpose, and a stand-alone microcomputer needs no telephone.

4C.4 Machine languages

Each operation that can be performed by a computer's CPU can be coded as a binary number and stored in the computer's memory. The computer executes these operations in sequence when a user runs a program. These binary instructions constitute the *machine language* of the computer, and they vary from one CPU to another.

Programming by entering machine-language instructions into a computer is very difficult but it did not take long for people who did this sort of thing to come up with the idea of writing a computer program that would translate expressions more understandable by humans into the zeros and ones required by the computer. The simplest type of translator is called an *assembler*. Assembly language provides mnemonic codes (such as ADD) for the control operations and can use decimal instead of binary numbers. However, there is nearly a one-to-one correspondence between statements in assembly language and control operations in the computer. This makes an assembly language specific to a given type of CPU. Moreover, since each control operation is very simple and not something humans are used to doing, a lot of program instructions are needed and the opportunity for error is great.

The drawbacks of assembly language can be alleviated to a considerable extent if either of the two kinds of *high-level language* is used. High-level refers to the fact that program statements bear a close resemblance to the words and mathematical formulae with which people are familiar. A *compiler* program, such as the one usually used with FORTRAN (short for *for*mula *tran*slator), reads through the entire set of statements provided by the programmer. The compiler notes any deviations from the syntax of the FORTRAN language by means of error messages, and generates machine-language instructions that will carry out the programmer's intent. The programmer's statements are referred to as the *source code*, and the machine-language program is called the *object code*. The number of object-code instructions greatly exceeds the number of source-code statements (often by a factor of 10 or more), because the source-code statements may specify rather complicated mathematical or logical operations. After the object-code program is complete, it can be executed.

An *interpreter* language, such as BASIC, also permits complex statements in near-natural language. The interpreter program, however, proceeds through the source code one statement at a time, checking, translating and immediately executing the statement before going on to the next. Interpreters are especially convenient in a timesharing environment. BASIC, for example, provides immediate feedback of readily understandable error messages, allowing a novice at a terminal to learn the language quickly. Usually BASIC also provides a simple means for editing and correcting program statements that are in error. When many similar calculations are to be done repetitively, however, a compiler has the advantage. Once the complete object-code program has been produced it can be saved and used over and over.

Also, because the compiler considers more than just one source-code statement at a time it can often produce fewer or more efficient machine-language instructions to solve the same problem. A high-level language makes it much easier to transfer programs from one machine to another. One FORTRAN compiler can operate on the same (or nearly the same) source code as another FORTRAN compiler, even though the object code produced is different for each different CPU.

Currently usage of BASIC and FORTRAN exceeds that of their nearest competitor by a large margin, and nearly every computing centre supports one or both of these languages. However, neither represents the current state of the programming art, and both are at least a decade old—ancient by computer standards.

Languages like PASCAL and ADA have been designed more recently to embody concepts of structured programming that make it far easier to produce error-free programs and to modify existing programs to meet changing needs. Also, for highly specialized programming tasks general-purpose languages like FORTRAN, BASIC or even PASCAL may become cumbersome. An example is a CAI program, which must recognize and respond to student inputs that may cover a broad range of words and symbols.

CAI languages like TUTOR (the language used by the PLATO system [45]) or common PILOT allow the programmer to cope with misspelled words and the same answer expressed in different units, situations that would require major programming efforts in BASIC or FORTRAN.

4C.5 Transportability of computer-based instructional materials

To use computer-based materials in chemistry instruction it is not necessary to be (or even to know) a computer programmer. A great many computer enthusiasts have already written programs that you might use, and most of those programmers are quite willing to share what they have done. The novice, therefore, could be excused for thinking that a request to, 'Send me all the programs you've got' is all that is needed to get started. Unfortunately it is not quite that simple.

The transportability of computer-based materials, that is, the ease with which they can be transferred to and used at institutions other than the one at which they were written, is still rather low. There are three factors in this transportability problem: hardware incompatibilities; software incompatibilities; and incomplete documentation.

4C.5.1 *Hardware incompatibilities*

Hardware incompatibilities include lack of sufficient memory or sufficient precision, the absence of appropriate I/O devices (such as a graphic terminal), the fact that a program may execute too slowly on a smaller, less expensive machine, or a mismatch in interchange devices (seven-track tape drive at one institution, nine-track at the other). The advent of low-cost microcomputers has improved transportability in one sense but exacerbated the problem in another.

By purchasing identical hardware you should be able to make use of a program without modifying it in any way. However, each brand of microcomputer, even those based on the same CPU, is an individual that will not run programs written for other brands, or even for other models of the same brand. For example, Apples come in regular and 'plus' versions, PETs may have old or new ROMs and forty or eighty columns on the screen. Newer models may or may not run all programs developed for older models, and if the older model is out of production you may not be able to purchase hardware identical to that of the person from whom you would like to obtain programs.

4C.5.2 *Software incompatibilities*

Software incompatibilities are often more subtle than simply the lack of a compiler or interpreter, should the program be written in a less-common language such as PASCAL. During the lifetimes of FORTRAN and BASIC there have been many significant developments in hardware, in the operation of computers and in the theory of programming. To capitalize on these developments, manufacturers have added new features to FORTRAN and BASIC, and such extensions often vary from one manufacturer to another.

Thus, although in principle a FORTRAN program should be machine-independent, in practice it often is not. Programs that include graphics or storage of data in disc files, for example, almost invariably require modification before they will run on a different computer. For large mainframes BASIC is much worse than FORTRAN in this regard, probably because there has been an accepted standard for FORTRAN for over a decade. When a minimal standard BASIC is constructed using only features that are common to almost all dialects of BASIC, so little remains that any program of reasonable size almost has to violate the standard. In the microcomputer field, however, Microsoft BASIC is somewhat standard, with many manufacturers providing essentially this same dialect of BASIC along with their computers.

4C.5.3 *Documentation needs*

Documentation refers to descriptive and explanatory material, both in the form of comments within a program and as instructor and/or student manuals. Technical documentation should specify the computing facilities (hardware, language, program size, machine dependencies) needed to run a program. It should include a printed *listing* of the program and at least one sample execution, showing the output to be expected from specified input. Good technical documentation can also minimize the effects of software incompatibilities by pinpointing them and explaining what to do about them.

Educational documentation should specify subject-matter, instructional objectives and pedagogical techniques. It should also explain how computer use is incorporated into coursework. Sufficient detail should be included so that a prospective user can judge the technical accuracy of the computer-based materials. Educational documentation should also reveal *embedded pedagogy*, an author's teaching method, approach to a topic, or strategy that is built into a computer program. If you disagree with the embedded pedagogy, then the program is worthless to you.

4C.6 **Computer-assisted instruction**

CAI refers to those applications of computers in instruction where a student engages in dialogue with a computer program to achieve a well-defined and measurable understanding or skill. CAI can profoundly alter the manner in which a course is presented if sufficient high-quality CAI material is available and an instructor is willing to adjust the roles of lectures and other class activities to accommodate it.

4C.6.1 *Types of CAI programs*

Some of the principal ways in which CAI has been incorporated into chemistry courses are: (i) problem-tutorial CAI, which provides built-in drill and remedial review on assigned homework problems; (ii) objective-directed tutorial and drill, which tutors and quizzes students on a specific learning objective; (iii) test-mode CAI, in which students at terminals are presented with unique quizzes that have been generated by random selection of questions; (iv) pre- and post-lab CAI, which simulates a laboratory exercise in advance or checks and gives assistance with analysis of data; (v) lab-extension and lab-substitute CAI, which simulates experiments that students have already done or that are too expensive or difficult to be run in the laboratory.

General strategies adopted in CAI include drill and practice, tutorial, simulation and gaming, with most CAI lessons involving all of these in various proportions. Drill and practice is appropriate for many rote memorization tasks, such as nomenclature. Tutorial CAI presents information of expository material in small segments, like the 'frames' of programmed instruction. The student's active participation is required by many requests for student responses. On the basis of these the program decides how rapidly material should be presented and whether review is needed.

Simulation of data, experimental results, apparatus that must be assembled, or situations that require that students apply their chemical knowledge can expand students' experience in many ways. Simulation is especially effective when combined with graphics and embedded in tutorial dialogue that checks whether the student understands the objective of the simulation and can interpret the results correctly.

Many programs can be made more interesting by keeping score or otherwise introducing a feature of competition. One example of this is a program that allows two students to compete at synthesizing organic compounds. One student chooses a substance for the other to make, and the computer evaluates the success of the synthetic method. Such games can also be played by a single student with the computer as opponent.

4C.6.2 *CAI programming and languages*

Programming a computer to carry out the strategies listed above is not a trivial task. Under the best of circumstances producing enough *courseware* (the instructional programs) for a course requires effort equivalent to writing a textbook. The CAI author must devote a great deal of time to revising and updating on the basis of experience gained as students interact with the courseware.

In principle the operations that are most fundamental to CAI can be expressed in nearly any programming language, but in practice general-purpose languages do not lend themselves to tasks such as interpreting student responses on the basis of keywords or detecting minor spelling errors. Consequently specialized high-level CAI author languages have been developed. At present there is no standard CAI language that, like BASIC or FORTRAN, is available for nearly every computer system. The most widespread CAI system is IBM's Coursewriter III. The best of the widely used languages is TUTOR, which is closely allied to the PLATO CAI system marketed by Control Data Corporation. Common PILOT is available for many different microcomputer systems. The next few paragraphs summarize features that are important in comparing CAI author languages. Several languages that have been used to create chemistry courseware are evaluated on the basis of these features in Table 1.

TABLE 1. Comparison of features of some commonly used CAI languages.[1]

Language	Coursewriter III IBM Program Product for System/360; also available on HP 3000	Extended BASIC (may not be applicable to all microcomputer implementations)	TUTOR The PLATO authoring language developed at University of Illinois	APL Originated by IBM, now more generally available.	Common PILOT Developed at Western Washington University
Who has developed chemistry materials?	S. Lower, Simon Fraser University	J. J. Lagowski, G. Culp, University of Texas; many others	S. Smith, University of Illinois	D. Macero, Syracuse University	G. Gerhold, Western Washington University
Language structure	Linear, global variables only, good implicit branching, moderate use of GOTOs	Subroutines, often no local variables; no implicit branching; moderate use of GOTOs. (Language structure varies from one implementa-tion to another)	Efficient block structure, but no local variables; good implicit branching and powerful CASE constructs	Procedure blocks with local and global variables; implicit branching difficult to simulate	Subroutines, no local variables; some implicit branching, moderate use of GOTOs but minimized by convenient conditional specification
Constructing and accessing data lists	Limited and complicated	Unlimited but complex	Unlimited and flexible; moderately complex	Varies with implementa-tion, but can always use arrays	Unlimited, moderately complex
Character data					
1. Answer judging	1. Powerful	1. Requires complex subroutines or system functions	1. Powerful	1. Powerful but inefficient, requires complex subroutines	1. Powerful
2. Manipulation	2. Very weak	2. Moderate–weak (good with system functions)	2. Powerful	2. Powerful	2. Moderate
Numeric data					
1. Answer judging	1. Very weak	1. Moderate; requires computation	1. Powerful; includes algebraic expressions and dimensions	1. Powerful	1. Moderate
2. Computation	2. Integers only in most versions	2. Powerful	2. Powerful; includes bit and byte manipula-tions	2. Extremely powerful	2. Powerful
Random selection					
1. Numbers	1. Awkward and limited	1. Computation required	1. Powerful and easy	1. Powerful and easy	1. Moderate computation required
2. Text and modules	2. By GOTOs	2. By subroutine call or CHAIN statement	2. By module call or CASE statement	2. By function call or array reference	2. By subroutine call

IDF Instructional Dialog Facility marketed by Hewlett-Packard written in BASIC	IPS Course design and documentation language under development at Simon Fraser University at Vancouver, B.C.	DIALOG CAI support system on SIGMA 7 at University of California, Irvine	CATALYST/PIL Available from University of Pittsburgh for DEC system-10	PASCAL/DIALOG Under development using University of California, San Diego PASCAL	DECAL DEC author language marketed by Digital Equipment Corp.
K. Gregg. Vancouver Commercial College; and others	S. Lower, Simon Fraser University	A. Bork, University of California, Irvine	K. J. Johnson, University of Pittsburgh	None yet (A. Bork, University of California, Irvine)	...
Subroutines, no local variables (except 12 counters), good implicit branching	Procedure blocks with local and global variables, good implicit branching and conditional modifiers with CASE constructs	About 200 assembly language macro calls; FORTRAN subroutines. APL-based screen design capabilities	Linear, subroutines allowed, global variables except with sub-routines, good implicit branch-ing and CASE constructs	PASCAL is a structured language with procedure blocks and full control structure	Segmented structure with ordered blocks, within seg-ments. Optional branches
Moderately complex	Easily done using DATA modules and GETDATA commands	Requires pro-gramming skill with macros; can use FORTRAN arrays	Unlimited, but requires indexing	Unlimited	
1. Moderate	1. Powerful	1. Powerful	1. Powerful	1. Powerful	1. Powerful
2. Weak	2. Powerful	2. Powerful	2. Powerful	2. Powerful	2. ...
1. Moderate; requires computation	1. Powerful	1. Powerful	1. Powerful	1. Powerful	1. Tolerance and ranges may be specified
2. Powerful	2. Powerful; uses APL primitives	2. Powerful	2. Powerful	2. Powerful	2. Powerful
Difficult	1. Powerful	1. Powerful	1. Powerful	1. Powerful	1. Computation required
	2. By subroutine call or CASE statement	2. Powerful	2. Powerful	2. Powerful	2. By branch random statement (similar to CASE)
Limited number, predefined but can access BASIC	Unlimited number, no dimensioning or specifications required	Unlimited number and names	Unlimited number, no dimensioning or specification required	Unlimited	Similar to BASIC

TABLE 1 (*continued*)

Language	Coursewriter III IBM Program Product for System/360; also available on HP 3000	Extended BASIC (may not be applicable to all microcomputer implementations)	TUTOR The PLATO authoring language developed at University of Illinois	APL Originated by IBM, now more generally available.	Common PILOT Developed at Western Washington University
Variables and data types	Limited number, predefined names, no floating point. (Can access BASIC on HP 3000)	Large number, restricted names, dimensioning required	Large number of predefined types, may be flexibly renamed	Unlimited number, no dimensioning or specification required	Large number, restricted names, dimensioning required
Extensibility	Difficult; requires assembly language functions. (Can access BASIC on HP 3000)	By BASIC functions	By function modules, but less necessary owing to large number of commands	Convenient, by use of APL functions	By BASIC-like functions.
Graphics	None	None, except in special dialects	Extremely powerful	None	Moderate; point, vector, and character graphics
Recording of student responses	Automatic	By user-constructed utility programs	Automatic	By user-constructed utility programs if file input/output is available	By user-constructed utility programs
Documentation					
1. Intelligibility of listings	1. Poor–fair	1. Poor	1. Good	1. Hopeless	1. Good
2. Commenting	2. Fair	2. Fair	2. Good	2. Fair	2. Fair
Major virtues	Large quantity of courseware, vendor support, easy to learn	Widely available, well known	Quality and amount of courseware, ease of using	Computational elegance, availability	Low-cost microprocessor implementations, ease of learning
Major drawbacks	Cost, inefficiency and obsolescence; limited computational facility	Difficulty in programming, too many dialects	Cost of system or CDC PLATO service	System limitations on ease of programming	Limited number of functions and author conveniences

Explanation to Table 1

Language structure. Modern computer programming practice places heavy emphasis on a style known as structured programming. Two structured-programming concepts that are especially important in CAI applications are top-down, modular design and minimization of GOTO commands. In the top-down approach a large problem or lesson is broken down successively into larger and larger collections of smaller and smaller program segments or modules. Each module is simple enough that it can be programmed and tested easily, and properly written modules can be reordered and reused to suit changing course requirements or teacher preferences. Modular programming is

IDF Instructional Dialog Facility marketed by Hewlett-Packard written in BASIC	IPS Course design and documentation language under development at Simon Fraser University at Vancouver, B.C.	DIALOG CAI support system on SIGMA 7 at University of California, Irvine	CATALYST/PIL Available from University of Pittsburgh for DEC system-10	PASCAL/DIALOG Under development using University of California, San Diego PASCAL	DECAL DEC author language marketed by Digital Equipment Corp.
By BASIC functions	By functions in command language, similar to APL	Difficult except by preparing new assembly language macros	By PIL functions (PIL is an interpreter like BASIC, but more powerful)	New procedures easy to prepare	Difficult; requires alteration of BASIC program
None	Not yet implemented	Powerful	None	Good, including limited animation	None
Automatic; integrated with HP Instructional Management Facility	Automatic	At author's discretion	Automatic	At author's discretion	Automatic
1. Fair	1. Good	1. Good	1. Fair	1. Excellent	1. Fair
2. Fair	2. Good	2. Fair	2. Variable	2. Good	2. Not available
Availability on H-P systems, ease of programming in author-prompting mode	Flexibility and power, efficiency in programming and documenting	Graphics capability and computational power	Two languages in one: CATALYST a CAI-author language, PIL an excellent interpreter	Powerful language becoming widely used. Available on many micro-computers.	Interactive programming in author mode; easy to learn and use
Limited programming options and flexibility	Lack of general availability; requires other systems for support	Several languages needed	Limited availability; written in assembly language for the DEC-system-10	Sophisticated language to learn.	Lack of flexibility, graphics, extensibility, and courseware

1. The authors acknowledge the co-operation and help of: Alfred Bork, University of California, Irvine; George Culp, University of Texas, Austin; Harry Keller, Digital Equipment Corporation; and Taylor Pohlman, Hewlett-Packard Corp. in providing information included in this table. One major CAI system, the TICCIT project, has not been described here because it includes no chemistry courseware.

best realized using an authoring language that has a procedure or block structure (as in ALGOL or PL/I. for example); failing this, there must at least be the facility for invoking subroutine calls as in FORTRAN or BASIC. Local variables are also necessary for full modularity. If *X* is a local variable in module 1 and in module 2, then changing the value of *X* in module 1 does not change the value of *X* in module 2. Many versions of BASIC, for example, permit only non-local (global) variables, and therefore modularity is limited.

Minimization of GOTO commands simplifies program logic and facilitates subsequent debugging or modification of programs. For the most part the sequence of program flow should be linear except for leapfrogging provided by conditional statements of the type: IF some condition is met THEN do one thing ELSE do something else. An example of such leapfrogging occurs when a student response is being checked: IF response is correct THEN increment correct answer count ELSE increment wrong answer count. Some CAI languages perform this kind of leapfrogging automatically, a function known as *implicit branching*. Even more elegant is the CASE construct, in which several possible actions are specified in parallel. Which one is executed depends on the value of some variable. Again using the example of chemical symbol drill, the statement

```
CASE student response OF
   SKIP: present next question
   HELP: call tutorial module
   Magnesium: print reinforcing message
   Manganese: print 'The symbol for manganese is Mn.'
```

would cause the indicated action, depending on whether the student response was 'SKIP' 'HELP', 'magnesium', or 'manganese'.

Access to data lists. An author should be able to construct, store, and access tables of data, such as element names, symbols, atomic numbers, and atomic weights. Preferably this should be possible without the usual formalities of memory allocation and disc-file management.

Character data. A student's response is always accepted by the computer as a string (sequence) of alphanumeric characters. For example, 'One mole of methyl chloride weighs 50.5 g.' The ability to manipulate such a character string and compare it with anticipated responses is the most important index of the power of a CAI language and of the sensitivity with which a program can communicate with students. All CAI languages use some type of keyword search. Powerful CAI languages also provide many other features including: alternative answers (such as choloromethane instead of methyl chloride); pattern matching to detect word fragments (chlor could be picked out of chloride); detection of spelling errors; phonetic matches (cloride for chloride); case editing (making Methyl Chloride equivalent to methyl chloride); and space editing (which, if desired, would make methyl chloride equivalent to methylchloride). BASIC, FORTRAN, and other computational languages are very weak in this area. This weakness can be overcome by writing sub-routines, but only at a significant cost in complexity and efficiency.

Numeric data. In addition to accepting numbers in standard and floating point (scientific notation) format, a CAI language should be able to extract numbers buried in character strings (such as 50.5 in the example given in the preceding paragraph). It should also recognize numeric matches independent of format (50.5, 5.05×10^1, and + 50.5 should be equivalent), and it should handle both absolute and relative ranges on acceptable answers and the dimensionality of scientific units. It is useful to be able to judge the equivalent algebraic expression and to compute the numerical result for expressions entered by the student. The language should be capable of the full range of scientific calculations and should be able to present numeric results to whatever degree of precision an author specifies.

Random selection. It should be possible to select numbers at random between lower and upper limits and to choose exponents at random within a given range. Random selection of text, parameters in problems, and even entire modules (preferably without excessive use of GOTO commands) allows for more variation in phrasing of problems and questions.

Variables. The language should handle a large number of numeric and string variables. Minimal restrictions on variable names permit mnemonic forms that make programs more intelligible. Numeric arrays and either string arrays or long strings with convenient string operators are essential. Programming is simpler when it is not necessary to allocate a specific number of memory locations for a variable (dimension the variable) or to specify its data type at the beginning of a program.

Extensibility. The possibility of modifying and extending a language in terms of its own command structure is important because the developers of a language never anticipate all the needs of those who eventually use it. A language like Coursewriter which is not easily extensible, eventually becomes obsolete. Extensibility requires that a language includes a rich assortment of elementary operators for manipulating numeric and string variables, together with the ability to construct author-defined functions from those operators.

Graphics. It should be convenient for the CAI author to position text or drawings anywhere on the output display, and true graphics ranging from graphs to complex animations should be possible. Elementary commands for drawing and erasing points and lines and for translating, rotating and scaling them are essential. Also valuable is some means for direct input of a graph or drawing (as opposed to input of the numbers or function that generate the diagram). Exactly what can be done in the area of graphics often depends more on the output terminal than on the CAI language, but a language that includes graphics will be preferable to one that does not when graphics terminals become more readily available.

Recording of student responses. The importance of analysing student responses as an aid to courseware development has already been mentioned. Most CAI languages provide for this automatically while some require special programming. Some CAI systems also provide for recording of student comments to the author. A related feature is the capability of recording the point at which a student stopped in the program so that the course can be resumed at a later session.

Documentation. Most CAI programs require continual revision and debugging as more experience is gained from student–computer interactions. Such revision is facilitated greatly if a source-code program listing clearly specifies the algorithm used by the original programmer. Some languages are inherently clear and easy to read. Others achieve clarity when comment lines are interspersed liberally among the executable program statements. Such documentation is essential to prevent the sad but common phenomenon of the program that has to be entirely rewritten because nobody (not even the original author) can remember or understand the algorithm.

Availability of chemistry courseware. The table indicates the names of some persons who have sizeable collections of CAI materials in chemistry. The existence of a body of usable courseware need not be the determining factor in selecting a CAI language or system, however. It is usually straightforward to write a program that translates CAI materials from a less powerful to a more powerful CAI language. Current information on translation services and programs usually can be obtained from the major users of any particular language.

4C.6.3 *CAI terminals*

CAI can be delivered to students via any interactive computer, including an inexpensive microcomputer that is dedicated to a single user. The I/O capabilities of such a microcomputer, or of a terminal connected to a timesharing system, define and limit the sophistication of the CAI that can be delivered. In the past much worthwhile CAI has been done using devices like teletype printers, which are limited to ten characters per second and have neither lower case, subscripts or superscripts, nor graphics. However, much better video data terminals are now becoming available for presentation of text, and microcomputers often provide excellent text and graphics output, usually at higher speeds than terminals linked by telephone to timeshared mainframes. Microcomputers can also be equipped with devices such as light pens or game paddles that permit touch or other non-character responses from a student.

For a terminal or microcomputer to deliver state-of-the-art CAI the following features are desirable: *Random-Access Write/Erase*, that is, the ability to write or erase one portion of a display without altering the rest. For character displays this can be accomplished with cursor controls. *Supplementary Character Set*, which can provide arrows, square brackets, Greek letters, subscripts and other non-standard characters that chemists use regularly. Combined with selective write/erase, special characters can also be used for animations. Some microcomputers have shape tables that can store often-used graphics or characters. *Touch Response* allows students to point to a display on a video screen rather than typing a word on a keyboard. This greatly increases the ease and speed of student–computer interactions. *Graphics* of at least medium resolution (roughly 256 × 256 dots on a screen) can add greatly to the richness of CAI by permitting display of spectra to be interpreted, complex molecular structures, plots of one variable versus another, drawings of apparatus and many other diagrams of the type that are found in textbooks. Combined with touch response, graphics capability is a powerful tool because visual stimuli can be presented and students can also respond in a visual, rather than verbal, way.

What about costs? The following are prices at the time of writing (1983) in the United States. There are, of course, fundamental differences in many countries where there are import taxes.

Terminals that offer all of the features listed above cost on the order of $6,000 to $8,000. Graphics terminals are available in the $3,000 price range, and character-only video terminals with cursor control cost from $300 to $1,500.

Probably the best bargains are personal microcomputers, which can provide excellent graphics, can often be equipped to operate as time-sharing remote terminals, and may be able to run the PILOT CAI author language. A microcomputer with many of these characteristics will cost from $600 to $3,000. Larger microcomputer systems with high-resolution graphics start at about $3,000, and such systems can often be set up to support several full-graphics terminals from a central computer and disc drive.

In some cases, notably PLATO [45], CAI programs can be written and developed on a large mainframe that has convenient editing and programming facilities and then transferred ('downloaded') to a microcomputer. Eventually sophisticated CAI systems that are entirely microcomputer-based can be expected to become commonplace.

4C.6.4 *Evaluation of CAI*

Experience with CAI shows that students appreciate the immediate feedback that CAI provides. They also like the opportunity to work at their own paces and at times that are convenient to them. They are enthusiastic supporters of high-quality lessons. CAI aids instructors in identifying students who need special help, and it relieves teachers of the need to cover routine material, freeing time that can be used to provide more creative help. A CAI-based course can be completed in 30 per cent less time than a conventional course, or conversely, in the same length of time a larger quantity of subject-matter can be taught. Finally, CAI enables a much larger fraction of students to master a given skill or skills than is practical using conventional instruction. For example, with CAI it becomes reasonable to require that all members of a class be able to assign for example oxidation numbers at a proficiency level of 90 per cent. Subsequent instruction becomes much easier, achieves greater success and can better address higher-level cognitive development when uniformly high mastery of basic skills such as these can be assumed.

Some very important generalizations can be drawn from the experience chemistry teachers have had with CAI to date. By far the most important component of good CAI is the skill and commitment of the author/teacher. The primary skill required is experience in working with students and coping with their difficulties. Probably a small number of skilled authors will write most of the very best CAI lessons, but many teachers will nevertheless prefer the personal touch of tailoring their own materials to the specific needs of their institutions and their students.

No matter whom the author may be, CAI materials must be tested continually and revised on the basis of insights gained. Writing, testing and revising CAI programs is much easier if a good author language and a state-of-the-art display terminal are available, and at least one of these components is almost a necessity. Personal microcomputers can, with appropriate options, provide high-quality graphics and other displays, and are capable of delivering state-of-the-art CAI.

If CAI materials of high quality are comprehensively incorporated into a course, CAI can be a powerful, cost-effective tool for facilitating learning and increasing teacher effectiveness. CAI is no longer an expensive toy for the few. In the near future, it will almost certainly become an important aspect of instruction in chemistry at all levels.

4C.7 Computer graphics

The advantages in CAI or other computer applications of using computer-drawn images of molecular structures of electron densities in atoms or molecules, of plots of one variable versus another, of spectra of various types, of laboratory apparatus and of many other items that chemists use regularly have already been pointed out. The advent of microcomputers that include graphics capabilities similar to those available only in much more expensive terminals until recently has made it possible for almost everyone to obtain hardware that will support graphics displays. Hence it is becoming more and more important for chemists to become familiar with the techniques of computer graphics.

4C.7.1 *Screen-oriented devices*

Computer graphics are most commonly displayed on a screen, and at present there are four main types of screen-oriented devices. On a *storage-tube terminal*, such as the Tektronix 4010 series devices, an image can be written on a special cathode ray tube (CRT) screen, where it will remain until the entire screen is erased. Such terminals are relatively inexpensive, starting at about $3,000, and they provide linear resolution of about 0.007 inch (0.0178 cm), permitting smooth curves to be drawn.

The resolution of a graphics device is most commonly expressed by dividing the screen into a rectangular grid, with each grid point being called a pixel (picture element). High resolution corresponds to at least 512 pixels vertically and 512 horizontally, medium resolution to at least 256×256 pixels, and low resolution to about 128×128. An illustration of the effect of various resolution levels on curved lines is shown in Figure 7. Generally at least medium resolution is needed to display effective graphics for CAI and other chemical applications.

The main advantage of a storage-tube CRT display is that the screen itself stores the image—no computer memory need be devoted to remembering whether a given pixel is illuminated or dark (or the intensity or colour of a pixel). Since a 512×512 display involves 262,144 pixels, it would require 32,768 (32K) eight-bit words to store the information needed to define every pixel of a simple black-and-white display, and colour or grey-shades would require even more memory. (Typical eight-bit microcomputers are limited to a total of 64K eight-bit words, so 32K of memory is a lot.)

Recently the cost of memory has fallen drastically and *non-storage-tube CRT displays* have become more readily available. These require that data representing the CRT image be stored in memory and the image be repeatedly written on the screen (refreshed). To avoid annoying flicker a refresh rate of 30 to 60 s^{-1} is necessary, and this places an added burden on the host computer (or on the graphics terminal, should it contain the image memory and refresh hardware).

There are two main types of refreshed CRT graphics displays. In a *directed-beam* or *vector-driven* device the end points of line segments (vectors) to be

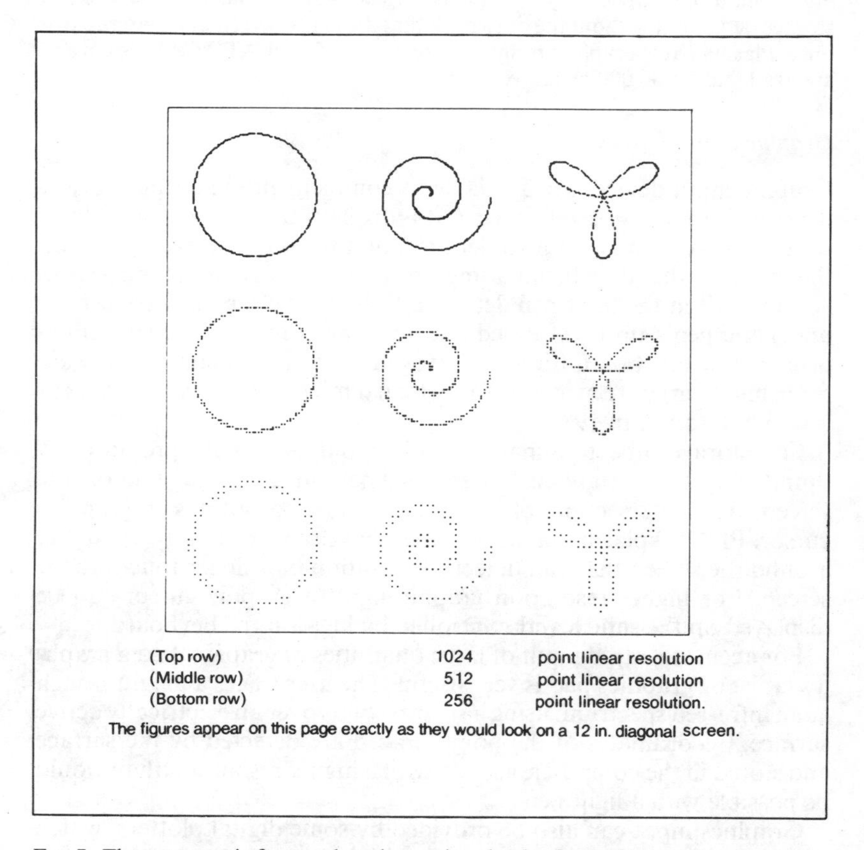

FIG. 7. Three geometric figures plotted at various levels of graphic resolution.

displayed on the screen are stored in memory and the electron beam of the CRT draws all vectors during each refresh cycle. This requires little memory and works well when there are only a few vectors on the screen but displaying too many vectors may result in lowered intensity or flicker. Most microcomputers and many graphics terminals use a *raster-scan display*, which is based on standard television technology. During each refresh cycle the electron beam of the CRT scans the entire screen in a raster pattern. In the United States this involves 525 horizontal lines on the screen, corresponding to a vertical resolution of 0.02 inch (0.058 cm) on a 12-inch (30-cm) screen. Both raster-scan and vector-driven displays have the advantage that portions of the displays can be selectively erased, and the entire display can be changed during each refresh cycle. Thus rotation of molecular structures and animation of, for example, an S_{N^2} mechanism, are possible with refreshed displays.

A fourth type of screen display is the *plasma panel*, which is not based on a CRT but rather on a layer of ionizable gas between two glass plates. A fine grid of 512 vertical and 512 horizontal electrodes allows each of the 262,144 pixels on the screen to be individually illuminated or extinguished, making the screen a storage device but without the drawback that the entire screen must be erased at once. Plasma panel graphics terminals are used by the PLATO CAI system and are available for $6,000 or more.

4C.7.2 *Graphic input devices*

Graphic input devices are available to complement the output devices described above. For refreshed CRT displays a light pen provides a simple, low-cost way for a student to point to something displayed on the screen (rather than formulating a response in words and typing it in). A photocell in the light pen detects a flash on the screen when a pixel under the pen's tip is refreshed and tells the computer the screen co-ordinates of that point. Light pens (together with rudimentary software for using them) were available for personal microcomputers at prices as low as $30 to $50 in 1982.

For storage-tube terminals graphic input is usually provided by thumbwheels that move a horizontal line and a vertical line on the screen. The intersection of these crosshairs constitutes a graphics cursor. PLATO plasma-panel terminals have sensors on a 16 × 16 grid around the screen that can detect the position of a finger touching the screen. For higher resolution graphic input a graphic cursor can be displayed on the screen and controlled by keys on the keyboard.

For accurate, rapid input of large quantities of graphic data a graphics tablet or graphics pad is very useful. The user traces a diagram such as an infra-red spectrum, using a special stylus over an electrically active surface. Co-ordinates of the points traced are detected by the surface and stored in the computer, usually with a higher resolution than would be possible with a light pen.

Graphics input can also be provided by some digital plotters, which have joysticks that can be used to guide the pen over an existing plot.

FIG. 8. Drawing of crystal lattice of ice produced on a wide-bed digital plotter. Hydrogen bonds appear as sticks, covalent bonds as intersections of spheres. Copyright © 1979 by John W. Moore and William G. Davies.

Some microcomputers have joysticks or game paddles that can be used to control a graphic cursor on a screen display.

4C.7.3 *Hard-copy graphics output*

Hard-copy graphics output is commonly provided by digital plotters, which have resolution on the order of 0.005 inch (0.0127 cm) and cost from $1,200 up, depending on the features available and the maximum plot size. Digital plotters can provide excellent reproducibility and can generate very detailed drawings. Figure 8 shows a drawing of the crystal lattice of ice that was produced on a Calcomp drum plotter with a 20-inch (50-cm) bed.

Hard-copy graphics can also be obtained from accessories that attach to graphics terminals, or from electrostatic plotters such as those made by Versatec. Often these devices require special paper, which can increase costs significantly.

There are now quite a few dot-matrix printers available that can reproduce graphics at medium resolution and can be used to print ordinary text as well. Examples are the DECwriter and the Integral Data Systems 'Paper Tiger' series. Dot-matrix printers usually have a vertical row of seven or nine wires, each of which can strike a typewriter-like ribbon against ordinary paper, producing a dot. The vertical row of wires scans horizontally across the page, and with an appropriate program it can be made to place dots at the desired locations to produce text or graphics.

4C.7.4 *Graphics software requirements*

Producing images on a screen or hard-copy graphics device imposes special software requirements. Given a set of x, y co-ordinates that define pixels or vectors to be drawn in a figure, it is necessary to determine what spatial region of the display medium the plot will occupy (a process known as *windowing*) and to determine what values of the variables to be plotted will correspond to the edges of the viewing window (to *scale* the image). In addition, the numerical co-ordinates must be coded into a form that makes the graphics device draw points or lines in the right places. For example, the Houston Instruments DMP-2 digital plotter requires that a lower-case 'p' be sent to move the pen +0.005 inch in the y direction. A lower-case 'q' moves the pen +0.005 inch in both the x and y directions (at a 45° angle) and a lower-case 'r' moves the pen +0.005 inch in the x direction. Other letters produce other pen movements. To lower the pen (so that it draws) a 'z' must be sent to the plotter. A 'y' raises the pen. These tasks are usually carried out by a graphics program, and the purchaser of any graphics device should inquire about the extent of software provided with the hardware.

The most fundamental level of graphics software consists of routines that generate the graphics codes mentioned above and perform special functions like clearing a screen. Such routines will be specific to a particular graphics device. Usually they are written in machine language (to provide a maximum speed of execution), which makes them specific to a particular CPU as well.

The next level of software is a graphics package that performs windowing, scaling, and perhaps rotation, perspective transformations, stereoscopic views, and so forth. Commonly such a package of subroutines is written in a high-level language such as FORTRAN, BASIC, or Pascal and is available from the supplier of graphics hardware. In some cases a graphics software package is into a high-level language so that graphics commands such as DRAW or SCALE are as much a part of the language as are FOR, NEXT, GOTO, DO or other common commands.

The applications programmer must make use of routines like those described above to write software that actually produces the desired graphic images on the hardware available. This can usually be accomplished without unreasonable effort by a programmer of average skill. Some graphics packages, notably the TUTOR language available with the PLATO CAI system, take optimum advantage of graphics equipment, providing convenient graphics input and other features that help a programmer to develop applications software. In other cases, such as where a FORTRAN program must be compiled and linked with graphics subroutines every time a trivial alteration or correction is made, program development is much slower.

Graphics hardware has continually increased in performance/cost ratio during the past few years, and this trend can be expected to continue over at least the near-term future. This is especially true in the case of microcomputers, many of which have graphics capabilities that exceed those of most of the reasonably priced graphics terminals.

Chemistry is a discipline whose practitioners cannot discuss for long without drawing a molecular structure or other diagram on a blackboard, napkin, tablecloth or whatever, and so any improvement in graphics capabilities will certainly provide corollary improvements in computer-aided instruction.

4C.8 Microcomputers

A number of applications of microcomputers and personal computers in chemical education have already been mentioned. This section will concentrate on features that are unique to microcomputers and that teachers or schools will need to be aware of if purchasing such computers for instructional use.

When a school purchases a microcomputer, the teachers become, in effect, the directors of a small-scale computer centre and no longer have the kind of support that most industry, government or university computer centres offer to users of their mainframes. Each will have to decide what hardware and software to purchase to carry out instructional or other tasks. In addition, should anything go wrong, it will be the teachers' job either to fix it or find some service facility that can carry out repairs. If the software has bugs or requires adaptation, they will have to see to the debugging or modification. These are real problems. Hardware repairs may require shipping a defective item to the manufacturer, waiting for an under-staffed facility to repair it, and then waiting for the return shipment. Software problems may require considerable programming time and effort to fix.

Nevertheless, microcomputers, often called personal computers, have so many advantageous features that many chemists are adopting them for instructional and other purposes. The biggest advantage of course is low cost per unit.

4C.8.1 *Microcomputer hardware*

In addition to low cost, personal computers provide high-speed display of text on their screens and considerable graphics capability, although each machine is unique in the way graphics displays are programmed and constructed on the screen. Many of these machines have powerful BASIC interpreters as standard software and are equipped with convenient editors that facilitate writing of programs. Some microcomputers can be equipped with letter-quality printers and word-processing software. Together with other business-oriented programs, this makes them attractive for record-keeping as well as instructional use.

A wide range of add-on peripheral devices and circuit cards are available for the most popular microcomputers, allowing them to be used as remote terminals to large mainframe computers, providing analog-to-digital and ditigal-to-analog conversion for instrument interfacing, allowing upgrades to higher-resolution graphics, providing audible outputs (including music), and generally permitting customization of the computer for a particular application. However, such customization may come at a high price, with add-on options sometimes costing as much as or more than the basic computer. It is also possible to configure systems in which several microcomputers are connected to a single high-priced peripheral device, such as a printer or disc drive. Such a multi-terminal system can provide excellent computer-aided instruction at a very low cost.

4C.8.2 *Microcomputer software*

Since microcomputer hardware is relatively inexpensive, it pays to purchase hardware on the basis of the software that is currently available and running on a particular model of microcomputer. Because each microcomputer is so much of an individual, things that can be done easily on, say, a PET might be well-nigh impossible on an Apple, and vice versa. Software for the more common microcomputers is already available from commercial suppliers, and there is a strong possibility that the tremendous resources of book publishers may soon be brought to bear on software distribution. However, in the near future there is little probability that the same software will be adapted to all models of all brands of microcomputers, and so it is best to look first at the applications programs you want to use and then choose a microcomputer that will run those programs.

The same rule applies to writing your own applications: choose hardware on the basis of the systems software—assemblers, compilers, interpreters, text editors, graphics packages, CAI author languages—that is available. You do not want to have to write (or supervise writing of) any more software than is absolutely necessary.

There is not space in this chapter to go into detail regarding the features of various microcomputer systems; they are evolving so rapidly that information today is not of much use next year.

4C.9 Other instructional applications

CMI and CATC systems often do not involve direct interaction between student and computer. Such systems are much less expensive per student than is tutorial CAI, and consequently they have been widely used and extensively developed during the past decade. Software for several of these systems is readily transportable from one large mainframe to another.

An example of computer-managed instruction in chemistry is CHEM TIPS [46]. Students take diagnostic tests (which do not count in determining grades) at weekly or bi-weekly intervals. The TIPS computer program grades each test and, based on which questions a student was unable to answer correctly, provides an individualized study guide for each student. This study guide suggests portions of the textbook that need to be restudied, directs the student towards audio-tutorial or other remedial lessons, and generally oversees the student's progress through the course. CMI systems can be designed so that study guides are printed in a large batch and distributed by hand, or students may obtain information about their progress by sitting down at a timesharing terminal and running an appropriate program. Faculty members can obtain summary data about individual students for advising or counselling, and all grade calculations can be done quickly by the CMI program. In some cases even the reporting of grades has been automated.

CATC involves using the computer to store or generate large numbers of examination questions. In question-retrieval (or item-banking) CATC systems, questions are typed into the computer and stored on magnetic tape or in some other machine-readable form. Also stored are attributes such as the subject-matter content of a question, the difficulty of the question, keywords on which retrieval of the question may be based, and so forth. The instructor then uses a computer program to select questions from the database by specifying subject-matter categories, number of questions desired and so on. The computer constructs one or more test forms as per the teacher's instructions and prints the tests. Since questions are selected at random from a large number of similar test items, a large number of different tests, all of equivalent difficulty, can be produced. These can be used for repeatable testing, where a student is tested more than once over the same subject-matter. In this way, a much greater percentage of students can be brought to a desired level of mastery. Computer-generated tests can also be graded by computer, and student scores can be handled by a CMI system.

An alternative to item banking is to write sub-routines, each of which can construct a specific type of question by filling data or test into a skeleton question. For example, a skeleton question might look like this:
Calculate the pH of a ____ M solution of ____. The pK_a of ____ is ____.
(a) (d)
(b) (e)
(c)
The computer would select one weak acid from a data file containing formulae and pK_as, and it would choose a concentration from within an appropriate range. It would calculate the correct answer and four distractors, basing some of the latter on common errors that students are known to make in solving such problems. Then it would print the question, as for example:
Calculate the pH of a 0.032 M solution of HCN. The pK_a of HCN is 9.3.
(a) 4.7 (d) 6.2
(b) 5.4 (e) 3.9
(c) 8.2
At the same time the computer would store the correct response to the question, to be used for later grading.

CATC in chemistry has been evaluated at several locations and has been found to contribute significantly to learning. Pedagogical advantages of CATC include: (a) relief of exam anxiety by repeatable testing; (b) advance notice of coverage, difficulty and style of questions; (c) opportunity and motivation for students to learn from their mistakes; (d) solution of the make-up and 'bad day' problems; (e) increase in study time; (f) release of class time for new material by specifying CATC questions that cover remedial or review material for self-study.

Future developments in CATC may be expected to involve refinement of the automated program generation technique to allow better questions to be generated by persons with little or no previous computer experience. A change that is already occurring and that will have considerable impact on CATC is improvement of printing technology. Quality of the text will improve during the next few years and prices should remain constant or decline. Finally, microcomputer systems for CATC will probably become available at prices that schools can afford since a microcomputer-based CATC system could be used by all departments in a secondary school, making it a real bargain on a per-student basis.

4C.10 Pocket calculators

When deciding which calculator to buy, one of the first decisions to be made concerns the order of entry of operators and operands. A majority of calculators employ the algebraic system, which corresponds with the way arithmetic expressions are written. Others use the Reverse Polish system, which normally includes a stack of perhaps four registers for temporary storage of intermediate results. The former is more natural but requires some planning for the evaluation of complicated algebraic expressions unless parentheses are available. The latter, after a somewhat longer learning period, allows complicated expressions to be handled with fewer key strokes.

What kind of calculator should a student in a chemistry class have? At the very least the calculator should have floating decimal point notation so that the very large and very small numbers of chemistry can be treated. The four arithmetic operations suffice for most work, but occasionally a logarithmic, exponential or square root function is needed. Because the cost of this additional capability is so small, a calculator with these functions is recommended. As a bonus, trigonometric and other functions will usually be included.

A final question concerns programmable calculators. Are they worth the money? They offer the advantages of an easy programming language, ready access to considerable calculating power and, for the expensive models, packaged programs on magnetic cards. Most of the methods of numerical analysis can be run on programmable calculators. On the other hand the output is rather primitive (one number at a time, no alphabetic output), there is no hard copy without additional expense and the program is lost when the calculator is turned off except on the most expensive calculators.

Some helpful simulations and games have been produced for use on various calculators [47].

4C.11 Conclusion

Computers and calculators can be applied to chemistry instruction in such a great variety of ways that a chapter like this one cannot but scratch the surface of the subject. There is no doubt that computer applications in chemical education are coming of age and that a great deal of development and refinement of the methods described here will occur during the next few years. Also there is no doubt that computers can be used in a variety of ways to enhance chemistry instruction and even to teach concepts that are very difficult to get across in any other way.

References

1. E. Apea, in A. Kornhauser (ed.), *Proceedings of the International Conference on Chemical Education*, p. 170, Ljubljana, DDU Univerzum, 1979.
2. *International Trade in Educational and Scientific Materials*, Geneva, United Nations Conference on Trade and Development (UNCTAD), 1976.

3. *Report on the Seventh Commonwealth Education Conference, Accra (Ghana), March 1977*, London, Commonwealth Secretariat, 1977.
4. C. Hakansson, in C. Y. Lim-Sylianco, D. F. Hernandez and G. R. Mendoza (eds.), *Proceedings of the Asian Symposium on Low-Cost Equipment for Integrated Science Education at all Levels*, p. 28, Quezon City, University of the Philippines, 1979.
5. *A Collection of Apparatus Assignments*, Penang (Malaysia), Regional Centre for Education in Science and Mathematics (RECSAM), 1976.
6. *New Unesco Sourcebook for Science Teaching*, Paris, The Unesco Press, 1979.
7. *Unesco Handbook for Science Teachers*, Paris/London/New York: Unesco/Heinemann/Unipub, 1980.
8. J. D. Lockard, *Guidebook to Constructing Inexpensive Science Teaching Equipment,* Vol. II: *Chemistry, Inexpensive Science Teaching Project*, College Park, Md., University of Maryland, 1972.
9. Jets of Zambia Production Technology, School of Engineering, University of Zambia, P.O. Box 31338, Lusaka (Zambia).
10. For details, write to the National Council of Educational Research and Training (NCERT), 924 Sri Aurobindo Marg, New Delhi 11016 (India).
11. For details, write to the Institute for the Promotion of Teaching of Science and Technology (IPTST), Sukhumvit Road, Bangkok 11 (Thailand).
12. S. Baholo, C. M. Chabane and P. J. Towse, *School Science Review*, Vol. 58, 1976, p. 78.
13. H. J. Arnikar and R. A. Kulkarni, *International Newsletter on Chemical Education*, Vol. 11, 1979, p. 16.
14. C. Y. Lim-Sylianco, D. F. Hernandez and G. R. Mendoza (eds.), *Proceedings of the Asian Symposium on Low-Cost Equipment for Integrated Science Education at all Levels*, Quezon City, University of the Philippines, 1979.
15. International Council of Associations for Science Education (ICASE), General Secretary, D. G. Chisman, 114 The Avenue, Sunbury on Thames, Middlesex (United Kingdom).
16. R. G. Sanchez and H. K. Schmidt, *Manual de Química Experimental*, published under the auspices of the Colegio Alemán de Sucre, Sucre (Bolivia), 1977.
17. H. K. Schmidt, private communication. Address: 6106 Erzhausen lei Darmstadt, Waldstrasse 5 (Federal Republic of Germany).
18. D. J. Waddington (ed.), *A Sourcebook of Chemical Experiments*, Vol. 1, Unesco/International Union of Pure and Applied Chemistry (IUPAC), 1975. (Also available in Spanish.)
19. S. I. Bayyuk, M. H. Freemantle and E. C. Watton (eds.), *A Sourcebook of Chemical Experiments*, Vol. 2, Paris, Unesco, 1977. (Also available in Spanish.)
20. For details, write to Dr K. V. Sane, Department of Chemistry, University of Delhi, Delhi 110007 (India).
21. L. Pataki and A. Hutter, *Félmikrokemiai Kisérletek*, Budapest, Tan-Könyvkiadó, 1974.
22. E. I. Alonge, *International Newsletter on Chemical Education*, Vol. 16, 1981, p. 8.
23. See ref. 14, p. 2.

24. K. Warren and N. K. Lowe, *The Production of School Science Equipment*, London, Commonwealth Secretariat, 1975.
25. For details, write to National Education Equipment Centre (NEEC), Wahdat Colony, Lahore (Pakistan).
26. C. J. Sturdy, private communication.
27. P. J. Towse, J. V. Binns, D. A. Futcher, S. Pardhan, J. S. Rank and J. W. Steward, *School Science Review*, Vol. 54, 1972, p. 331; M. Carroll, S. Pardhan and J. W. Steward, *School Science Review*, Vol. 57, 1975, p. 254.
28. N. K. Lowe (ed.), *Directory of Educational Equipment Production Centres*, London, British Council, 1979.
29. *The Role of Laboratory Teaching in Chemistry: A Report of the International Conference*, Adelaide, South Australia, South Australian Institute of Technology, February 1978.
30. *The Role of Laboratory Teaching in University Chemistry Courses; A Report on the Unesco International Chemistry Congress, Perth, Western Australia, 1978*, Canberra City, Australian Government Publishing Service, 1978.
31. *International Newsletter on Science Technician Training*, Vol. 1, 1981. Obtainable from Dr D. G. Rivers, Centre for Overseas Studies, Huddersfield Polytechnic, Holly Bank, Huddersfield (United Kingdom).
32. D. J. Waddington (ed.), *ALECSO Conference on Chemical Education, Alexandria (Egypt), 1976, Plenary Lectures.*
33. First Laboratory Workshop in Chemistry at University Level, Department of Chemistry, Seoul National University, Seoul (Republic of Korea), August 1975.
34. Second Laboratory Workshop in Chemistry at University Level, Department of Chemistry, University of Jordan, Amman (Jordan), February 1976.
35. *The Role of Laboratory and its Influence on University Chemistry Courses: A Regional Seminar*, Mexico City, Department of Chemistry, Universidad Nacional Autónoma de México, November 1977.
36. W. T. Lippincott (ed.), *Proceedings of the Sixth International Conference on Chemical Education*, p. 219, College Park, Md., University of Maryland, 1982; A. M. Sarquis, 'Secondary Chemistry Teachers Leave Their Mark on the 6th International Chemical Education Conference', *Journal of Chemical Education*, Vol. 59, 1982, p. 116.
37. A Kornhauser, C. N. R. Rao and D. J. Waddington (eds.), *Chemical Education in the Seventies*, 2nd ed., Oxford, Pergamon Press, 1982.
38. *Safety in the Lab*. A booklet written by the Oxford and District Section of the Association of Science Education. Association of Science Education, College Lane, Hatfield, Hertfordshire, AL10 9AA (United Kingdom).
39. *Education in Science*, Vol. 87, 1980, p. 16.
40. W. T. Lippincott (ed.), *Sourcebook for Chemistry Teachers*, p. 41, Washington, D.C., Division of Chemical Education, American Chemical Society, 1981.
41. *Chemical Nomenclature, Symbols and Terminology*, Hatfield, Hertfordshire, Association of Science Education.
42. *Education in Science*, Vol. 87, 1980, p. 19.
43. Published by the Royal Society for the Prevention of Accidents, Royal Oak Centre, Brighton Road, Purley, Surrey CR2 2UR (United Kingdom). (Publication No. SE. 39b.)

44. See Section 6.3. In *Art of the Science Teacher*, one of the STEP publications, there is an excellent article by E. W. Jenkins on Safety (pp. 127–36). The lesson is suggested in 'Theory into Practice', pp. 54–5.
45. See, for example, R. Chabay and S. G. Smith, *Journal of Chemical Education*, Vol. 54, 1977, p. 745.
46. CHEM TIPS. Professor Bassam Shakhashri, Department of Chemistry, University of Wisconsin-Madison, Madison, WI 53706 (United States).
47. See, for example, D. K. Holdsworth, 'Use of Pocket Calculators in Chemical Education', *New Trends in Chemistry Teaching*, Vol. V, p. 245, Paris, Unesco, 1981.

Bibliography

HOLDSWORTH, D. K. Use of Pocket Calculators in Chemical Education. *New Trends in Chemistry Teaching*, Vol. V, pp. 245–8. Paris, Unesco, 1981. (Although he describes the use of one simple electronic calculator, the method can be transferred to other calculators. Further examples of the work are given in the *Australian Science Teachers Journal*, Vol. 23, 1977, pp. 74 and 113 and in several issues of *School Science Review*.)

Journal of Chemical Education. (Published by the Division of Chemical Education, American Chemical Society. The journal publishes a series of articles entitled 'Computer Series'. The first was published in 1979 (Vol. 56, p. 140).)

MOORE, J. W. (ed.). *Iterations: Computing in the Journal of Chemical Education, Articles 1979–1981. Bibliography 1959–1980*. Washington, D.C., Division of Chemical Education, American Chemical Society.

SLEDGE, D. *Microcomputers in Education*, London W1N 2BA, Council for Education Technology, 3 Devonshire Street, 1979. (This gives a series of articles for teachers who are taking their first steps in the field of microcomputers in education. There are no specific articles for chemistry but the articles give a general introduction and provide background information.)

5 Assessment of students

J. C. Mathews

It has long been recognized that for curriculum development to be successful there first must be thorough pre-service and in-service training and, second, the assessment of students must be sensitive to the aims and objectives of the curriculum. In Chapter 2 , some of the curricula developed over the last twenty years or so were described; in Chapter 4, the training of teachers was discussed. This chapter centres on the assessment of students.

Hand-in-hand with the development of curricula, much work has been devoted to assessment techniques. In this chapter, John Mathews outlines both the functions and techniques of assessment. He describes the various types of fixed response assessment that have gained currency and which in some countries play at present a dominant role in both the classroom and in the examination room. However, other techniques are discussed, some of which are older and have been refined and adapted to meet the needs of the new curricula. These include short-answer and structured questions, information-giving questions and essay questions. Among the newer techniques discussed is oral assessment. Mathews also discusses methods by which practical work can be assessed, an important aspect if we are to encourage effective use of experimentally based curricula.

Employed judiciously, assessment results can be used to evaluate curricula, particularly to determine difficulties (references to their use are given in this chapter and in Chapter 2). The chapter ends with a suggested treatment of in-service training for teachers in assessment techniques, an essential as we progress to more internal assessment.

5.1 Introduction

'Exams: powerful agents for good and evil in teaching.' Few will argue with these words [1] and in this section we shall explore the way these 'powerful agents' can affect the teaching and learning of chemistry.

The moves to change the chemistry curriculum and style of teaching, which began in the late 1950s, gathered momentum in many countries in the next two decades [2]. The changes in the curriculum were facilitated in some countries which did not have the constraints of a centralized examination system. Furthermore, the changes were sometimes inspired by a desire to be as free as possible from these constraints.

The term 'public examination' in this section is taken to mean a large-scale system of assessment at regional or national level, the results of which are intended to have general currency as a measure of attainment. In the 1950s and 1960s, many teachers (for example, in Australia, Canada, the United Kingdom and the United States, felt that the examinations as then constructed encouraged the memorization and repetition of factual chemistry at the expense of the understanding and use of the concepts of chemistry and scientific methods in general [3]. Thus if curriculum change was to be effective, the examinations had to reflect and then reinforce the style and content of the course.

The statement by the International Baccalaureate: 'The criteria of evaluation in the examining of chemistry ... are inseparable from the aims of the course' [4] reflects the often repeated assertion that the aims, content and learning processes of chemical education should determine the nature of the examination; the examination was to be given a subordinate, not a determinant, function. The backwash effect of examinations on the chemistry curriculum was acknowledged; but, it was argued, if the examinations truly reflected what was desirable, they would themselves encourage good teaching and learning: 'Any educational scheme will be made or broken according to, whether or not, at times of assessment, in public examinations or at school, the demands met by the types of questions set encourage the intentions behind the scheme' [5].

Thus, 'assessment is not merely a part of teaching; it is central to teaching. It is an activity that goes most quickly and directly to the heart of teaching' [6].

In a recent detailed survey in which correspondents in forty countries described changes that had occurred in chemical education in their countries during the 1970s, it became plain that the assessment techniques developed and used with the curricula of the 1950s and 1960s were being more widely adopted in the 1970s [7].

5.2 The functions of assessment

Students are subjected to day-to-day internal assessment which teachers use to diagnose their strengths and weaknesses and to find out the strengths and weaknesses of their teaching, and to public assessment which is applied generally and has general currency. Of course, the two are not separable, for teachers will give considerable weight in their courses to the type of questions which their students will eventually take in public examinations. The principles and techniques are essentially the same whether the assessment is public or internal, and we hope that the material in this section will be as useful to the teacher in his school as it is to those concerned with public examinations.

Two main functions of assessment can be perceived. One is *norm-referenced* in which the main purpose is to discriminate among students and to place them in an order of merit; it is essentially competitive, the results being used for selection purposes either internally, within a school, or externally, by employers or teachers in tertiary education. The second is *criterion-referenced* in which the main purpose is descriptive; it is essentially diagnostic, describing the knowledge and skills which students have or have not attained.

A good deal that has been written about chemistry assessment in recent years indicates at least an attempt to emphasize the *criterion-referenced* function [8]. In particular, much thought has been given to the quantitative specification of knowledge and skills as a basis for the design of specific tests (Section 5.4). But it is wrong to assume that these two functions are completely distinct. Every so-called norm-referenced test makes statements about the attainment of knowledge and skills as well as putting the students in an order of merit. And every so-called criterion-referenced test, when operated on a large scale, will put the students in an order of merit as well as make statements about their performance [9].

Evidence indicates that public examinations, although they are sometimes more closely specified in terms of student performance than they used to be, are still predominantly norm-referenced. It follows that their main function is still competitive and this has strong implications for their effect on education in chemistry. Although there has been little direct research on internal assessment, all the indications are that it is still essentially diagnostic; its bias is towards criterion-referencing. But the question must be asked, 'What criteria do teachers use in devising their own tests?' If teachers' tests simply demand the recall of information, then teachers have much to learn from the criteria or specifications of performance which some examiners now use in constructing public examinations in chemistry. However, it is likely that teachers will be influenced by the external examination and will imitate and use the style of questions their pupils will face eventually (if not use questions from past papers). It is more often a question of whether they can imitate questions successfully and for the right reasons.

It should be said in passing that the move towards a closer specification of the knowledge, skills and content of chemistry assessments allows the possibility of using the results not only to evaluate the students but to evaluate the curriculum and the assessment system itself [10]. The evidence from the tests of what students can *actually do* at the end of a chemistry course may be a more objective basis for the development of the course than subjective opinion of what they *ought* to be able to do.

Other secondary functions of public assessment include that of motivation both for teachers and students. No matter how much the actual process of being assessed is disliked, there is every reason to suppose that it provides a tangible and accepted goal towards which teachers and students can work, often in a common purpose. It could be argued that internal motivation would be more productive than this external motivation, but few teachers or students appear to want to rely on it entirely. The degree of this motivational pressure varies from country to country. In some parts of the world, this pressure is at a level which gives cause for concern. For example, Shimozawa has pointed out [11] that examinations for entrance to university in his country are becoming more and more difficult. The Daigaku Nyushi Centre (National Centre for University Entrance Examination) has now been formed, implementing a new procedure whereby 'the chemistry questions were written by professors with acknowledged experience in examination preparation and with knowledge of the "Course of Study"'. Shimozawa is expressing a concern that many others have but do not like to commit to paper.

In addition public assessment, particularly if it is a single, centralized system, can be used to control and change the curriculum. This is less possible in those countries in which public assessment is diversified and not centralized. It is worth noting, however, that in the United Kingdom [12], for example, there are now moves towards the centralization of curricula. Closer co-ordination of the different chemistry curricula has been suggested in Australia [13] and Canada [14].

The incentive to achieving success, together with the competitive, selection function of examinations, will inevitably cause teachers and students to adopt those methods of teaching and learning which will best prepare them to gain credit in the eyes of the examiner. Given these circumstances, it is idle to suppose that aims and learning strategies proclaimed in chemistry curricula will be adopted unless they can be shown to enhance success in the examinations. With this in mind, it behoves curriculum developers to be realistic in their published intentions no matter how much they may deplore this formative function of examinations.

5.3 The agency of assessment

This contribution is primarily concerned with the assessment of attainment in chemistry on a large scale, usually nationally and sometimes internationally. The problems arising from assessing large numbers of students, and the use of the results for selection purposes, have tended to give emphasis to the production of consistently reliable grades serving as a general currency for attainment in chemistry. In most countries there is a substantial degree of uniformity and centralization in the administration of assessment, and a single agency is responsible. It can be said, therefore, that the agency for the assessment of chemistry is normally *external*; that it is administered from outside the schools and not from within the schools. This is not to say that teachers of chemistry have *no* function in public assessment, and in this section a brief analysis of this function will be attempted. But complete control by teachers within a single school is rare. In the United Kingdom a school may choose for its pupils taking the Certificate of Secondary Education (taken by the 20 to 60 percentile at 16 years old), a style of examination known as Mode III. If so, the teachers in that school not only construct the syllabus but also the examinations. However, even here, the syllabus and examination are moderated by an external examination board [15].

The argument for an increasing element of teacher control of public assessment has been given force by the contribution to curriculum development which in recent years chemistry teachers have made throughout the world—in so many countries that it would be invidious to pick on any one as an example. Teachers have been used in policy-making, in design teams, in writing curriculum schemes and in school trials. It is natural and proper, therefore, that they should be used in the design and operation of assessment processes and there is now considerable experience of how this function has been exercised in various curriculum development projects. Most teachers of chemistry would argue that, since it is they who guide the learning, they should have at least some control of the assessment of the outcome of that learning. Few—we hope—would deny this; the questions are: at what level, in what form and on what scale should the teachers' contribution be made?

Four levels of participation of teachers in the public assessment of attainment can be perceived:

5.3.1 *The policy level*

Two kinds of teacher participation are apparent in the composition of those bodies which decide assessment policy on a national scale: political and professional. The political decisions need not concern us here since they deal with assessment systems as a whole rather than with the assessment of attainment in particular subjects. Furthermore, it is not particularly profitable to pursue this line. Using the term described by P. J. Fensham, the assessment system is not usually within an individual teacher's 'frame' [16], although in some countries, where there is a strong Science Teachers' Association or strong teacher representation in the education division of the learned and professional chemistry (or science) society, it may be within the frame of that body in the sense that it might influence political decisions about assessment.

5.3.2 *The planning level*

This level is concerned with the general structure of assessment schemes in chemistry. Decisions are made about the specification of content and objectives, the types of question which are to be used, the degree of choice to be allowed to teachers and candidates, the relative weighting of the various parts, the relative proportions to be assessed externally and internally and the criteria for grading. This function usually rests with the subject committees of examining boards, which may or may not have substantial teacher representation. The advent of the large-scale curriculum projects in chemistry in some countries widened the opportunity for teachers to participate in it.

5.3.3 *The operational level*

The main functions at the operational level are: writing questions and devising marking schemes, constructing tests or examination papers and marking the answers of the students. Teachers are often associated with the last of those and in some countries teachers are in the majority. Until recently, however, the construction of questions and tests has remained with a very few people, who may or may not be teachers.

The participation of a much larger number of teachers in the construction part of the operation arises from the growth of three techniques in the assessment of attainment in chemistry: objective tests, structured questions and the internal assessment of practical work. Each of these will be described in more detail later; the point to be made here is that each of them requires a considerable number of people with those particular skills and experience which are to be found in practising teachers. In fact, the third technique, if universally adopted, requires the participation of nearly all chemistry teachers.

5.3.4 *The feedback or monitoring level*

If any system is to remain healthy, it requires feedback of information from its productive end: in this case, chemistry teachers and students. Even if teachers and students are given no constructional function, it is self-evident that effective monitoring of their opinion and performance within the assessment system is essential. Yet there is little evidence that this is happening on a large scale. In one of the 'A' level chemistry examinations (taken at 18 years by the more academic children) in the United Kingdom, teachers meet examiners after the examination, on a regional basis, to discuss the paper. This helps to dispel the natural antipathy and perhaps resentment that grows up between examiners and teachers.

Worldwide, there does seem to be a move towards the greater involvement of chemistry teachers in public assessment systems. It varies widely from country to country; in some, it has not got much further than the marking function at Level 3. This section has not been written with the intention of prescribing massive participation of chemistry teachers at all levels: even if that were feasible, it may not be desirable. But there is a strong case for much greater participation, especially at the operational level, in the writing of questions and the construction of tests. The constraining factor here is sometimes the lack of expertise in teachers themselves, particularly in those techniques mentioned above, and, in some countries, difficulties of communication. Not all countries face such problems. For example, in the United States there is no shortage of teachers able to help in the construction of the tests organized by the American Chemical Society [17]. However, where there is a problem, the training of chemistry teachers in the techniques of assessment is essential; this is dealt with in Section 5.10.

5.4 The substance of assessment-specifications

In the past two decades there has been a conscious attempt to make assessment match the curriculum. The concept of *curriculum validity* of a scheme of assessment is an important one and transcends the more limited, traditional definitions of validity. In short, it means the degree to which the assessment system reflects the curriculum system in all the latter's essential dimensions: *purpose*, *content* and *processes*. '... the situation in which a student finds himself in an examination should not be significantly different from that which he has experienced in his learning situation' [18].

Complete curriculum validity is an ideal, not a reality. One has only to read the intentions and design of many chemistry curriculum projects to realize that it is not possible to reflect them fully in a reliable assessment scheme on a large scale. The most we can expect is a modest correlation between the two or, to put it negatively: the assessment system should not significantly distort accepted educational processes. In view of the strong formative effect of assessment on the curriculum discussed above, this requirement places a heavy burden on those responsible for the design and operation of assessment systems.

In the 1960s and 1970s curriculum development in chemistry placed much emphasis on *understanding* chemistry, rather than remembering chemistry, and on the process of *investigation* in chemistry. There were other features, of course: a move to bring the content of chemistry up to date and to give greater consideration to social, environmental and industrial aspects of chemistry. Kenya provides an interesting example.

At the lower secondary school level, the School Science Project (SSP), introduced in 1973, emphasized the importance of thinking about why changes take place. It also emphasized the discovery of ideas and principles, as well as the interplay between observation and explanation of facts. The examination was reconstructed so that it reinforced the aims of the curriculum developers. The examination [19] had multiple choice, structured, semi-structured and essay questions (Section 5.5). The curriculum was then revised in 1980 in the light of nearly ten years' experience and one of the principal reasons was because 'it was felt necessary that a greater emphasis was needed to be placed on chemistry and society'. To test this aspect, the examination contains 'thinking type questions' in which

> most of the information required has been given in the stem and pupils are asked to explain or interpret the data in a particular way. By a wider use of this type of question it is hoped that teachers will encourage their pupils to think more deeply about problems in general (that may or may not be related to chemistry in particular) and thereby increase their awareness of the environment around them [20].

This is an example of how the two dimensions of understanding and investigation dominated developments in curriculum design and teaching methods. The schemes for the assessment of attainment in chemistry were then redesigned to attempt to reflect the new curricula and the new teaching approaches. There is no doubt of the influence of the Bloom taxonomy in the cognitive domain [21]. The main categories of that taxonomy—knowledge, comprehension, application, analysis, synthesis and evaluation—have featured increasingly in test specification, with the almost universal trend to include more in the higher (so-called) skills. This system of classification was, and is, applied not only to objective questions but to short answer, structured and open-ended questions. There can be little doubt that the application of the Bloom cognitive taxonomy has been found to be helpful in the classification of questions and in communication between examiners and teachers, but only if confined to the six—or fewer—main categories. Even then, there is room for much overlap between them, leading to disagreement in classification; and there are some performances, specific to the subject, which do not fit conveniently into the taxonomy. Curriculum developers and evaluators may well feel the need for generalizable, theoretical constructs to give a kind of scientific respectability to their work: but if their work is more a craft than a science, their search may be in vain.

At the same time, there has been a move to specify the relative weighting of content in tests of chemistry in forms other than a detailed examination syllabus. The two-way grid, with abilities in one dimension and content in the other, is now commonplace, and with the declining emphasis on the recall of specific facts the content has often been specified in terms of the broad topics and themes of chemistry.

Decisions about the quantitative weighting of specifications have been reached largely by a consensus of examiners and curriculum developers rather than by empirical means; and on the whole only slight acknowledgement has been made to theories of learning. Table 1 is an *example* of a two-way grid that can be drawn up.

A worrying aspect of assessment is our present lack of understanding what makes a topic difficult for a pupil to grasp and what makes a question difficult to answer. This would not be so important if chemistry were to remain a subject for those of highest ability only. But if chemistry in some form is to be part of the education of the majority, and if attainment in it is to be tested, some of the present assumptions about the levels of difficulty must be challenged. Is it possible that we shall see a reversal of the trend towards the higher Bloom categories and a return to a common core of 'knowledge'? This is also discussed in Chapter 2 [2].

The other influence on assessment has been that of changing modes of teaching and learning chemistry over the past twenty years. How far 'guided discovery', 'open-ended investigation' and 'learning by doing' have become realities could be debated. But certainly in many countries the intention was there, and this influenced assessment design. The influence shows itself partly in the forms of written question; for example, in the development of information giving, structured questions and free-response questions (Section 5.5). Especially, however, the influence was felt in the approach to assessment of attainment in practical and project work (Section 5.6). There are many examples of specifications of attainment in practical work—one is given on p. 251: the ability to observe, to deduce from observations, to plan a practical scheme and to manipulate occur frequently. Of particular interest is the introduction of a modest element of assessment of attitudes.

It seems, however, that no taxonomy model in the practical domain has had the same influence as Bloom's in the cognitive. Some theoretical work has been done and that of Alles, in Sri Lanka, may be of particular interest to chemists [23].

Despite these moves, however, the overwhelming impression, obtained from a survey of assessment schemes around the world is that the cognitive domain still dominates. Furthermore, even within that domain, the emphasis on recall and routine comprehension has not declined as much as the intentions of test specifications would lead one to support. These statements are not intended to be value judgements, nor are the undoubted improvements in assessment denied; but it would be idle to suppose that present schemes of assessment at the operational level in chemistry are as revolutionary as some of the published intentions of curriculum developers. Nevertheless, it is probably true to say that a more conscious effort has been made in recent years to improve the curriculum validity of chemistry examinations and test reliability has not suffered in that time; in fact, reliability may be a good deal higher

TABLE 1. An example of a two-way grid (22)

Content / Behaviour objectives	*Area 1* Chemical change, element compound; laws of chemical combination, kinetic molecular theory, gas laws, atomic theory, chemical symbolism, formulae, equations, calculations	*Area 2* Solutions, ionization acids, bases, salts, neutralization, hydrolysis, activity series, electrochemistry	*Area 3* Oxidation – reduction, heats of reaction velocity, chemical equilibrium	*Area 4* Atomic structure and chemical behaviour, periodic system	*Area 5* Chemistry of the common elements and their compounds. Physical and chemical properties of hydrogen, oxygen, halogens, sulphur, nitrogen family, typical metals
I. Ability to recall important information Knowledge of important facts Knowledge of definition of important terms Acquaintance with important concepts Verbal understanding of theories and principles General knowledge of the physical and chemical properties of the more important elements and their compounds					
II. Ability to apply principles in making simple predictions Functional understanding of principles and theories of chemistry and their interrelationships Application of a definition Application of principles in situations similar to those encountered in a typical course Application of principles in new situations taken from everyday life Interpretation of a set of data and drawing conclusions from them					

III. Ability to apply principles quantitatively by carrying out calculations Quantitative meaning of chemical symbolism Balancing chemical equations					
IV. Ability to use the scientific method. Includes the following aspects of the method: Distinction between observed phenomena and their theoretical explanation Explaining phenomena in terms of theory Giving the experimental evidence for a theory Factors to be controlled in an experiment Statements which are true merely by definition *Source* [22].					

than it was. The point being made here is that we should be realistic in what we can expect from our assessment systems in chemistry and not claim that they can carry all that we would wish in our more idealistic moments.

5.5 The techniques of assessment

Although this will be the longest section, space will not permit more than an outline of the techniques of assessment which have been developed over the past twenty years. To understand them fully and to become proficient in their use requires more of the reader than mere reading: training and experience are necessary. It is for this reason that some discussion on the training of teachers in assessment techniques will be undertaken in Section 5.10. However, it is hoped that this brief review, coupled with a few examples, will at least indicate recent trends and the present situation.

Written questions and answers, the most predominant mode of assessment, will be described first, followed by other modes of assessment. The classification of types of questions and answers is in some confusion, so we start with some theoretical considerations in an attempt to clarify it.

Perhaps the most important criterion for classification is *the degree to which the answer is predetermined by the examiner*. Using this, questions may be classified on a simple scale ranging from *fixed response* to *free response* according to the type of answer which is required.

5.5.1 *Fixed response*

Fixed response questions, more usually called objective test items, are now widely used in written examinations in chemistry. The most familiar is the multiple choice question consisting of a stem, which asks the question and provides whatever data may be required, followed by four or five responses. The candidate either chooses the correct response and scores a mark or one of the other responses (distractors) and scores zero or (more rarely) a penalty mark. An example is as follows:

Given the following standard electrode potentials:

$E^0 Zn^{2+}/Zn = -0.76$ volt
$E^0 H^+/H_2 = 0.00$ volt
$E^0 Cu^{2+}/Cu = +0.34$ volt
$E^0 Ag^+/Ag = +0.80$ volt

The metal ion(s) which would be reduced by hydrogen is (are)

(a) Zn^{2+} (b) Zn^{2+} and Cu^{2+} (c) Cu^{2+} and Ag^+ (d) Zn^{2+} and Ag^+
(e) Zn^{2+}, Cu^{2+} and Ag^+

(New Zealand University Entrance Examinations)

The classification set of fixed response items is another form of objective testing which is particularly useful in chemistry, where the ability to classify and the ability to use classifications is central to the subject. For example:

For each of the underlined species in Questions A13 to A16, choose, from the list A to E, the most appropriate description of its function.

A acting as an acid
B acting as a base
C acting as an oxidising agent
D acting as a reducing agent
E acting as none of the above (A to D)

A13 $\underline{Cl_2} + 2FeCl_2 \rightarrow 2FeCl_3$

A14 $\underline{NH_3} + HCl \rightarrow NH_4Cl$

A15 $\underline{C} + CO_2 \rightarrow 2CO$

A16 $\underline{Fe^{3+}} + e^- \rightarrow Fe^{2+}$

(Joint Matriculation Board, GCE 'O' level)

Two other forms, multiple completion and relationship analysis, have fairly wide use:

For each of the questions below, one or more of the respones given are correct. Decide whether each of the responses is or is not correct. Then choose

A if only I, II and III are correct
B if only I and III are correct
C if only II and IV are correct
D if only IV is correct
E if I, II, III and IV are correct

Directions summarised

A	B	C	D	E
I, II, III only	I, II only	II, IV only	IV only	I, II, III, IV (all four)

26. Lead can be obtained from lead(II) oxide by
    I passing hydrogen over heated lead(II) oxide.
    II electrolysing molten lead(II) oxide.
    III heating lead(II) oxide with carbon.
    IV heating lead(II) oxide in the absence of air.

(Malysian Certificate of Education)

For the 10 questions 41 to 50 two statements are given. From the table given below, select the best description 1, 2, 3, 4 or 5 that fits with the two statements for each of the questions and mark appropriately.

First Statement	Second Statement
(1) True	True and is a correct explanation of the first statement.
(2) True	True but does not explain the first statement correctly.
(3) True	False.
(4) False	True.
(5) False	False.

	First Statement	Second Statement
41.	The vapour pressure of a solution is always less than that of the pure solvent.	There is interaction between solvent and solute molecules.
42.	The PCl_3 molecule is planar.	There are three identical P—Cl bonds in the molecule.
43.	An element in the colloidal state is much more reactive than when it is in its normal state.	The element in its finely divided colloidal state has more surface area per unit mass than when it is in its normal state.

(Sri Lanka GCE 'A' level)

It is doubtful whether any advantages which the last two types of fixed response questions may have outweigh the problems arising from the complexity of their structure, and the uncertainties which arise when deciding whether a statement is completely true or completely false.

The reasons for the wide adoption of objective tests at school level are well known, but it may be useful to go over them:

They can make a substantial contribution to curriculum validity; they provide in a short period a large number of questions giving good coverage of abilities and syllabus content.

They are cheaply and objectively scored, often by automatic scanner; and this, coupled with the large number of items, is a contributory factor to their high reliability.

They can be pretested so that anomalies in draft questions can be removed and the characteristics of facility and discrimination predetermined.

They help candidates who find difficulty in giving open-ended responses requiring extended answers.

They can be used diagnostically by the application of post-test analysis. This analysis can serve as a guide to student progress, particularly in self-instructional and computerized learning. It can also be used as an instrument of curriculum evaluation [10].

The items can be classified and banked, giving test constructors access to large numbers of items from which they can construct tests of predetermined characteristics (at least in theory).

The writing of objective test items is usually undertaken by teams of trained writers. If these writers are teachers, both the training and the writing can provide insights into the teaching of chemistry which otherwise would not have been acquired.

In several countries it has been worthwhile to produce 'banks' of questions. Pre-eminent undoubtedly are those in the United States [24], but there are other examples. Thus, in 1972 work began on an Australian Item Bank Project at the Australian Council for Educational Research on the assumption that teachers lacked the time and expertise to develop good objective items and that these would be needed as schools accepted greater responsibility for curriculum development and evaluation. Due to changed economic circumstances, computerization is no longer possible on a national basis, but a bank of over 2,500 items in science has been published [25]. Similarly, a major project in India, aimed to help students and teachers get adjusted to new curricula and assessment, has been undertaken by the Association of Indian Universities. This relates to the formation of a centralized Question Bank. By conducting intensive workshops all over the country, this organization has been able to produce a comprehensive Question Bank Book Series. The volume dealing with chemistry lists over 7,000 sample questions with answers. All the entries are classified into objective (e.g. multiple choice, true–false, short answer and other types of questions). This kind of book is likely to prove invaluable in changing the pattern of the examination papers, which in the past have emphasized learning by rote, thus discouraging students from making an effort to understand chemistry [26].

In Yugoslavia, objective testing represents approximately half of the school assessment. A computer test bank is being developed which should serve teachers and students of elementary and secondary level as well as freshmen courses [27].

The many advantages of objective tests should not lead us to overlook there *disadvantages*. They are essentially conformist: unless the candidate answers exactly as the examiner has predetermined, he can get no credit—hence their classification as fixed response; so they give no opportunity for imaginative discussion or free response to the candidate. 'The answer is suggested; the tests look for passive, not active, understanding' [17]. They are susceptible to misinterpretation for linguistic reasons. It is wrong to assume that they are objectively constructed, apart from the use of pre-test data. The writing and selection of the questions, and the responses, are subjective processes and must remain so.

The questions are required in large numbers and they are expensive to produce. The use of item banks may help in this, but security of the items can be a problem, so can the production of large numbers of items each year if an item bank is not used.

To produce a large-scale national examination takes considerable time and no little administrative skill. As an appendix to the section, a timetable is shown for the construction of the test by the American Chemical Society.

5.5.2 *Free response questions*

It is an experience common to chemistry teachers that many of their pupils, after an initial enthusiasm for the subject—reacting to some of the more spectacular aspects of chemistry—either reject it or learn it only because they must. Among the many reasons advanced—somewhat speculatively—for this is that there is no room for open-ended discussion within the subject particularly on its social, industrial and environmental aspects; that there is no room for opinion, or speculation, or value judgement, which characterize study of the humanities and the arts.

Some curriculum development projects attempted to demonstrate that this was not so: 'In all new situations one gropes and fumbles and is likely to make mistakes. This, however, is the exercise by which judgement develops. A pupil must have graded opportunities to be right or wrong' [28]. To reflect in an examination this open-ended approach to learning chemistry is a matter of considerable difficulty. Although the attempt has been made, examples are rare of examination questions which deliberately invite open-ended responses, the marking of which would give credit for divergence of opinion from that of the examiner. Two examples give the flavour of a free-response question.

(i) 'Scientists, far from wanting to impose their ideas upon other people, are often too little interested in the social implications of their work ...' (W. R. Brain). Discuss some of the social implications of some of the chemistry you have studied in the Nuffield course, and, in particular, refer to the importance of chemistry in helping to solve some of mankind's most pressing problems.

(ii) Classify in an appropriate form the various hazards to safety and health which may arise in your Sixth Form practical chemistry. State the basis of your classification and illustrate it with examples of the hazards together with the means whereby they can be avoided.

(Nuffield A-level examinations)

The growing emphasis on the social, political, economic and technological aspects in school chemistry curricula [2] has also led to changes in assessment so that many questions invite free response [29]:

Either (a) Explain what it is that you find attractive in the prospect of a career in some branch of science or technology.

Or (b) What is it that you fail to find attractive about the prospect of a career in any branch of science or technology?

The characteristics of free response lie not so much in the question itself but in the way in which the answer is marked. If a detailed, rigid and predetermined mark scheme is applied to the answers, then our examples to are be placed towards the fixed-response end of our scale. But, if a genuine free response *is* intended and allowed, how may answers to such questions be marked?

The limited experience of free-response examining [30] indicates that such marking is possible with an acceptable level of reliability, provided that the following conditions apply.

The scale used for marking should carry few intervals; for example, 5—4—3—2—1—0, the process being one of grading rather than marking.

The criteria of performance are few and expressed in general terms. In response to the questions quoted above, an example of guidance to markers could be that they should look for: (a) an appropriate selection of evidence; (b) a logical development of argument on the basis of the evidence; (c) an imaginative discussion of the issues involved; and (d) coherent and effective communication.

To increase reliability, the answers should be marked by two examiners and, if they differ by more than one grade interval, by a third examiner. Trials to determine the compatibility of markets are essential. Even so, one must be aware that the marking is still subjective, and the conclusions are arrived at by a small group of examiners.

However, some difficulties remain. The marking and standardizing process is costly and time consuming. An efficient and rapid postal service is necessary; even then, organizational problems may arise. That double marking of free-response answers can work on a modest scale has been demonstrated [30]; but it would be unwise to assume that it could be universally applied. Furthermore, it may be argued that such questions and answers are not 'chemistry' and that they give advantage to those who write well while knowing no 'chemistry'. That is a point which is considered later in this section and it is hoped that the whole of this book will encourage such debate amongst those who are considering the future of the teaching of the subject.

5.5.3 *Short answer and structured questions*

A chemistry examination consisting entirely of objective tests may be thought to demand too fixed and convergent a response; while, at the other end of the scale, an examination consisting entirely of open-ended questions may be thought to allow too free and divergent a response. Perhaps the most interesting development over the past twenty years is that of the structured question of which the following is an example:

5.0 cm^3 of 2.0 M aqueous solution of Y chloride (where Y is a metal) was placed in each of eight similar test-tubes. Different volumes of 2.0 M aqueous solution of silver nitrate were then added to the solution in each of the test-tubes. The resulting mixtures were shaken and allowed to settle. The heights of the precipitates obtained in each test-tube were plotted against the volumes of the silver nitrate solution added. The graph obtained is as shown in Figure 1.

(*a*) (i) Name the precipitate formed.

....

(ii) What is the initial colour of the precipitate?

....

(*b*) How many moles of Y chloride is present in 5.0 cm^3 of 2.0 M solution of Y chloride?

(*c*) What is the volume of the silver nitrate solution that will be just sufficient to react completely with 5.0 cm^3 of the Y chloride solution?

....

(*d*) Calculate the number of moles of silver nitrate that will react with one mole of Y chloride.

(*e*) If the volume of the aqueous solution of Y chloride used is slightly more than 5.0 cm^3, the maximum height of the precipitate obtained will be different. Sketch the graph you would expect to obtain on Fig. 1.

(*f*) 14.0 cm^3 of the silver nitrate solution is mixed with 5.0 cm^3 of the Y chloride solution. The precipitate formed is filtered.

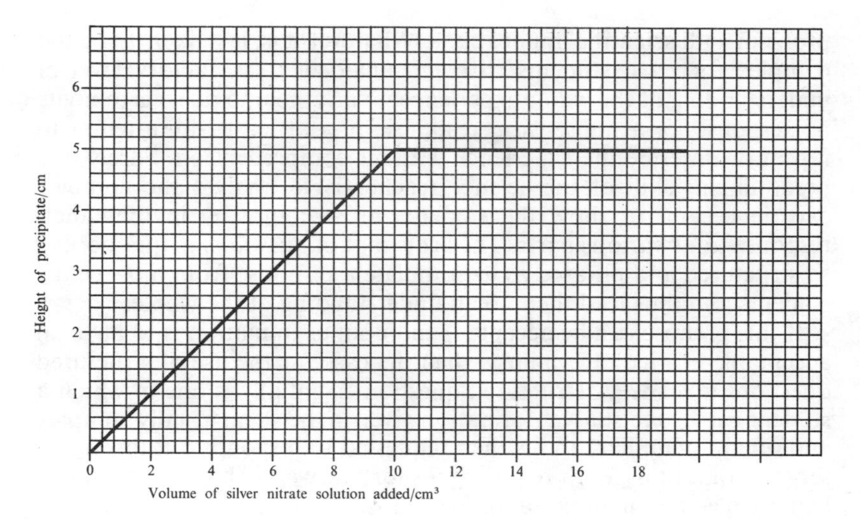

FIG. 1.

(i) What will be observed if the precipitate obtained is exposed to sunlight for a few hours?

....

(ii) Metal Y is above copper in the electrochemical series. Describe what will be observed when a piece of copper foil is placed in the filtrate.

....

....

(Malaysian Certificate of Education).

The justification for the use of structured questions is, or should be, that at their best they reflect good teaching and learning techniques. Many teachers may still see their role as givers of information to be received, remembered and repeated by their pupils; but, whatever reservations one may have of recent trends in science education, there seems to be growing recognition of the desirability of a more active participation by pupils in the learning process. Increasingly teachers see their role, not as mere providers of information, but as guides to understanding. In science this often takes the form first of introducing their pupils to some form of experience—perhaps a class assignment, or a teacher demonstration, or quantitative data, or an industrial process, or a social problem with scientific implications. Whatever it is, it is likely to be too complex for most students to reach unaided a satisfactory level of understanding of the whole experience; to ask them to do so is to invite no response, or a nonsensical response, or—at least—an attempt to repeat or describe the experience without comprehension. Surely, the predominant skill of the science teacher is to analyse the complex experience and to show the students its structure by inviting their response to its various parts. The art is that of *questioning*, questioning so structured and phrased as to stimulate a response from most of the students and thus lead them to an understanding of the whole.

If this premise can be accepted, and if an examination is to have an acceptable curriculum validity, then the case for the use of structured questions is made. A structured question essentially is one in which a student is asked to study information given in the stem, usually complex and often unfamiliar, and is then requested to respond to it through a series of questions, each requiring a short answer. (This type would be better called a structured set of questions.)

Although there is evidence that the use of structured questions is growing, it seems likely that training in their construction is less readily available than for objective questions. It may help teachers, therefore, to make use of the following guidelines for the writing of structured questions. At the same time the guidelines will serve as a summary of the characteristics of this form of assessment:

1. The *stem* provides information for the student and acts as a focus for the set of questions which follow it. It may take various forms: descriptions of experiments or industrial processes, diagrams, graphs, histograms, tabulated data, photographs, etc.: in short, the stem can consist of any chemical information on which the candidates can work and apply their knowledge.
2. The questions which follow should relate to the stem and should be in a sequence according to one or more of the following principles:
   (a) a teaching sequence through which the information normally would be studied;
   (b) a logical sequence of operations such as steps of a calculation;
   (c) increasing difficulty or hierarchy of skills (often linked to increasing mark allocation).
3. When formulating the *questions* it is useful to think first of what *response* they are intended to elicit. (This is what many teachers continually do in class.) In order to get a precise answer, the question must be equally precise.
4. It follows that a draft *mark scheme* should be written at the same time as the questions. The number of marks allocated to each question usually ranges from 1 to 5 and the mark allocation may be shown on the question paper, as shown in the following examples.

3. The sectional diagram (from a manufacturer's catalogue) shows a bomb calorimeter.

3 marks (*a*) What substance would be put into:

(i) the crucible,

....

(ii) the bomb,

....

(iii) the calorimeter vessel?

....

1 mark (*b*) How could the combustion be initiated?

....

1 mark (*c*) Where would you place the bulb of a thermometer in the apparatus?

....

2 marks (*d*) What is the purpose of:

(i) the calorimeter vessel,

....

(ii) the stirrer?

....

2 marks (*e*) How would the heat capacity of the apparatus best be determined?

....

....

....

2 marks (*f*) Why is the number of joules produced by the complete combustion of 1 mole of the substance in this apparatus not the same as the enthalpy change of combustion of the substance?

....

....

....

4. Liquid ammonia, which boils at 240 K, is an ionizing solvent. Salts are less ionized in liquid ammonia than they are in water but, owing to the lower viscosity, the movement of ions through liquid ammonia is much more rapid for a given potential gradient. The ionization of liquid ammonia

$$2NH_3 \rightleftharpoons NH_4^+ + NH_2^-$$

is very slight. The ionic product $[NH_4^+][NH_2^-] = 10^{-28}$ mol^2 dm^{-6} at the boiling point. Definitions of an acid and a base similar to those used for aqueous solutions can be used for solutes in liquid ammonia. This question is mainly about acid–base reactions in liquid ammonia as solvent.

1 mark (*a*) Write the formula of the solvated proton in the ammonia system.

....

3 marks (*b*) In the ammonia system state the bases corresponding to each of the following species in the water system.

(i) H_2O ....

(ii) OH^- ....

(iii) O^{2-} ....

4 marks (*c*) Write equations for the reactions in liquid ammonia of:

(i) sodium to give a base and hydrogen,

....

(ii) the neutralization reaction corresponding to:

$$HCl(aq) + NaOH(aq) \rightarrow NaCl(aq) + H_2O(l)$$

....

2 marks (*d*) What would the concentration be of NH_2^- (in mol dm^{-3}) in a solution of liquid ammonia containing 0.01 mol dm^{-3} of ammonium ions?

....

....

....

....

2 marks (*e*) The dissociation constant of acetic acid in liquid ammonia is greater than it is in water. Suggest a reason for the difference.

....

....

....

....

....

....

(Nuffield A-level Chemistry examination)

Each mark should be related to a specific point in the anticipated answer, i.e. marking is positive rather than negative.

5. It is common practice to allow sufficiently large space between the questions to contain the anticipated answer, but not so big as to invite too long an answer. Markers find that this alternation of question and answer, coupled with a tightly specified mark scheme, leads to rapid and accurate marking.
6. The number of questions in a set ranges from about 3 to 10 with an average of 5 or 6. This gives an average total score for the set of about 15.
7. The draft mark scheme is unlikely to cover all the responses which may arise during an examination. It is important, therefore, that examiners should meet after marking a sample of scripts to revise, and usually extend, the mark scheme into its final form. Even then, some discretion has to be allowed to the judgement of individual examiners.
8. Usually choice is not allowed either between the questions within a set or between sets of questions. This is because their length and complexity is such that the exercise of choice by a student under the stress and shortness of time of an examination is too much to expect. (The average time allowed for each structured question is about 15 minutes.)
9. A set of questions may have either a *linear* structure or a *branched* structure. In the former, each question depends on a preceding one; in the latter, the questions do not depend on each other although they all relate to a common stem. The branched type is more common in chemistry and is to be preferred. Sometimes the linear type is necessary; for example, if the sequence is that of logical steps in a calculation. In this case, the mark scheme should allow for the effect that a mistake early in the sequence will have on the subsequent questions.
10. Pre-testing and pre-test analysis of structured questions is not common practice. However, post-test analysis is useful and will produce facility and discrimination indices [31]. Like objective tests, therefore, structured questions can be used for formative evaluation and diagnosis, although they cannot be expected to provide as much coverage of content as an objective test in the same time.
11. Experience has shown that training and practice are necessary for examiners who intend to use this technique. On those occasions, the most important advice is that they should try to view the stem and the questions *through the eyes of the students*. If follows that practising teachers should be deeply involved in the construction of these questions.

5.5.4 *Information-demanding or information-giving questions*

A noticeable trend in chemistry assessment is a move away from the short, information-*demanding* type of question: 'Give an account of the laboratory preparation and properties of ... or the industrial manufacture of', etc. There are now many more examples of questions which *give* students information within the question itself on which they are to work. This has been mentioned in structured questions, but it can apply equally well to objective items and to free-response questions. This trend reflects the general move in chemical education towards less reliance on remembered facts and an increased demand for the ability to *apply* knowledge to unfamiliar situations. The examples which follow show the two contrasting styles of question on the same topic. It should be noted, however, that the 'information-giving' style does not necessarily make the question easier.

13 The figure shows a simplified flow sheet of the manufacture of nitric acid from ammonia.

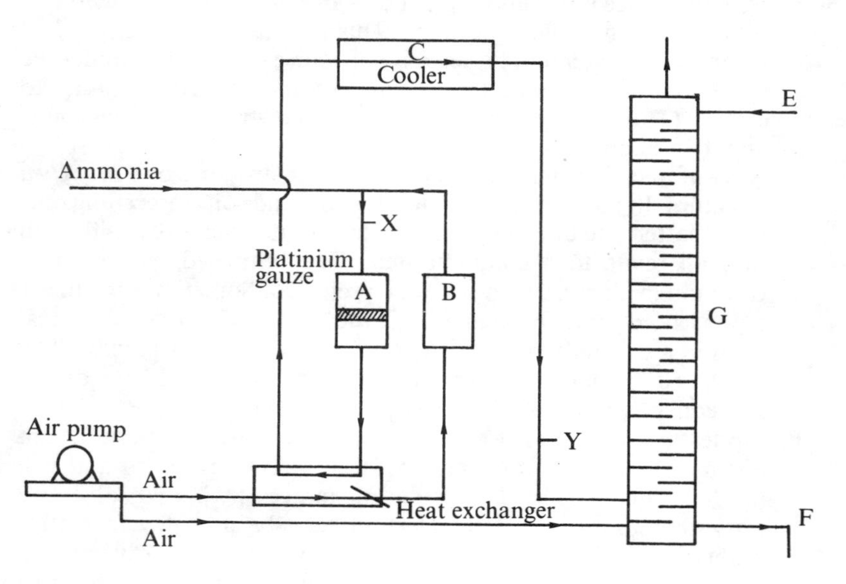

The following are the basic reactions.

1 $4NH_3(g) + 5O_2(g) \rightarrow 4NO(g) + 6H_2O(g)$ $\Delta H = -907$ kJ (217 kcal)

2 $2NO(g) + O_2(g) \rightarrow 2NO_2(g)$ $\Delta H = -113$ kJ (27 kcal)

3 $3NO_2(g) + H_2O(l) \rightarrow 2HNO_3(aq) + NO(g)$ $\Delta H = -134$ kJ (32 kcal)

The quickest reaction is 1 and the slowest 2

a In which part of the plant does reaction 1 take place (state the letter)?
b In which part of the plant does reaction 2 take place?
c In which part of the plant does reaction 3 take place?
d At which part of the plant is the air filtered before it reacts with the ammonia?
e The platinum gauze is kept at a temperature of about 900 °C. How is this temperature maintained?
f What is the purpose of the heat exchanger?
g Reaction 2 is reversible and high temperatures favour the decomposition of nitrogen dioxide. Therefore the reaction is allowed to take place at a low temperature to increase the yield of nitrogen dioxide. State one *disadvantage* of using a lower temperature.
h What enters the tower at E?
i What comes out of the tower at F?
j What is the main component of the waste gas which leaves from the top of the tower?
k What is the purpose of the plates in the tower?
l Suppose a leak occurred at the point X. What substance would cause the main toxic danger to the surroundings?
m Suppose a leak occurred at the point Y. What substance would cause the main toxic danger to the surroundings?

(Reproduced from J. C. Mathews, *Problems in Modern Chemistry*, London, Hutchinson.)

This contrasts with a question on the same topic:

Describe the industrial manufacture of nitric acid from ammonia.

The 'information-giving' question raises problems for chemistry examiners:

The search for appropriate information to serve as the stem of a question is difficult enough, particularly if a teacher does not have access to literature and other resources. Furthermore, the question writer has to go to considerable trouble to check the accuracy and appropriate presentation of the material—a problem which does not arise in the 'Give an account of ...' type of question. Equally important is the amount of space which long stems require, coupled with the ever increasing costs of paper and printing.

One possible alternative is the use of an open-book examination. An example of such a question is:

6. Using your Book of Chemical Data where possible, but otherwise making your own estimate, give a value for the energy required in ($kJ\ mol^{-1}$) to remove one electron from each of the species shown below, and then arrange each group of three in order of increasing energy. (Estimated values should be shown with an asterisk.)

Attempt to explain the reason for the increasing energy in each case.

(a) Na, Mg, Al;
(b) K, Ca, Sc;
(c) K^+, Ca^+, Sc^+;
(d) K, Rb, Cs;
(e) Cu, Ag, Au;
(f) C, N, O;
(g) Na^+, Ne, F^-;
(h) Fe, Fe^{2+}, Fe^{3+}.

(Nuffield A level Chemistry examination)

But it must be admitted that the use of a data book of mainly quantitative information has limited application to examination questions. All candidates must have access to the same data, and data books are expensive to produce and distribute. Nevertheless, the growth of microprocessors as data storage and retrieval systems may add another dimension to this problem. These devices in miniaturized, hand-held form are likely to make qualitative and quantitative data readily available to students, and now present both problems and opportunities to chemistry examiners.

5.6 *The assessment of practical work*

It is disappointing that one of the main features of recent curriculum development in chemistry, the emphasis on practical investigations, should prove to be one of the most difficult operational problems in assessment. Perhaps one reason is that curriculum designers and teachers do not have clear objectives in mind [32]. In the past it was clear that the principal aim was mastery of manipulative skill; now, there is a wide range of other aims. In one discussion [33], twenty aims were suggested, ranging from being 'a way to promote a logical, reasoning method of thought', to the more mundane 'to prepare the student for practical examinations'.

There are three reponses to the problem: to continue with externally devised practical examinations, or to introduce a form of internal assessment by teachers, or to have no assessment of practical work.

The externally devised practical examination is still widely used, but not always in the traditional form of qualitative and volumetric analysis as the following example illustrates:

Exercise 2 Chromatographic identification of a transition metal ion.

Identify the transition metal in the solution PB using the following chromatographic technique. Solutions containing approximately equal concentrations of cobalt(II), nickel(II) and copper(II) ions are provided.

(a) Pour sufficient of the solvent supplied (which contains 90 parts propanone (*acetone*): 5 parts water: 5 parts concentrated hydrochloric acid) to form a pool approximately 0.5 cm deep in the beaker provided. Place the glass cover on the beaker.

(b) Mark the slotted chromatography paper supplied, in pencil, Co, Ni, Cu and PB at the top of convenient strips, and also put your name at the top of the paper. Spot the appropriate solutions approximately 1 cm above the bottom of the marked strip, and allow the paper to dry for a few minutes.

(c) Roll the paper into a cylinder, secure it top and bottom with the stapler provided, and place the paper in the beaker containing solvent. Replace the cover on the beaker.

(d) Allow the chromatogram to develop until the solvent front is near the top of the strips: this should take 10–15 minutes.

While the chromatogram is developing, prepare the apparatus and solutions for Exercise 3.

(e) When development is complete, CAREFULLY remove the paper from the beaker (the wet paper is easily torn). Mark the position of the solvent front and hang the paper up to dry.

(f) Place the dried paper in the ammonia tank provided for about two minutes (do not allow the paper to dip into the small vessel of concentrated ammonia solution). Remove the paper, hold it vertically in the fume cupboard and immediately spray both sides lightly with the dithiooxamide solution provided. Allow the paper to dry once more.

(g) Identify the unknown ion in solution PB.

Submit the dried chromatogram for inspection in the polythene bag provided, which must be stapled to page 8 of the answer book.
(JMB A level Chemistry examination)

Despite efforts to introduce more variety into practical examinations, most of the disadvantages of this form of assessment remain.

The form and substance of the assessment must allow uniform administration by an external agency to a large number of candidates on a single occasion. It follows that apparatus, materials and laboratories must be commonly available to many candidates during a single examining session. Inevitably the scope of the examination must be restricted, particularly in countries where resources do not allow diversity. Even so, the large-scale external organization and administration of practical examinations give rise to many difficulties concerned with resources and security, not to mention the many technical problems which frequently arise during the examination itself. One obvious example is security, particularly if only a few different questions are available.

It seems inevitable, where examination pressure is strong, that limited choice of experiments in the practical examination will be reflected in similarly limited experiments during the course of study.

The reliability of the single occasion practical examination must be suspect [34]. This, and the limited range of experiments, cast doubts on the validity of practical examinations and may explain why their contribution to the total assessment, in terms of nominated weighting, is never high, despite exhortations in curriculum design about the need for practical investigation to be central to the teaching of chemistry.

External assessment of practical work necessarily requires that the marking is applied to the *outcome*—often in written form—rather than to the *process*. While acknowledging that a correct result is important, the means by which it is obtained are, in an educational context, equally if not more important.

These disadvantages of external practical examination are well known. Why, then, does the practice persist? Mainly because it is thought that any practical assessment is better than none; that it will prevent the absence of practical assessment being used as an excuse for depriving schools of laboratory resources for practical work; and at least it is the same for all candidates, although this last assumption is not always justified.

Internal assessment of practical work has been practised on a much smaller scale, but for a sufficient time for some tentative evaluation to be made [35]. It is based on the assumption that assessment of practical work by teachers of their own students on several occasions during the course of study can be more reliable and valid than an external practical examination, an assumption that has the backing of some research [36]. The following advantages are perceived:

The frequency of assessment increases reliability because the incidence of chance failure or success is decreased. In view of this, most systems of internal assessment accept without question the order of merit of the students produced by the teacher.

The medium of assessment is not restricted to written evidence; direct observation and oral assessment are both possible and are encouraged. It follows that the range of attributes of students is extended to include those which are displayed *during* the work as well as at the end of it.

Another consequence is that the range of experiments and the types of work can be extended. In fact, most of the practical work in a modern chemistry course could be used for assessment purposes.

There is a minority of teachers who are enthusiastic about internal assessment. They remain a minority partly because there is little opportunity to participate in such a system. But even where the alternative exists there are sufficient problems to make some teachers cautious:

There is some concern that the dual role of teacher and assessor could adversely affect the relationship between teachers and students. There is some evidence that this is not necessarily so [37]; but it would be unwise to generalize. It may be that this form of assessment is acceptable only if the weighting is low. (This is probably also true for practical examinations.) Furthermore, internal assessment may not be equally acceptable in countries with different educational traditions.

A fundamental issue is that of the moderation of teachers' scores. No system has come to our notice in which the new scores given by teachers are accepted without question. In other words, it is recognized that teachers have differing degrees of toughness or tenderness in marking. No system of moderation can be completely valid; nevertheless, systems have been devised which have proved to be acceptable to all parties, including those who use the results.

There are two main methods of moderating scores obtained by internal assessment. One is by *inspection*. In this, a sample of the work of a group of students is remarked by an external moderator who then decides whether all the marks awarded by a teacher should be allowed to stand, or should be adjusted. Where the number of candidates is small and where the external moderator can see a representative sample of the work, this can be effective. It is, however, time-consuming and expensive and raises problems of standardization when more than one moderator is involved. The second method involves a *statistical* moderation using an external written paper as a moderating instrument. The assumption is that the performance of a group of students on the internal assessment should be similar to their performance in the external assessment. These statistical processes can be applied at various levels of exactness; however, the relationship between the internal and external tests is rarely close enough to allow a rigid statistical moderation. This method is useful in that it can be applied quickly to a large number of candidates, but it is probably at its best when using a system of tolerances rather than an exact scaling procedure. The detailed procedures for both kinds of moderation are complex and to explain them fully would require a much longer treatment than is possible here.

There may be a problem in ensuring comparability between schools, not only in marks but in the objectives, content and method of assessment. The problem lies in the need to have sufficient similarity between schools in these three dimensions while still allowing the flexibility, variety and an element of choice which are dominant reasons for introducing internal assessment. There is considerable similarity in the regulations and guidelines of those schemes of internal assessment which have come to our notice [35]; an abbreviated version of the notes for guidance published for the Hong Kong Advanced Level Examination (for 18-year-olds) will serve as an example:

Means of assessment

A. Laboratory Performance (40 per cent)—assess the candidate's manipulative skills, observation and general bench performance.

B. Written Reports (40 per cent)—assess the candidate's presentation of data, interpretation of results and planning of experiment.
C. Overall impression (20 per cent)—to be assessed at the end of each year of the Advanced Level course, and is an assessment of the candidate's attitude towards practical chemistry.

It is intended that internal assessment of practical chemistry should be part of the teaching programme, and made as unobtrusively as possible. (Author's Note: Detailed criteria for assessment under each of the three means then follows. Of particular interest may be the criteria for the highest score under C: 'Self-reliant and resourceful, able to work with little supervision, interested, eager and curious. Willing to tackle problems and persistent in approach. Works well independently or in a small group. Conscious of safety in the laboratory.')

Making assessments

Assessment scale
10–9 very good
8–7 good
6–5 average (refers to the average 'A' level candidates)
4–3 weak
2–1 very weak

Number of assessment occasions over two years
For each year a minimum of two assessments should be made for each of the abilities A and B. For each year, at least three assessment occasions should be made.

Over a period of two years, there should be a total minimum of eight assessment occasions. Assessment C should be made once at the end of each year.

Recording assessments
A sample record card is given on page 253.

Areas and types of experiments
Teachers are encouraged to assess their students on a variety of laboratory experience covering:
Areas: Changes in substances and patterns in changes in substances; Equilibria; Kinetics; Energetics.
and Types: Quantitative exercises; Qualitative exercises; Preparative exercises. (Details are given for all three types.)

Allocation of marks
Three methods are suggested:
1. A mark scheme—particularly for written reports.
2. By impression on a single occasion—particularly during direct observation during an experiment.
3. By overall impression over a period of time.

HONG KONG EXAMINATIONS AUTHORITY
HONG KONG ADVANCED LEVEL EXAMINATION
TEACHER ASSESSMENT OF PRACTICAL WORK IN A-LEVEL CHEMISTRY
STUDENT'S INDIVIDUAL RECORD CARD

Year of Exam.: 1 9 8 1

Candidate No.: 0 0 0 2 3 5

Name of School: Alpha College

Name of Student: CHAN Siu-ming, Peter

1st year 1 9 7 9 – 8 0

Expt. No.	Date of Exercise	Marks awarded to abilities tested				
		A	B	C	D	E
3	2/10/79	6	4			
5	16/10/79		5			
8	14/11/79	8				
12	18/12/79	7				
15	5/2/80		5			
19	6/3/80	7				
23	6/5/80	9	6			
				6		
1st year Average mark (maximum 10)		14.8/20	10.0/20	6/10		

2nd year 1 9 8 0 – 8 1

Expt. No.	Date of Exercise	Marks awarded to abilities tested				
		A	B	C	D	E
2	16/ 9/80		6			
6	4/11/80	8				
10	9/12/80		7			
15	6/1/81	9	7			
19	3/3/81	7				
21	20/3/81	8	8			
				8		
2nd year Average mark (maximum 10)		16.0/20	14.0/20	8/10		
Average of 2 Years		31/40	24/40	14/20		

= Total 69

Comment: Give relevant comment where it is considered that this student's performance has been affected by illness or disability.

(1st year) — NIL —

(2nd year) — NIL —

Signature of Teacher: [signature]

Date: 25/6/1980

Signature of Teacher: [signature]

Date: 15/4/1981

FIG. 2. Sample of a completed record card.

Moderation

'Moderation of internally assessed marks for practical work will be by a statistical procedure based on the performance of each school on the theory papers in the written examination. The rank order of candidates within a school will be preserved but the moderation procedure will take account of the differing standards of assessment and mark distributions between schools. Where more than one teacher in a school is involved in internal assessment, it will be necessary for the teachers concerned to discuss and agree upon the standards of assessment they will be adopting. The moderation procedure will not take into account variation is standards of assessment within a school.'

The Hong Kong scheme for internal assessment is not open to all schools; a satisfactory period of one year is required before the Examination Authority will grant permission for a school to join the scheme. Teachers in these schools must attend the meetings or conferences organized by the supervisors of their group of schools.

It will be clear that internal assessment of practical work involves both teachers and administrators in a good deal of work and it demands a high level of competence and professional integrity. It requires co-operation at national and local level and it may be costly. In the author's experience, it is something to be aimed at rather than to be adopted precipitately. But the benefits are considerable, not only in an improved assessment of the students, but also—and equally important—in the form of valuable in-service teacher training at the operational level.

5.7 The assessment of project work

Although many curriculum schemes emphasize the role of independent thought in experimental chemistry, few allow for any extensive use of project work. For example, only one major chemistry curriculum development in the United Kingdom has a project as part of the assessed work. The development has been left to the curricula developed for the less-academic student (the Certificate of Secondary Education—CSE).

'In project work, pupils are expected to assume some level of personal responsibility for their work and to organize their time for constructive study' [38].

Three types of project may be identified [39].

Report type in which pupils collected information from books, journals and other secondary sources.

Discovery type in which pupils used 'the results of their own experimentation, observation and practical exercises to answer a specific question or test a specific hypothesis, the findings being written up in the form of a simple research paper'.

Combination type in which 'an attempt was made to marry the theoretical and experimental aspects of a topic by combining the reporting and discovery aspects of it'.

It would seem on the face of it that projects should allow the exercise and development of those skills of planning, design, investigation and interpretation so frequently claimed as central to chemical education. Whether they do and whether they can be assessed is less certain. Sadler [39] reports

> the evidence presented in this paper indicates that much of the project work undertaken in the CSE science subjects falls short of these ideals, involving highly structured, teacher directed reporting exercises and demanding little of the pupils in terms of the skills outlined above. When viewed in this light, the value of project work as a teaching strategy is without doubt open to suspicion *and its validity as an assessment tool is immediately questionable.* (Author's italics.)

However, Sadler goes on to suggest that many of the problems associated with project work and its assessment are organizational and that 'rejection of project work at this stage would be rash'.

The very fact that so much is sometimes claimed for projects as an educational activity makes assessment of them difficult. In particular, affective objectives, such as confidence, initiative, self-reliance, are often associated with them. This is understandable if part of chemical education is to 'work as chemists work', but it does raise all the problems associated with the formal and public assessment of affective attributes, particularly that of social acceptability. In addition, there are the problems of assessing work which is done by groups of students rather than by individuals. It can be argued that projects in the real world of chemistry are nowadays rarely, if ever, the work of individuals and, therefore, educational projects in groups should develop 'communal and cooperative attitudes' [40]. In such projects, the difficulty of differentiating between the performances and attributes of members of a group is self-evident.

Despite these problems, some experience is now available of the formal and public assessment of projects at school level. (There is also experience in Higher Education, but this is outside the scope of this publication.) One example is the project work in the Scottish Certificate of Sixth Form Studies [41]; the aims are the development of:

> (i) skill in devising an appropriate scheme for studying a problem in chemistry; (ii) skill in handling, classifying, interpreting and presenting information (including graphs and quantitative results); (iii) ability to apply previously acquired understanding to new situations and to show creative thought; (iv) resourcefulness on the part of the pupil with a corresponding lessening of dependence on the class teacher.

Handy and Johnstone [42] summarize the form of assessment of these projects as follows:

> A dissertation marked mainly on the basis of five point scales. An oral discussion of the dissertation (with an external examiner) carried out in the presence of the class teacher, partly to reassure the candidate and partly to supply information not readily available to the examiner.
>
> A discussion between the examiner and the class teacher to arrive at an agreed mark.

Handy and Johnstone demonstrate an acceptable validity and reliability for this assessment scheme. It is noteworthy that the affective aspects get special treatment, particularly in the teacher's function and in the oral examination. There remains, of course, the problem of economy; in the Scottish scheme the assessment was feasible, allowing no more than ten candidates each day for each examiner. But under different circumstances—numbers of candidates, distances of travel and available finance—the feasibility of such a scheme could be less, and for large numbers over longer distances, probably impossible.

This typifies the dilemma in which educators and examiners in chemistry find themselves: the more the educators move towards the study of chemistry as a process rather than as a body of knowledge, the more difficult it is for the examiners to design a valid and acceptable form of assessment. The question must therefore be asked: should the whole area of affective and creative objectives be transferred to a more informal mode of education and a more informal mode of assessment? This is perhaps best illustrated by the exhibited work of pupils in meetings of associations for science education and similar organizations. An example is afforded by the Zambian organization, JETS [43]. Much of the project work exhibited at science fairs is remarkable in its originality and resourcefulness. One is tempted to think that the quality of the work arises, in part at least, from the fact that, although it is assessed and is competitive, it is not part of a norm-referenced system of assessment and selection. Of course, the objection may be raised that not all students participate as they would if project work was an obligatory part of formal assessment. This is so, but at least it may be preferable to the rejection of all project work in chemical education on the grounds that it cannot be formally assessed.

5.8 Oral assessment

Oral examinations have long been accepted as a normal element in student assessment, particularly at university level in many European countries, as well as in schools. Ashford has pointed out that oral examinations vary widely with respect to such characteristics as length, form, procedure, purpose, level, effectiveness and reliability and that this form has not yet yielded to critical analysis [6]. This is still true—years after Ashford's comment. Nevertheless, there are strong advocates for this procedure, often because it encourages verbal communication skills. Oral examination for university entrance is practised, for example, in Norway [44]. In Yugoslavia, Kornhauser points out [45] that

> written tests have never been accepted as the only form of assessment. The student has the right and the teacher the duty to assess orally at least two to three times each term and to consider here as many parameters mentioned earlier as possible. There were some attempts to introduce only written tests, but mainly such suggestions were not accepted by school councils.

Those interested in oral examinations may find the booklet *CSE: An Experiment in the Oral Examining of Chemistry* particularly helpful since the researchers based their tests firmly on the content of an existing syllabus [46]. The findings of the research were:

> The oral examinations used in the investigation had a reasonably high reliability and a reasonably high validity.
>
> The provision of an oral examination for each pupil at CSE level presents considerable problems of administration. Nevertheless, the type of examination described could serve two very important functions: (a) as an instrument of external moderation; (b) as a satisfactory method of assessment for those schools choosing the Mode II or Mode III forms of examination (i.e. where the school has chosen to construct its own curricula).
>
> The 'backwash' effect of an oral examination could be healthy for two reasons: (a) the discussion that would become a necessary part of preparation could promote the effective use of spoken English; (b) the oral test is sufficiently flexible to make excessive restriction of the practical work done in a course unlikely.
>
> CSE candidates throughout the full ability range expressed a liking for and an interest in practical and oral examinations. Expressed interest in and liking for written examinations declined sharply as one moved down the ability range.
>
> In this investigation it did not appear that the nervousness of the candidate was a limiting factor.
>
> The examiners found the oral examination interesting and potentially valuable.
>
> The examiner/candidate relationship was demonstrably an important factor in the oral examination (for example, 62 per cent of one set of pupils examined by one pair of examiners chose the oral method as the method they liked best; another pair of examiners were very unpopular—only 7 per cent of this set of pupils chose oral examinations as the best method).

5.9 Reporting and using the results of assessment

From the evidence we have received, it appears that public examination results in chemistry are usually published in the form of a single grade.

This grade may be numerical or literal, often on a seven-point scale and with something close to a normal distribution. Although the design of a scheme of assessment, as a whole and in parts, may be based on a quantitative specification showing the weightings for various skills and subject areas, the scores for the various parts are usually added together to give a single total score. After various standardizing procedures, the award of grades is made by establishing grade boundaries in terms of total scores.

The information we received contained no examples of differentiated or profile reporting in which performance on different skills or different areas of work was reported separately. No doubt, profile reporting does occur but, if so, it cannot be common. Analysis of profiles has been undertaken for purposes of research [10], but in public reporting the information received by those who will use the results is largely in the form of a single grade.

The single grade reporting has advantages of course. It provides a simple, easy-to-use measure for those who rely on examination results to select students, particularly for further and higher education. It is usually made up of several sub-tests or modes of assessment and this tends to increase the reliability of the grade at least in the sense of internal consistency; to put it another way, the reliability of the whole examination is greater than any of its constituent parts.

Both issues, ease of use and reliability, are important; but it can be argued that neither should be elevated to the extent that other equally important aspects of chemistry assessment are diminished or excluded. If it is accepted that the assessment of chemistry is essentially a form of competition with the highest possible reliability, that it is strongly biased towards norm-reference, then a single grade as the only form of reporting might be equally acceptable. But, if it is thought that the reporting of assessment results should say something about the *kind* of performance of the students as well as their relative performance, then some move towards criterion referencing and profile reporting will be necessary. And it should not be overlooked that the reliability of chemistry examinations in recent years is likely to have improved. Although there is little evidence about the reliability of the whole examination there is a good deal to suggest that the parts now commonly used—objective tests, structured questions, double impression marking and concurrent assessment of practical work—all give increased reliability to the whole. (We should distinguish between the profile reporting of individual students and profile reporting of groups of students. This will be discussed in more detail in the next section.)

At present profiling is more a matter for debate than for prescription. The techniques certainly exist for its implementation, and their reliability is acceptable at least for the profile reporting of the attainment of *groups* of students; whether they are sufficiently reliable for the profile reporting of *individual* attainment is less certain. The question depends to a large extent on the structure of profiling and this in turn depends on the use to which it is to be put. The latter is all important; it may be interesting to know the different levels of attainment of a candidate in organic, physical and inorganic chemistry, for example; but if the information is not to be used, profiling is a waste of time.

Perhaps one of the most significant moves on this issue is that towards the identification of a common core in chemical education. That there is a good deal of common content and common skills in various chemistry curricula around the world cannot be denied. But, under our present system of reporting it does not follow that two students, both with, say, Grade C, have the same knowledge and the same skills; in fact, they may have substantially different knowledge and different skills even if they have taken the same examination. In a purely norm-referenced system this may not matter so much; but, even there, it is surely useful to those who are concerned with the next stage in the chemical education of a group of students to have some idea of *what common knowledge and skills he can assume them to have*. To take the argument a little further: the first step towards profile reporting should be to distinguish between that chemistry which should be learned in common by all students and that which is open to variation and choice; the second step would be to report attainment in each part separately as well as in a combined score. And, since co-operation and discussion between chemical educators has been so widespread in recent years, there seems no reason why this should not be investigated internationally, as well as nationally.

5.9.1 *The use of results for formative evaluation*

Understandably the emphasis in the reporting and use of attainment tests is to evaluate student performance. However, it should be recognized that the data available in the outcomes of assessment in its various forms can be used to evaluate the educational process and the assessment process which gave rise to these outcomes. In other words, public assessments can be used to evaluate not only students but chemical education itself.

The problems of unreliability which arise in reporting a profile of attainment of an individual student give less cause for concern when considering the profile of attainment of a group of students, provided that the group is large enough to give adequate significance to the results. Certainly, if one analyses the attainment of students in various areas of content, skills, types of work, and types of examination question, it is possible to identify those areas which gave rise to performance which is higher or lower than the norm. For example, a profile analysis of attainment in Nuffield 'A' level chemistry revealed a relative weakness in performance in carbon chemistry, while performance in the study of oxidation states was relatively high [31]. A similar analysis can reveal differences in attainment between different categories of students, male and female for example.

A profile analysis of attainment in chemistry at a national level requires certain conditions to be effective:

There must be high curriculum validity in the assessment system with an explicit specification of each part of it in terms of content, skills, types of work and types of question. Second, a substantial part (at least 50 per cent) of the assessment must be common to all students, usually related to the common element in the curriculum. In addition, analysis is easier if the common part of the assessment is in the form of a large number of questions, such as fixed response and short answer structured questions, which require brief answers and can be marked objectively.

One of the criticisms of recent curriculum development projects in chemistry is that they have tended to be *batch* processes: usually a period of trials followed by publication and implementation with no mechanisms for concurrent evaluation and adjustment once the project has been put generally into effect. While continuous tinkering with the chemistry curriculum and its assessment system would not be tolerated, an evolutionary process of evaluation and change arising from an objective analysis of student attainment is surely both practicable and desirable; the alternative is to proceed on a basis of subjective hunches or half-baked learning theories.

There is another aspect of formative evaluation which should be considered. The many development projects in chemical education around the world in recent years have necessarily involved the active participation of teachers and other educators in the development and trials of materials. Prestt [47] makes the point:

> Rather than continuing to attempt to hand curriculum projects to teachers as products we must seek a strategy for curriculum development and renewal which will clearly involve teachers in the process of development.

Many of those teachers continue to be involved in the design and construction and marking of tests. Although this will be considered in more detail in the next section, it is relevant here to point to many groups of teachers with substantial skills in course design and assessment. In addition there are all those teachers who have a direct influence on the assessment system through their work on internal assessment. It seems that there is a growing feeling, at least in some countries, that chemistry teachers as a whole now have a more direct influence on the design and operation of assessment, and the traditional perception of remotely and externally administered examinations is declining. As discussed earlier, teachers in some countries now have an opportunity to influence the system and are more willing to provide evidence, when asked, about the effectiveness of the assessment which is applied to their students. But to do so, in an informed and objective way, requires that teachers be drawn into parts of the assessment system other than simply marking scripts, bringing them to an understanding of how attainment tests are designed and operated. This in turn requires training, which brings us to the next section.

5.10 The training of teachers in assessment techniques

It is probably true to say that in recent years training in assessment techniques has formed an increasing part of the pre-service education of teachers, but it is doubtful whether it is adequate to meet the increasing involvement of teachers in assessment and the complexity of modern assessment in chemistry. Whether it is so or not, it cannot be doubted that older generations of teachers have received no training in the techniques of assessing attainment in chemistry. Many of the teachers involved in assessment have picked up what expertise they have by experience rather than by formal training.

There are, of course, teachers who claim that assessment is no concern of theirs. However, even if teachers do not wish to become involved in the public assessment of their students, it is surely taking a narrow view of the profession to say that assessment of student progress has no part in the professional skills of a teacher. And it is not sufficient simply to have access to tests written by someone else: to understand some of the theory and actually to undertake some of the practice of test construction provides insights into chemical education and into the subject itself which cannot be attained entirely by other means. Assessment is part of chemical education; and, if we accept that the training of teachers is necessary, we must also accept that training in assessment techniques should be part of it.

Although the intention of this section is to concentrate on the design and operation of *in-service* courses for practising teachers, the essential content of the course will be the same whether it is pre-service or in-service, although the way it is organized will be different.

From those courses in which the author has been involved some common principles and practices have emerged [48].

The term 'course' does not denote the kind of training required; it implies that teachers can be told what to do, but this is far removed from what is required. A workshop in which tutors and participants work together to produce actual tests is much more effective and much more enjoyable.

At this point a distinction must be made between simply training teachers to act as assistant examiners in the marking of scripts and training teachers in the principles, construction and use of all the important aspects of testing as part of their professional competence. The first function, of course, is important, but it is narrow and serves mainly to recruit examiners who can mark reliably. It is with the second function that this section is concerned and it will take the form of setting out guidelines for those who wish to run workshops in assessment for practising teachers.

5.10.1 *The purpose of the workshop*

In part the workshop is to introduce teachers to the principles of question writing and test construction. But the main aim of the workshop must be to make the teachers effective *operationally*. That is, they must actually participate in the construction of tests. Of course, it is unlikely that a teacher working alone would be able to produce a complete assessment scheme; nor would it be desirable, because it is the co-operative nature of the work which should be a central feature of the workshop. The joint production of a complete range of tests, which each member of the workshop can take back and administer to his own institution, proves to be a tangible and strongly motivating goal.

5.10.2 *The workshop participants and tutors*

The optimum number of participants probably lies between fifteen and twenty. This is small enough to give the group coherence and a sense of common purpose, while being large enough to produce the necessary amount of test material (inevitably much material is discarded). It also allows the formation of three sub-groups of five or six members each, which is an effective number for the review, rewriting and editing of questions.

A workshop of this size can be run by one tutor provided that he is experienced in all forms of testing in chemistry and provided that he has three assistants who can head each of the three sub-groups. These assistants are an important feature of the workshops, because, clearly, a single tutor cannot alone head each sub-group, and their presence allows the tutor to move from one group to another. Ideally these three assistants should be people who have attended a previous workshop and are attending a second one with a view to becoming tutors and running their own workshops elsewhere.

5.10.3 *Workshop activities*

Lectures

Assuming that most of the participants have had no previous training, they will require some basic information. Informal lectures or seminars are convenient ways of doing so provided that they are accompanied by written materials. The following may serve as an indication of what may be required, although it should not be taken as definitive:

	Lecture title	*Materials*
1.	Specification of cognitive performance: curriculum validity linked to recent trends in chemical education	Examples of various categories of cognitive objectives. Examples of test specifications in the form of two-way grids.
2	Types of tests: their advantages and disadvantages	Examples of test items. Glossary of terms.
3.	Writing objective test items: item reviewing	Outlines of guidance for writing and reviewing common faults.
4.	Writing and marking structured short answer questions and free response questions	Outlines of guidance for writing and marking—examples of mark schemes.
5.	Assessment of practical work	Examples of a scheme for teacher assessment of practical work.
6.	The construction and administration of an objective test. *Simple* statistical treatment of results (item and test analysis). Pre- and post-testing.	Guidelines and examples. Simple concepts and devices: distribution, reliability, validity, item analysis, regression, prediction, correlation.
7.	Construction and administration of tests of structured and free response questions	Guidelines and examples.

In addition to the lecture sessions, there should be allowance for other plenary sessions at which the workshop tutor can discuss matters of general concern as they arise. It is not possible to predetermine what these may be since they usually arise from problems which have been perceived in the group or individual work.

Individual work

Since the main aim of a workshop is to give the participants experience in producing a scheme of assessment, it is necessary that each one of them gets experience in every aspect of test production. The starting-point is the writing of questions and, in the first instance, that is usually a solitary occupation. A substantial part of the time of the workshop is spent in participants working alone, writing or refining questions. It is only by extensive practice, comment, discussion and refinement that anyone can obtain facility in writing these questions; it is an art that is difficult to master.

Group work

The co-operative nature of the work has already been stressed. In particular, the process of reviewing (shredding) questions is necessarily a group activity. For those who have not previously experienced it, this can be a somewhat disturbing experience, but it is probably the most rewarding way of teaching people to distinguish between good and bad items (and, sometimes, between good and bad chemistry!). It is here that the assistant tutors play their most important role. They should ensure that criticism and response is constructive and good-humoured. They must be sufficiently expert themselves to be able to guide the group in the refining of questions and to know when to advise and when to stay silent. The tutor himself can be a silent observer on those occasions except when asked to arbitrate.

Later in the workshop each group can turn itself into a test construction unit. In this they will undertake the final editing, test assembly, agree on mark schedules and supervision of reproduction.

Practical work

Too often, assessment workshops are entirely concerned with written work. While it must be recognized that practical work is time-consuming and that it may be difficult to fit formal sessions into the workshop schedule, it is useful if the participants can have access to a laboratory for two reasons. First, they may wish to try out experiments which are to serve as the basis for the stem of a written question. (In such instances it is very important that the question writer should have actually done the experiment so that the information given to the examination candidate is authentic.) Second, they may wish to carry out experiments in order to decide how they may be used as part of a scheme for the assessment of practical work. In the first case, the practical work could be done individually; but in the second it is more productive if the participants could work in groups of two or three.

5.10.4 *Workshop schedule*

A predominant consideration in the workshop as a whole, and in the group sessions in a particular, is to ensure that every participant is *actively* involved. These should be very little time for passive reception of information; even in the lectures the general tone should be informal and co-operative so that questions and discussion arise frequently and spontaneously. (One could say that there are similarities here with some of the more recent styles of chemistry teaching itself.)

Starting with largely untrained participants, a full-time workshop of fourteen working days is desirable if complete, edited tests in all the main forms are to be produced. In the early days of the workshop there will be two conflicting requirements: on the one hand, there is the need to start the actual task of writing questions as soon as possible: and, on the other, there is the need to provide sufficient information and guidance without which the writing by participants would be unproductive. This can be partly overcome by providing some pre-course reading, especially in the form of sample questions and notes on writing questions.

Thereafter, although the general sequence of events can be largely predetermined, the amount of time required for each cannot, and a suitable flexibility has to be built into the timetable. There are several related problems: the groups may get out of phase with each other, some working quicker than others; the reprographic service may cause delays; there may be a deficiency in the output of a certain type of question in sufficient quantity. It is useful, therefore, if each day starts with a short plenary session (if a formal lecture has not been arranged). This gives the tutor an opportunity to give supplementary guidance on the basis of his observations of the group work.

Another principle of scheduling is that of diversity of activity. Individual item writing day after day becomes tedious; so does listening to lectures; so do group shredding sessions. No one day should be entirely devoted to a single kind of activity. One of the main concerns of the tutor and the organizer is to keep up the momentum of the work and they should be finely tuned to the explicit and implicit messages which come from the participants. Singleness of purpose is not to be confused with rigidity.

It is important, if the workshop is to have a suitably high degree of common purpose and coherence, that there is some uniformity in the chemistry topics and objectives which are to form the substance of the tests. Particular care must be taken if the workshop is to be regional or international in character.

5.10.5 *Workshop administration and equipment*

While some degree of charisma on the part of the tutor and his assistants is helpful, it will not alleviate the problem (indeed chaos!) which can arise from poor administration and inadequate equipment. An organizer who is a member of the institution which houses the workshop is an essential member of the team, if not as tutor, then certainly as an assistant tutor.

Next, if not equally important, is quick and accurate reprographic work. Once the workshop is fully into its stride there should be more than enough work to keep a typist fully occupied. A good deal of revision and editing and subsequent retyping has to be done. (Of course, having access to a word processor has enormous advantages, for small, but important, changes can be done immediately and with very little extra work.)

Photocopying facilities, although they may be expensive in view of the great amount of material produced at these workshops, are almost essential. This is particularly so in the early days of the workshop when groups are shredding questions and mark schemes in their draft, manuscript form.

5.10.6 *Workshop output*

The output of a workshop depends, of course, on many factors, but a target for a three-week workshop of twenty participants could be: eighty fixed response items forming two objective tests; twenty structured questions, complete with mark schemes, forming two or three tests; forty free response questions; and a scheme for the assessment of practical work. (The latter will usually take the form of an existing scheme adapted to meet the circumstances of the participants.)

It may seem a trivial point, but there is much satisfaction to participants to return home with tests in their luggage, ready for immediate application in their own institutions.

5.11 **Conclusion**

Although the students of chemistry may not think so at the time, the proper assessment of attainment should be primarily for their benefit. No one can pretend that students will *like* the process, whether it be fixed response or free response, internal or external, criterion-referenced or norm-referenced, etc. Nevertheless, it seems inevitable in modern society, as it is at present constituted, that they will have to subject themselves to assessment in some form or other and that this assessment will often be used to select students for further study or employment.

Having said that, students of chemistry have legitimate grievance if their assessment is merely competitive without thought to the validity of the performance which they are asked to display. In this sense the purpose, content and techniques of chemistry assessment cannot be divorced from a consideration of the purpose, content and technique of chemical education itself, just as the latter cannot be divorced from the nature and needs of society at large. Nothing can do chemical education and its standing in society more harm than the thoughtless inclusion in chemistry tests of demands for information and skills which have little or no relevance either to the student, to chemistry or to society.

It is only too easy for a teacher working alone, or for an examination board proceeding down the tramlines of past precedents, to fall into an habitual, routine construction of assessment instruments. Thus one must be aware of the dangers that a test bank has; it can lead to rigidity. Such a process must be continually challenged; the inclusion of every test and every question for an examination should be debated until the author(s) can convince his (their) fellow examiners that it is necessary. Item banks must be revised regularly. It is for this reason that the central theme of this contribution has been that of consultation between teachers and external examining authorities and, through the teachers, or even directly, with the students themselves. Hence, the function of examination systems to monitor and interact with the curriculum both in its outcomes and its processes, and the use of workshops for the co-operative production of tests by teachers in the process of their professional training, have been emphasized. Those who have participated in a workshop which has reviewed all aspects of assessment (not just objective testing as so many do) will know how salutary an experience it can be. One is forced to consider the relevance of what one is doing in setting questions, and indirectly of chemical education itself. Given the gregarious nature of most chemical educators, these meetings can be not only fruitful but also enjoyable.

In some respects the workshops in assessment typify a critical issue in chemical education: what constitutes the *fixed* curriculum, that is the curriculum prescribed by higher authority, and what constitutes the *free* curriculum, that is the topics and activities in which institutions, teachers and students have an element of choice? Workshops on assessment force attention on this debate because the answers to these questions determine the structure, the content and the techniques of the assessment system which is devised to serve the curriculum.

When teachers meet to construct tests of attainment, they need to know the degrees of freedom which are available to them. Some readers of this section may have gained the impression that a shift to more teacher participation and freedom in assessment is being advocated. This is only partly true: complete teacher autonomy in assessment would result in a chaotic situation in chemical education. The suggestion is here that the limits of the fixed and free curriculum, and hence of the fixed and free assessment, in chemistry be more clearly defined. There is a trend in some chemistry examinations towards a polarization between *fixed responses* (objective tests and structured questions) on the one hand, and *free responses* (open-ended questions and internal assessment) on the other. This is a more satisfactory situation than the one which has traditionally left these boundaries blurred. Freedom both in choosing what chemistry is to be taught and what chemistry is to be assessed can be spread too thinly; the polarization to which we have referred could lead, in the long run, to greater autonomy for particular institutions, teachers and students, while not losing the central core of chemical skills and knowledge.

The danger, of course, is that the central core of fixed response assessment will dominate the free response and indirectly have a reflected effect on chemistry teaching itself. This is a realistic, if somewhat pessimistic, view which can only be countered by enhancing the professional status of chemistry teachers. If teachers see themselves simply as executives of a centralized assessment authority, motivation for development at school level will be lost. If, on the other hand, the assessment system allows a large area for diversity, and, equally important, continually monitors the composition and effectiveness of the core component, there is every hope that chemical education will not reach a state of fossilization which an authoritarian assessment system could bring about.

At the same time there is a need for equal clarity in determining those qualities and performances in chemical education which can and cannot be effectively assessed. Far-reaching claims are sometimes made for the goals of chemical education, particularly with regard to affective attributes. Perhaps the time has come for us to recognize not only the difficulty of assessing some of these attributes, but also the undesirable effect on chemical education which attempts at assessment of them may have. There comes a point where the assessment, if any, of some aspects of chemical education must be left in the hands and to the good sense of the participants, the teachers and students. For example, where the formal assessment of practical work in school chemistry in certificates for the 16- (approximately) year-old group has declined, can we say that the place of practical work in chemical education has declined as a result? It may well be that a greater variety, if not a greater quantity, of practical work has arisen directly as a result of the liberation from a formal assessment of it. It could be equally so in other equally desirable areas of chemical education.

It seems that much more thought and resources now go into the assessment of attainment in chemistry than did twenty years ago. At the same time the tests have become more complex and expensive and this gives rise to problems of resources of skill and finance in many countries. There comes a time when additional funds to produce only marginal improvements in the validity of assessment techniques may be thought to be better allocated elsewhere in the education system.

All the forms and modes of assessment mentioned in this section are in operation somewhere in the world and it seems unlikely that any revolutionary change in them will arise in their essential structure in the near future. There remains, however, the application of microprocessors. Some mention has already been made of the use of word processors in the writing of items and their collation into tests and item books. It seems that this aspect of testing, which is essentially one of the storage, retrieval and communication of test information, will grow rapidly with the use of microprocessors and could well lighten our present burden of a cumbersome test processing and administration system. Furthermore, electronic systems of this type could facilitate international test banking and application. Some early planning in this direction must be encouraged.

There is another aspect of the use of microprocessors in this field which should be considered. Mention has already been made of the use of data books in chemistry examinations. The problem there lies in the difficulty in storing sufficient data of the right kind in a printed form from which particular information can be quickly abstracted and at a cost which allows equal access to all students. A large capacity and, ultimately, inexpensive electronic information source is now feasible in a miniaturized form, which could be readily used by a student during a chemistry examination. This opens up the possibility of a revolutionary open book type of assessment within the next decade. Its appropriate function will need careful thought on the part of chemical educators and assessors and it is not too early for them to start to think about its implications.

Appendix. Procedure for the construction of ACS tests ('flow sheet')

	first academic year
1. *Organization of the subcommittee* Selection of subcommittee chairman. Designation of other officers. Selection of members. Indication of collaborators and critics.	Winter or Spring
2. *Preliminary meeting* Specification of test. —Nature and composition of areas to be covered —Kinds of knowledge —Levels of attainment —Types of abilities —Level of difficulty —Parts and length of test (Preferably from survey, examination of textbooks, etc., or may be based on combined judgement of subcommittee members and other collaborators.) Depending on the experience of the members of the subcommittee the following may require extensive discussion at this meeting or at the beginning of the first working session (Step 4): a. Content objectives: Areas to be covered from analysis of texts, courses of study, or by conducting a survey. b. Behaviour objectives: Kinds of knowledge; levels of attainment; types of abilities to be tested—from psychological theory. c. Techniques in writing good test items—from test theory. (Literature and reference materials are distributed at this time.)	Winter or Spring ACS Meeting
3. *First circularization* The skeleton form of the test—containing a clear designation and specification of each part and representative items—is sent to subcommittee members and collaborators. Members and collaborators are asked to write items for the test and submit them to the subcommittee chair-	Spring and Summer

Appendix (*continued*)

man for preliminary editing and collating for the first working session (Step 4).

4. *First working session*
Two-day working session during which the items are criticized and improved upon by the group. Two preliminary tests are assembled, each containing three-fourths of the total number of items expected in the final form. The two tests taken together are balanced for coverage, level of difficulty and kinds of items (using 'two-dimensional analysis'—content versus behaviour objectives in Step 2).
second academic year
Fall ACS Meeting

5. *Second circularization*
Preliminary forms are mailed to subcommittee members and collaborators for final criticisms prior to pre-testing. This step may require considerable time, depending on the need for editing and refining. A second working session may be needed (Step 6). Arrangements for administration of the experimental forms are begun at this time.
Fall

6. *Second working session* (optional)
This second working session usually includes the subcommittee chairman and two or three persons who live within a few hours of driving time. In cases of a 'difficult' test, a full working session may be necessary at the spring meeting although this may cut the time too short to complete the pre-testing during the spring season. The two pretests are put into final form. Each pre-test should not be longer than three-fourths of the final form so that accurate data may be obtained on all items.
Winter or Spring ACS Meeting

7. *Printing of experimental forms*
Mimeograph duplication—preferably in the office of the chairman of the subcommittee—after review by committee editor.
April or earlier

8. *Administration of experimental forms*
The pre-test is to be given to 400–800 students (200–400 students for each form) from eight to ten schools or more. At this time, copies of the tests, answer sheets and scoring stencil are sent to collaborators with instructions for administration. Collaborators should return answer sheets to the subcommittee chairman together with examiner's reports and criticisms and comments. The collaborators should return or *destroy* all pre-test copies and scoring stencils.
May and June

9. *Analysis of experimental forms*
When all pre-testing is completed, the subcommittee chair-

man forwards the answer sheets and examiner's report to the Distribution Centre. Statistical analysis is carried out to determine the difficulty and validity of items and the overall validity, reliability and difficulty of the test.

Summer

---

third academic year

10. *Third working session—final form*
One- or two-day session to consider and weigh the statistical characteristics of the items, the face validity of the items, and the overall coverage and to assemble the final form of the test.

Fall ACS Meeting

---

11. *Third circularization—final form*
The final form is sent to a selected list of subcommittee members and/or collaborators for final criticism.

September

---

12. *Preparation of final copy for printing*
Final manuscript is sent to Editor (of main committee) and then to Chairman (of main committee). Final form is typed for printing at the Distribution Centre. Preparation of key, other accessories and publicity materials is also completed at this time. All mailings of original test materials are to be registered and insured for $100. Xeroxed copies are sent by ordinary mail.

October

---

13. *Proofreading of final form*
Final form and publicity materials are proofread by Chairman, Editor and subcommittee chairman.

December

---

14. *Printing of final form and advertising*
At the Distribution Centre (University of South Florida).

January

---

15. *Distribution of tests*
By Distribution Centre. Copies of new tests are sent to various journals for review.
A personal copy is sent to all committee and subcommittee members and collaborators, with a letter of thanks. A xeroxed copy of the test and a letter is also sent to the superiors of those who worked on the test.
The test is now offered for sale, by advertising in various journals and by direct mailings.

Spring

---

16. *Norms*
After the first year testing programme for the new test, norms are calculated based on reports of scores sent in by users. If possible, several meaningful groups are selected and separate norms are calculated. After calculation, norms are sent to all users who have contributed data. Thereafter norms are sent routinely with all new orders. They are available to all other users upon request. Con-

Summer and Fall

densed norms are sent to the *Journal of Chemical Education* for publication in a spring issue.

Detailed analysis of Steps 12–14

Final manuscript prepared by subcommittee chairman (September and October).
→ Mailed to Editor (of main Committee).
→ Mailed to Chairman (of main Committee).
→ If concurrence of opinion on editorial changes, to Programme Manager for typing.
→ Proofread by Chairman.
→ Xerox copies mailed to subcommittee chairman and Editor.
→ Xerox copies mailed to Chairman for collation of changes.
→ To Programme Manager for incorporation of changes.
→ Final proofreading of manuscript by Chairman.
→ To printer (mid-December).
→ Printer's proofs ('brown copies') received and studied by Chairman.
→ Mailed to subcommittee chairman and Editor for proofreading.
→ Mailed to Chairman for collation of changes.
→ To Programme Manager if additional typing is needed.
→ To printer in final form (mid-January).

Note: Thus there are fourteen sub-steps in Steps 12–14. In many of these sub-steps there is a mailing, and time is required for study and for typing. Any of the sub-steps may take a week. More than a week should be allowed for typing, and certainly the printer must be allowed more time. Further, if there is no concurrence, it may require another 'merry-go-round.' Thus two and one-half months is the *very minimum* time required from the time the final manuscript is sent to the Editor.

References

1. E. M. Rogers, *American Journal of Physics*, Vol. 37, 1969, p. 954.
2. R. B. Ingle and A. M. Ranaweera, *Teaching School Chemistry*, Chapter 2.
3. For references, see Chapter 2, Section 2.2.
4. International Baccalaureate Office (IBO), *Criteria of Evaluation*. Available from the International Baccalaureate Office, Palais Wilson, CH 1211–Geneva 14, Switzerland.
5. Nuffield Foundation, *Introduction and Guide, Chemistry*, p. 8, London/Harmondsworth, Longmans/Penguin Books, 1966.
6. T. A. Ashford, 'Improved Techniques of Assessing Student Achievement', *New Trends in Chemistry Teaching*, Vol. IV, p. 107, Paris, Unesco, 1975.
7. A. Kornhauser, C. N. R. Rao and D. J. Waddington (eds.), *Chemical Education in the Seventies*, 2nd ed., Oxford, Pergamon Press, 1982.
8. See, for example, S. N. Imenda, *Science Education in Zambia*, Vol. 10, 1980, p. 13.
9. Much has been written about these two much misused terms, norm- and criterion-references tests. For fairly simple expositions, readers are referred

to: J. M. Thyne, *Principles of Examining*, London, University of London Press, 1974; and R. J. Montgomery, *A New Examination of Examinations*, London, Routledge & Kegan Paul, 1978.

10. J. C. Mathews and J. R. Leece, *Examinations: Their Use in Curriculum Evaluation and Development*. London, Evans Bros/Methuen, 1976. (Schools Council Examinations Bulletins, 33.)
11. J. T. Shimozawa, in A. Kornhauser, C. N. R. Rao and D. J. Waddington (eds.), *Chemical Education in the Seventies*, 2nd ed., p. 166, Oxford, Pergamon Press, 1982.
12. See, for example, *A View of the Curriculum*, London, HMSO, 1980 (HMI Series: Matters for Discussion 11); *A Framework for the School Curriculum*, London, HMSO, 1980.
13. D. Watts and N. Bayless, *School Chemistry Project: A Draft of a Secondary School Chemistry Syllabus for Comment*, Canberra City, Australian Academy of Science, 1978.
14. *Chem. 13 News*, No. 66, pp. 2–6; No. 68, p. 10; No. 70, pp. 2 and 19; No. 71, pp. 7–10 (1975); No. 76, p. 6; No. 78, p. 2 (1976); No. 85, p. 16 (1977).
15. Certificate of Secondary Education is for the less academic student taking examinations at 16+. There are three different forms of syllabus and examination. Mode I is externally controlled. The syllabus in Mode II is constructed by the school and moderated externally; the examination is externally controlled. Mode III syllabus and examination are both constructed at school and externally moderated.
16. P. J. Fensham, *Teaching School Chemistry*, Chapter 7.
17. T. A. Ashford, Chairman, Examinations Committee, Division of Chemical Education, American Chemical Society, private communication.
18. J. C. Mathews, *The Use of Objective Tests*, Bailrigg, Lancaster, University of Lancaster, 1978 (Teaching in Higher Education Series, 9.)
19. D. Odhiambo, in A. Kornhauser, C. N. R. Rao and D. J. Waddington (eds.), *Chemical Education in the Seventies*, 2nd ed., p. 175, Oxford, Pergamon Press, 1982.
20. M. Sinclair, *International Newsletter in Chemical Education*, Vol. 15, 1981, p. 17.
21. B. S. Bloom, *Taxonomy of Educational Objectives I*, London, Longmans, 1966.
22. T. A. Ashford, 'Improved Techniques of Assessing Student Achievement', *New Trends in Chemistry Teaching*, Vol. IV, p. 111, Paris, Unesco, 1975.
23. J. Alles et al., *Theoretical Constructs in Curriculum Development and Evaluation*, Colombo (Sri Lanka), Ministry of Education, 1967.
24. For further details: Division of Chemical Education, American Chemical Society, 1155 16th Street, N. W. Washington, D.C. 20036.
25. M. Cropley, *A Sample Collection of Multiple Choice Items from the Science Item Bank*, Melbourne, Australian Council for Educational Research (ACER), 1976.
26. K. V. Sane, in A. Kornhauser, C. N. R. Rao and D. J. Waddington (eds.), *Chemical Education in the Seventies*, 2nd ed., p. 132, Oxford, Pergamon Press, 1982.
27. S. Glavar and J. Urbanc, *Computer Test Bank for Chemistry*, Department of Chemistry—RCPU, University of Ljubljana (Yugoslavia).
28. *Nuffield Chemistry 'O' level. Revised Teachers' Guide I*, p. 4, Longman, 1975.

29. 'Science in Society' course, Oxford and Cambridge School Examinations Board. J. L. Lewis (ed.), *Science and Society. Teachers Guide.* Hatfield/London, Association for Science Education/Heinemann, 1981.
30. J. C. Mathews and J. R. Leece, *School Science Review*, Vol. 57, 1975, p. 362.
31. J. R. Leece and J. C. Mathews, *School Science Review*, Vol. 57, 1975, p. 148. *Facility Index*: 'the mean of scores actually attained on each question divided by the maximum possible score'. *Discrimination Index*: 'the product moment correlation of performance on each question with performance on the whole examination'.
32. D. J. Waddington, *The Role of Laboratory Teaching in Chemistry, Conference Proceedings*, Royal Australian Chemical Institute, 1978.
33. J. J. Thompson, *Practical Work in Sixth-form Science*, University of Oxford, 1975.
34. *Research in Assessment*, London, The Chemical Society, 1978.
35. The following chemistry examinations are among those that use internal assessment for practical work: Joint Matriculation Board, 'A' level Chemistry, Manchester, United Kingdom; University of London Schools Examination Board, 'A' level Chemistry, London, United Kingdom (who also administer the examinations based on the Nuffield 'A' level Chemistry course); Hong Kong Examinations Authority 'A' level Examination, Hong Kong.
36. R. Wood and C. M. Ferguson, *School Science Review*, Vol. 56, 1975, p. 605.
37. G. Sleightholme, *Nuffield Chemistry: An Investigation into the Operation of Teacher Assessment of Practical Work*, University of Lancaster, 1972.
38. J. Sadler, *The Nature and Management of Pupil Project Work in CSE Science: The Teachers' Perspective*, University of Leicester, 1979.
39. Ibid., p. 28.
40. M. J. Tomlinson, *Education in Chemistry*, Vol. 11, 1974, p. 40.
41. *Project Work in Chemistry for the Certificate of Sixth From Studies*, Scottish Centre for Mathematics, Science and Technical Education, 1976 (Memorandum No. 23); J. McGuire and A. H. Johnstone, *Education in Chemistry*, Vol. 11, 1974, p. 5.
42. J. Handy and A. H. Johnstone, *Education in Chemistry*, Vol. 11, 1974, p. 56.
43. *Junior Engineers, Technicians, Scientists (JETS), Journal and Annual Reports.* Ministry of Education, Lusaka (Zambia).
44. B. Pedersen, V. Ringnes and K. H. Holm, in A. Kornhauser, C. N. R. Rao and D. J. Waddington (eds.), *Chemical Education in the Seventies*, 2nd ed., p. 222, Oxford, Pergamon Press, 1982.
45. A. Kornhauser, in A. Kornhauser, C. N. R. Rao and D. J. Waddington (eds.), *Chemical Education in the Seventies*, 2nd ed., p. 292, Oxford, Pergamon Press, 1982.
46. P. Brown, P. J. Hitchman and G. D. Yeoman, *CSE: An Experiment in the Oral Examining of Chemistry*, London, Evans Bros/Methuen, 1971 (Schools Council Examinations Bulletins, 21.)
47. B. Prestt, *Take Your Time: Curriculum Development at Your Own Pace*, Manchester, Manchester Polytechnic.
48. Among others, a series of courses were organised by the Regional Centre for Education in Science and Mathematics (RECSAM), Penang (Malaysia).

6 Education and training of teachers

A. V. Bogatski, D. Cros and J. N. Lazonby

Our entry into new stages of the scientific and technological revolution makes it all the more important that all our students learn and understand the essential elements of science and scientific thought. This requires that there be, in any system of education, an adequate supply of 'good' teachers. Effective pre-service training of teachers is of cardinal importance and there is now a growing awareness that, however difficult they might be to implement, in-service courses for teachers are also fundamental to the well-being of chemical education in schools.

In the introduction to this chapter, a number of interesting problems are presented, for example:

Should aspiring chemistry teachers be taught at university as chemists and then undertake a course of teacher training consecutively, or should they be taught chemistry and receive teacher training concurrently?

Should chemistry teachers be trained as chemistry specialists or should they be prepared to teach school curricula which are wider in perspective (for example, integrated science, environmental science)?

The relative importance of educational theory in pre-service teacher training courses.

What are the most effective ways of keeping teachers up to date in their subject and in methods of teaching?

With the generous assistance of many chemical educators across the world, the authors try to see whether there are general answers to any of these questions.

6.1 Introduction

What makes a 'good' teacher?

As F. Diesterweg wrote, 'A bad teacher teaches the truth; a good teacher teaches how to find it' [1]. 'Good' teachers may be defined as those who, by the force of their example, by thorough knowledge of their subject and broad general background, coupled with the desire to keep on improving and augmenting their knowledge, by their integrity, high-mindedness and clear-cut philosophy of education, are able to kindle in their students an interest in and love for a subject to which they have dedicated their lives and inspire them to follow in their footsteps and emulate their qualities of character. Another view of a 'good' teacher is a congenial and conscientious person who leads an ordinary normal life. He is respected, intelligent, has a sense of humour and an aptitude for teaching; he is cultivated [2]. This fails, however, to recognize that a good teacher must be a person with a high sense of principle, an aptitude for creative work and scientific curiosity. However, enough has been said to make clear, in essence, what qualities a 'good' teacher must possess. This being so, let us give some thought to the question of what distinguishes a 'good' chemistry teacher from a 'good' teacher in general, and then consider the kind of training required to produce one.

The training of a 'good' chemistry teacher depends largely on the four following factors: (a) the careful selection of candidates; (b) the educational process; (c) appropriate working conditions; and (d) the efficacy of retraining programmes.

Selection of candidates for pre-service training

Those who contemplate a teaching career must be dedicated, and sincerely interested in communicating knowledge to students and willing to assume the arduous task of educating the younger generation. The search for qualified candidates among young men and women is a noble undertaking. It should begin while future candidates are at school. If universities and colleges are really serious about training, they must work in close collaboration with schoolteachers, helping to identify future generations of teachers. A very useful role can be played by university and college teachers in taking part, with schoolteachers and pupils, in chemistry competitions, evening get-togethers, science clubs, weekend field and laboratory excursions and similar gatherings [2].

However, one must be careful, for success in a competition may reveal a student's aptitude for chemistry as a science but not necessarily his aptitude for a future teaching career in this field. Only personal contacts and close acquaintance with potential teacher-training candidates can ensure success in the search for boys and girls who are sufficiently talented and motivated to become 'good' teachers [3].

The process of selection by field of interest should not, however, be confined to the period preceding a student's entrance into the university, institute or college. It should continue throughout his whole course of study in an institution of higher education. University and college teachers should identify among their students those who are genuinely interested in becoming chemistry teachers. The process of selection does not end there. Workshops, refresher courses and practical work experience also provide excellent opportunities for identifying especially promising teacher candidates. Many of us have had the experience to know and teach many people who, earlier in life, evinced no particular interest in a teaching career, and yet later, as they experience the joy that comes to a teacher who sees his students, acquire a taste for teaching and develop into 'good' teachers.

Thus, taking into account the importance of selecting those who are drawn to the ideals of teaching, and basing selection on the students' particular fields of interest, the search for those who will make 'good' teachers should be carried out by teachers at every level, among secondary pupils, undergraduates and those who have graduated.

The choice of the teaching profession and, in this instance, choosing to be a chemistry teacher, is of great social significance. The number of teachers (and the number of 'good' teachers) depends directly on the number of young people choosing this difficult career. It is hard to describe this social problem in general terms as it is bound up in the complexities of the social, political and economic circumstances in the country. But overall, the society will only recruit 'good' teachers into the profession if it is held in high esteem.

In what institutions should chemistry teachers be trained?

In some countries, all those who wish to study chemistry in any of the professional and vocational fields take the same undergraduate course. After obtaining their degree in chemistry or subjects in which chemistry plays an important role, those who wish to teach are then trained for, say, a year in an institution which may be attached to a university or be separate. In others, all chemistry teachers are trained in special institutions (e.g. colleges of education) which may or may not be part of the university. Some countries have both systems. Aspiring chemistry teachers (particularly those who want to teach chemistry at the more elementary levels in a secondary school) are often trained in two subjects; for example, in chemistry and biology or chemistry and physics. Those who are trained in colleges often have a reduced number of hours allotted to their chemistry education. It is said that this is compensated for by the ample time they now have for the study of aspects of educational theory and chemistry teaching methods [4]. This is a crucial question which will be raised several times in this chapter. Many believe that formal and abstract knowledge of 'teaching techniques', in the absence of thorough knowledge of the subject, does not produce good results and that a teacher must, first and foremost, be a specialist in his field; in the given instance he must be a good chemist.

This raises the interesting question of what combination of science subjects is most appropriate for a teacher of chemistry, if the students are taking a course in which two sciences are studied [4]. As chemistry develops as a subject, as Campbell has shown in Chapter 1, there is little doubt that chemistry teachers should become acquainted with those aspects of physics, biology and other natural sciences which chemists need and use. But how can this be effected in their training?

What should a 'good' chemistry teacher know?

Teacher-training programmes in general should be designed in such a way as to ensure the training of 'good' teachers and not just teachers, and must, therefore, be highly imaginative and creative.

Thus, as described earlier, the future teacher of chemistry must have a thorough knowledge of the subject, not only the core of organic, inorganic and physical chemistry; he should be acquainted with chemical technology, together with some of the newer areas of the subject in which other natural sciences impinge. For example, at present, both bio-inorganic and bio-organic chemistry are advancing very rapidly and are centres of great interest and importance. Similarly, the use of computer techniques will become steadily more important to the chemistry teacher, both in his role as a teacher and as a chemist.

However, above all we must always remember that, if the objective is the formation of 'good' teachers, it can be attained only if students are given the maximum opportunities to reveal their creative talents, together with optimum conditions for the bold and free deployment of their aptitudes.

What are the appropriate working conditions for a good chemistry teacher?

Young teachers, if they are to develop into 'good' teachers, must be ensured of appropriate working conditions which are conducive to creative growth and provide an opportunity for continuing professional improvement. As noted earlier, it often happens that the talents of a student are only apparent after graduation from an institution of higher education. Young graduates rarely possess those traits that immediately place them in the category of a 'good' teacher; that is too much to expect. They can only become one in their work of the process of training young people to be chemists.

Accordingly, appropriate working conditions should include the following: (a) provision of graduates with certainty of employment; (b) encouragement of society by giving them the esteem they deserve, for much of our development depends eventually on good science teaching; (c) furnishing of young teachers with the material conditions necessary for their work. It is especially important for chemistry teachers to have a laboratory where they can teach students to carry out experiments; and (d) making it possible for them to build on the education and training they have received in their pre-service course.

What in-service courses are needed?

There are several possible ways in which schoolteachers can and do receive further training. The first is by self-improvement, which not only includes, but goes hand in hand, with other available means. It entails reading books, pamphlets and journals (both specialized and of a popular science character), consulting specialists, keeping in contact with one's *alma mater*, going to lectures, and the like. Opportunities are also offered by television programmes. Educational programmes are particularly useful and popular science programmes can also be valuable. There are many examples to show that people who genuinely wish to build on their existing attainment can accomplish a great deal through self-improvement, which is, of course, possible only if the requisite material conditions for it exist: a person must have the time and money to buy books and subscribe to journals (or else the time to use them in libraries). It must not be forgotten that in many countries chemistry textbooks are often far behind the times and of little or no use to teachers who are interested in improving their own qualifications. Furthermore, if schoolteachers have the opportunity to attend lectures and maintain contacts with the university, the lectures and links must be of mutual interest, giving the teachers opportunities to share their experiences and aspirations.

There are, however, some secondary-school teachers who, for a variety of professional, psychological or material reasons, are simply not inclined to use this method for improving their knowledge and qualifications.

Another way in which the further training of secondary-school teachers can be carried out is through the development of periodic or regular refresher courses under university or other auspices. Various systems of this kind are in operation in many countries, to help secondary-school teachers to establish working links with scientific groups, obtain first-hand information and become immersed in the mainstream of modern scientific thought.

Another way in which teachers can obtain retraining or continued training to enhance their qualifications, is through systematic participation in the revision and improvement of textbooks and syllabuses. Bringing teachers together regularly to discuss these matters and elicit their concrete suggestions is an important and integral feature of further training.

Four questions have been raised so far: (a) Should chemistry teachers be trained with other chemistry students or in separate courses? (b) Should they be trained to teach chemistry as a specialist subject or as part of a wider science curriculum? (c) What content should be provided in pre-service courses—is there a consensus about the relative importance of chemistry-based courses and educationally based courses? (d) What in-service courses should be provided?

Although the first three are concerned principally with pre-service training and the last with in-service training, some educationalists also see the importance of training in the first few years of teaching as cardinal to the success of teacher education.

One can recognize four stages in the training of teachers:

1. Acquisition of a mastery of the specialist curriculum.
2. Pre-service teacher training.
3. Induction of the newly trained teacher.
4. In-service training for experienced teachers.

Much of this chapter will be devoted to examples of teacher training which are offered in many countries illustrating the decisions which have to be taken and the factors which might influence these decisions. Because of their close interrelationship, we will deal with 1 to 3, above, in a single section (Section 6.2). In subsequent sections (Section 6.4 and 6.5), we will consider in-service training. In Section 6.3 we will look at some curriculum development in the area of teacher training.

6.2 Pre-service training

6.2.1 *Consecutive and concurrent causes in chemistry and teacher training*

On looking at the systems of teacher education worldwide, one can see one important distinction between those countries where teacher training occurs 'end-on' to the training in the curriculum specialism and those where the training in the science subjects and in education is concurrent. Whatever the systems, there will be three elements to the overall course: (a) the acquisition of knowledge in one or more sciences; (b) foundations in education: courses may be given in aspects of sociology, psychology, history and philosophy related to education; and (c) teaching methods and teaching practice.

A secondary-school teacher must acquire a good knowledge of his own specialist subject(s). Whether or not this should be in chemistry only or in chemistry and a second science subject depends on the secondary school science curriculum.

Of course, the decisions taken on the secondary school curriculum depend on both educational factors (part of the overall decisions made for the curriculum patterns in both primary and secondary schools) and economic factors (whether there are enough specialist teachers and the overall manpower aims of the education system). These factors are themselves interrelated and depend on political and social factors within the country.

In many countries chemistry as a single subject is only taught at the senior secondary level and most science teaching below that is carried out as a combination of subjects, sometimes in small units of biology, chemistry and physics or, more recently, as an integrated science curriculum [5]. Thus, there is a growth of training courses where a broader science curriculum is studied.

For example, in the Netherlands the changes in both the school curriculum and educational structure are having a profound effect on teacher training. Since the beginning of the 1970s, the old existing training programmes for the teaching profession in the MAVO (intermediate general education) school system and in the lower and intermediate vocational schools are being replaced. In the so-called new colleges for teacher training (Nieuwe Leraren Opleiding), teachers are trained for the third and second degree certificates (which are teaching qualifications in intermediate and lower level general and vocational secondary education). Instead of specializing in only one subject such as, for example, the teaching of chemistry, they are now required to take two separate specializations simultaneously and, consequently, third or second degree teaching certificates for each of these two specializations are being awarded [6]. Another example is in Austria, where secondary-school teachers in the Allgemeinbildende Höhere Schule (AHS—15 per cent of the 14 to 18 age group) are trained in the universities and technical universities. Two main science subjects are compulsory, so chemistry must be fitted into four and a half to five of the nine semesters. This is probably the general trend [7].

However, in Finland, the opposite is observed [8]. Chemistry teachers in the comprehensive and senior secondary schools usually take degrees in which there are components of chemistry, physics and mathematics. Only a minority of those teaching chemistry have taken it as a major subject. Because chemistry is to become a compulsory subject in the senior secondary school, teacher training is now being remodelled to produce more specialist teachers. Perhaps the systems are moving from different starting-points to similar positions.

As described above, there are two distinct ways in which teacher training is organized. Teachers for primary (and sometimes lower secondary) classes are usually trained in colleges of education, which may or may not be attached to universities. Teachers for senior secondary classes have followed a science course in a university. However, one of the controversies surrounding the training of science teachers is whether prospective science teachers attend the same science classes as those who are to become professional scientists. In many countries teachers are taught separately from those who are to practise chemistry, often in separate departments even if they are in the same university or institution.

Indeed, as Lourdusamy has pointed out [9], there is an increasing number of university courses devoted to chemistry and education, and students have had to choose *before* going to university whether or not they wish to teach.

At university level, teacher education is giving way to concurrent degree programmes and special integrated degree programmes. Those who enter universities have to make up their mind right from the start whether they want to be teachers or not and follow a programme that will eventually qualify them to teach in schools.

Such a system has the potential of breeding two classes of student and two classes of university teacher. Those who teach pre-service courses are often denied facilities for research and status enjoyed by those who teach potential scientists. The system tends then to fail to attract really good teacher-educators.

Secondary-school chemistry teachers are in short supply in most countries. In the developed countries, it is a relatively marginal difficulty, with some schools finding it difficult to obtain a specialist, i.e. someone who has majored in that subject. However, in many developing countries, the situation is chronic. As Bajah has pointed out for his own country, Nigeria [10]:

The truth is that the rate of expansion of secondary schools in which chemistry is taught far exceeds the rate at which chemistry teachers are being trained.

Indeed, in Malaysia, the acute shortage of graduate science and mathematics teachers led to the introduction of science with education (B.Sc.Ed.) degree [11, 12] at four universities. The education component is taught concurrently with the science components and the duration of study can be reduced, overall, by one year. Other countries in which teachers may take similar types of courses include India and Australia.

Many countries, though, train their senior secondary-school teachers by a different route within the university. This entails the student taking a chemistry degree or, as we have seen, chemistry and another science subject, followed by a one-year postgraduate course leading to a separate qualification.

Many countries still adopt this system. In Czechoslovakia, for example, many teachers of chemistry for the upper grades of secondary schools have to complete their studies in a faculty of natural sciences, as does anyone else wishing to use chemistry in his or her career. Teachers for the lower grade of secondary schools study in a pedagogical faculty of a university for four years [13]. In the German Democratic Republic, the latter take a four-year course in a college of education [14]. Variations of this system can be seen in a wide range of developed and developing countries [15].

Yugoslavia provides an example of a bridge between the methods for training senior secondary science teachers at university. In some of the universities [16], an inter-linked study scheme has been introduced. All chemistry students at these universities, whether they wish to become research chemists, chemistry teachers or chemical engineers, have a common curriculum for the first two years. In the third year, about 20 to 40 per cent of the curriculum is specific for one of the three professional orientations and in the fourth this is increased to 50 to 60 per cent. This enables students to change their professional orientation during their studies:

> They are living together till the end of their studies and so are more prepared for mutual co-operation during their working period in schools or industry or research organizations.

This type of structure is also seen in the United Kingdom. For example, over the last fifteen years, one university has had a Chemistry-with-Education course in which about 65 per cent of the time is spent by the student working alongside the chemists, taking the same courses and examinations. The other 35 per cent is used for educational studies. Students still have to take the fourth year postgraduate course [17]. The course is taken mostly by those who are committed at a relatively early age to the teaching profession, but the vast majority of teachers at secondary level have studied chemistry (and perhaps another science) and then taken the fourth year course for teachers.

An interesting educational development occurring in some chemistry courses is illustrated by two examples. In Sri Lanka, some elements of chemical education have been introduced into university chemistry courses:

> This step was taken in view of the fact that many graduates who offered chemistry as a subject became teachers [18].

Chemical education is an option given to students in the chemistry courses at a university in the United Kingdom. In this course, which is for general education rather than for vocational training, most of the students take the option out of interest rather than because they necessarily intend to become teachers [19].

In the United States, four years of concurrent study of chemistry and other sciences in *science* departments and education in *curriculum* and *instruction* departments is the common pattern. This leads to courses of approximately 60 per cent science, 20 per cent education and 20 per cent general education (liberal arts).

6.2.2 *Teacher training: balance between theory and practice*

As described earlier, one can see three elements to this part of the training which must be interrelated: the acquisition of knowledge in the sciences; the foundation in education; and teaching methods and practice.

The relative importance attached to the three parts and the degree of integration between them varies from country to country; much of the decision-making has occurred when deciding in what type of institutional system teacher education and training will take place.

Difficult choices have to be made in the training of chemistry teachers. If the 'end-on' method is being used, there has to be a balance in the postgraduate course, between methodology and teaching practice, on the one hand, and appropriate educational theory and wider educational issues on the other. If the course in science and education is concurrent, even more fundamental decisions of balance must be made.

One important factor which influences decisions to be made about the initial training course is whether in-service training is being established on a wide scale. Many initial training courses spend what seems to students to be a disproportionate time on elements of theoretical education, such as sociology and history, which might appear to have no bearing on their work in school at the start of their career. Yet the reason for including these subjects is to try to develop a framework to enable teachers to evaluate critically educational issues which they will face. Those responsible for courses are trying to balance teachers' long-term career needs with their short-term needs. The elements of theoretical education may indeed give more help to, and be understood better by, experienced teachers. Thus, there is a movement throughout the world to make teacher-training courses more responsive to the immediate rather than long-term needs of the young teacher. A greater emphasis is being given to methodology rather than to basic educational theories.

Thus instead of attempting to introduce students to disciplines such as sociology and psychology, aspects of these disciplines are used to add additional dimensions to considerations of issues which are more closely related to the teaching of science. For example, when considering criteria for selection of science curricula, it seems important to include references to relevant theories of cognitive development. Similarly, a consideration of language in science education may involve reference to some sociological aspects, whereas a systematic study of these disciplines may not be appropriate for a pre-service teacher training course.

One aspect of moving the balance in favour of methodology is the need to arrange as much teaching practice as possible. This depends on whether there is a well-established science teaching profession already in existence. For example, there is considerable advantage of having a substantial element of school-based training in which experienced teachers look after the training, either during the initial training period or immediately after it—or indeed, both. Not only does this need an already established teaching force but also effective communication between school and university/college. If this is possible, then it can lead to the school assisting in the assessment of competence of the teacher-student. Such a scheme alters the whole balance of the training programme and makes the training 'real' for students.

If it is not possible to arrange extensive school-based training, every effort should still be made to give some teaching practice in a school. The visits should be arranged to allow discussions between the student-teacher, their university/college teacher and the experienced school-teacher. The latter's experience and expertise can be of inestimable value to the trainee. The teacher's attitudes too make an impression on a student which is impossible to exaggerate.

In the Netherlands, for example, new teacher training programmes were introduced in the 1970s, in which a total of eighty hours is devoted to practice-teaching in secondary schools under the supervision of specially assigned schoolteachers [6]. In the United Kingdom and the United States, a student must spend at least three months in schools during the course in addition to earlier observation and participation experience.

However, giving students practice in schools may not only be expensive but sometimes impossible. The university or training college may just not have enough schools near by in which to place its many students. It is too easy for those of us working in relatively densely populated countries with effective means of communication (both for travelling and for the large amount of essential correspondence between the trainer and the schools) to forget the difficulties faced by teacher-educators in many developing countries.

There are various institution-based activities which can complement school experience or compensate for the lack of it. Peer-group teaching where a student teaches a group of fellow students can be a worthwhile experience and useful as a focus for the consideration of specific skills. It may even be possible for small groups of pupils to be brought into the institution for micro-teaching exercises. Such exercises are often felt to be very effective and save much time. They will never be able to totally replace the experience gained in the classroom but they encourage team work, give immediate feedback to teachers, enable prospective teachers to evaluate themselves, and enable teacher-educators to link theory and practice [15]. If video recording facilities are available, the subsequent analysis of such activities can be all the more effective [20].

6.2.3 *Content of methodology courses*

Methodology courses include not only methods of teaching but also a study and evaluation of curricula being studied at school, and, with advantage, an examination of curricula being developed in different parts of the world to meet needs similar to the teacher's own country.

Science curriculum components of pre-service courses may involve, as a minimum, familiarization with recommended school science curricula, but depending on the curricular freedom there may be a need to consider the processes of curriculum development and certainly the conversion of examination syllabuses into teaching syllabuses.

Most chemical educators believe that one of the most important aspects of chemistry, if it is to be taught in schools as part of a general education, is that it is an experimental subject, in which the importance of experimentation and observation can be emphasized. The following page taken from *Guidelines and Recommendations for the Preparation and Continuing Education of Secondary School Teachers of Chemistry* summarizes this [21]:

The laboratory

Laboratory work is an indispensable part of chemistry instruction. In the hands of a good teacher it is in itself an excellent vehicle for instruction

It gives an insight into authentic science.

It is intellectually exacting. It requires careful attention to planning, execution, and observation. It demands interpretation and abstraction.

Most students enjoy it. It couples cognitive activities with 'hands-on,' action-oriented, concrete operational activities. It is a stimulating change from lectures and other forms of instruction.

It is interdisciplinary. It sharpens students' skills in: reading and writing; applying mathematics to real problems; classifying, measuring, interpolating, extrapolating, drawing inferences, framing hypotheses, and building mathematical, verbal, and physical models.

It helps students to develop manual dexterity, patience, dependability, and a spirit of co-operation.

It helps to enhance and strengthen student–student and student–teacher interpersonal relations.

A learning laboratory is a busy laboratory. Although it is time-consuming for students, and mentally, emotionally and physically taxing for teachers, laboratory work is indispensable—if the spirit of science is to be conveyed to the younger generation and if they are to develop the capacity for independent learning.

This is the lesson of the laboratory method, and the lesson which all education has to learn: The laboratory is a discovery of the conditions under which labour may become intellectually fruitful and not merely externally productive.

John Dewey (1859–1952), educator and philosopher.

However, many curricular projects built around experimental work have run into difficulties. This is due in part to lack of training for laboratory instruction, partly because of the lack of facilities and supplies in schools and partly because of lack of motivation. Teachers who are ill-prepared to teach experimental chemistry, even under good conditions, find it a hopeless task given the conditions under which many work at school.

In some countries the reasons are not hard to find. Teacher-training courses may take place entirely without laboratory provision, with the only practical work done in the chemistry content of the course. Little is done to prepare teachers for practical work in schools.

Another danger looms in the opposite direction: namely that the teacher is trained in circumstances far more ideal than those he will find in school. Indeed, where experimental work plays an important part in the chemistry curriculum, a situation of crisis proportions is developing. The number of students taking science at secondary school is increasing rapidly in many countries; there is also strong pressure to increase the proportion of time given to practical work. But neither trend is compatible with the effect the economic climate has on school budgets in the majority of countries.

Not only must teachers be trained in the techniques of teaching chemistry as an experimental subject, but they must be trained to make, use and maintain simple equipment and to use inexpensive, locally available chemicals. There have been valiant attempts to introduce these ideas but a concerted effort is necessary at the training stages before real progress can be made. Curricula based on experiments are doomed if teachers are not trained to do these in the milieu of the current situation in the school and become confident in their use. Further, it is important for teachers to receive thorough training in the techniques required in the teaching of experimental chemistry; for example, how to control a class effectively, fulfil the objectives of a practical, manage stores maintain safety and so on. It is vital that maximum use is made of time and the resources being used.

The content of the methodology part of the course must also include an appreciation of assessment techniques. After all, assessment will be a crucial part of their pupils' work and a thorough training in setting questions and marking answers is needed. There is a very wide range of techniques, as described, for example, by Mathews (Section 5.5). The format of the in-service course described in Section 5.10 could be adopted for use in a pre-service course.

There is a significant gap between the chemical educators engaged in research and those engaged in teaching in the classroom and laboratory. An appreciation both of the value or otherwise—in other words a *critical* appreciation of present-day chemical education research—would be very useful. (Fensham, in Chapter 7, outlines some of the interesting current research in chemical education and links this to the real world of the teacher.) This would also provide a valuable insight during pre-service training.

Teaching strategies can range along a continuum from a formal lecture to the student-centred unprescribed inquiry one might encounter during project work. The approach advocated during the teacher-training process may well depend on the extent to which the training course is seen as an agent of reform rather than a means of preparing students to be proficient in the existing practices found in schools. In reality, it is likely to be a balance of these two extremes. The cultural background of the pupils and what they perceive as their own needs may determine how far or how quickly it is possible to move along this continuum towards the inquiry type of teaching strategies. It must be remembered that student teachers are usually the relatively successful products of a country's education system and so any attempt to encourage reform via the teacher-training process has to be deliberately planned and convincing. The secret of success lies in teacher trainers using the approaches with student teachers which they hope that the students (when qualified) will use with their pupils.

6.2.4 *Induction of teachers*

The first one or two years of schoolteaching, with its many varied calls on a young teacher in teaching and management skills and in displaying a competence of his ability as a scientist and a communicator, are daunting. Some see this period as one of training, as vital as the training that has taken place at the university or college.

The key to the success of this period 'as one of training' is whether the schools to which the teachers go have staff who are able to guide them and whether there are resources to enable them to have time for study, either in the school itself or a teacher's centre (which may be based in a nearby university or college or which may be specially set up by central or regional authorities for teachers).

6.3 Curriculum development in teacher education and training

The type and extent of the school experience that training institutions are able to offer their students may well be predetermined by factors which are beyond their control. Whether the experience is a short period of teaching practice or an extensive programme of structured activities within the schools, there is clearly a need to identify the skill areas which the trainee teacher ought to develop.

It may be possible to draw up a list of skill areas associated with science teaching which are requisities for teachers in all countries, but the order of priority of importance will vary according to the circumstances in a particular country or region of the country. In a recent United Kingdom survey [22] of teacher trainers and new and experienced secondary-school science teachers, there was considerable agreement that the seven most important skills out of a list of twenty-seven in which trainee science teachers should gain competence are:

1. Lesson planning and preparation.
2. Lessons presentations.
3. Practical work organization.
4. Teacher demonstrations.
5. Safety in the laboratory.
6. Discipline and class control.
7. Class questioning skills.

Clearly, depending on such factors as the esteem of education, social relationships between teachers and pupils, curricular freedom and availability of resources, the order of priority will vary in different countries. Thus each country will determine its own priorities, but whatever they are and whatever the balance is between school and institution-based training, these skill areas must be systematically studied and developed.

The areas listed above are concerned with the short-term aims of pre-service training, that is, they aim to prepare and equip the student for the first few years of classroom teaching. They assume competence in the subject (biology, chemistry, physics, etc.) and they ignore what the wider, more long-term aspects of the teacher's job may be.

A more long-term aim may be that the teacher would be prepared to think beyond the day-to-day activities of classroom teaching and contribute to the debate on educational issues which might lead to innovation at the departmental school level and indeed, where this is feasible, at national level. A preparation concerned with such long-term aims must acquaint students with, for example, the history, philosophy, sociology and economics of the educational system. Consideration of such areas is not aimed at influencing the students' performance in the classroom directly, and should not be judged according to this criterion.

However, in many countries, the first priority has to be the short-term aim, to train teachers so that they teach their subjects competently. Thus this is the primary concern of the rest of the section and in fact, the whole chapter.

Various sources exist which suggest guidelines for teacher training. One is *Guidelines and Standards for the Education of Secondary School Teachers of Science and Mathematics* [23]. Published in 1971, it is a series of guidelines in terms of humaneness, societal issues, nature of science and mathematics, science competencies, communication of science and mathematics, learning conditions, materials and strategies, and others. Another booklet, *Guidelines and Recommendations for the Preparation and Continuing Education of Secondary School Teachers of Chemistry* [21] published by the American Chemical Society, gives a pithy set of guidelines for courses and recommendations addressed not only to school, college and university chemistry teachers but also to officials and groups who administer the system and to professional teacher organizations.

A number of curriculum development projects in teacher education have been commissioned in different parts of the world with two overall aims: (a) to identify those aspects of science teaching methods which ought to be covered in pre-service training; (b) to pool the experience and expertise of leading teacher trainers and to share them with others.

At the University of Monash in Australia, the Australian Science Teacher Project (1976) was co-ordinated with science teacher-educators across Australia participating. ASTEP introduced forty-seven units of activities and experiences, in six sections [24]:

1. Understanding science (7 units).
2. Understanding pupils (6 units).
3. Models of teaching (12 units).
4. Considering the curriculum (8 units).
5. The laboratory as a teaching resource (9 units).
6. The Australian context (5 units).

The Thai Science Teaching Project, Thai-STEP, is one, if not the first, such project in a developing country and is aimed at improving the pre-service training in all of the higher education institutions with teacher-training responsibilities across Thailand [25].

In the United Kingdom, the Nuffield Foundation funded the Science Teacher Education Project (STEP) [26]. STEP pooled the ideas of over fifty science tutors in training institutions and developed and tested materials. Examples of the materials for laboratory safety are given in Section 4B.6.

More recently, the British government funded the Teacher Education Project [27]. This project is not confined to science teaching and has undertaken research, and development in the areas of mixed ability teaching, class management and control, language across the curriculum, the education of slow learners and bright pupils, teaching skills such as questioning and explaining, and job analysis of the teacher's role.

These projects are of value beyond their country of origin as they provide, on the one hand, a range of materials and activities which can be selectively used or modified and, on the other, guidelines for curriculum development in teacher education.

For such curriculum development work to be successful not only must it identify the areas to be covered, but it must devise effective student activities rather than a list of points to be covered in lectures. The formulation of an agreed list of skills does not imply a desire to produce a stereotyped teacher, but that each teacher in training should be provided with the stimulus, activities and experiences which will allow him or her to formulate what will be the most effective policies for each of the agreed skill areas.

Each stage—the identification of skill areas and the production of student activities—should involve a partnership between research and development. Thus, for example, during the Teacher Education Project, class management and control was investigated by a number of research procedures including case study research of students said to be 'good at handling classes' and students said to be 'poor at handling classes' [27]. The findings of research such as this can be influential in the design of activities for students in training. It is indicative of the trend to relate developments to research findings that this project has produced an extensive collection of research papers and reports in addition to the materials for use in the training courses [28].

Similarly in Thai-STEP the following objectives were set: (a) to correlate research findings as guidelines for designing structures of science and mathematics teacher curricula; (b) to produce printed materials as well as experimental kits and audio-visual aids for developing skills and competencies of science and mathematics teachers which will be used in teacher training institutions throughout the country.

This project has so far produced a number of research instruments such as a 'Science Teaching Profession Competency Test' as well as instructional packages for enhancing important teacher competencies [29].

The projects mention student activities and resources. The important trend is to broaden the concept of student activity and consequently that of resources. Listening to a lecture is a student activity and the lecture is a resource, but the outcomes of such an activity are likely to be mostly cognitive, whereas those aimed for in teacher education are likely to be more sophisticated and involved. Competency-based teacher-training programmes in the United States aim to include performance, consequence and affective as well as cognitive objectives. Thus, for example, when we are considering the safety aspects of science teaching, as well as wishing pre-service students to acquire the knowledge of potential hazards in a school laboratory, we will also wish to influence their attitudes towards safety and for them to develop the skills required to perform a demonstration or direct a class practical safely. The activities which might be considered appropriate for these objectives could be quite different. Thus, for example, the acquiring knowledge objective could involve prospective teachers working individually or in small groups in a safety quiz with the aid of suitable reference material (Section 4B.6). In this way, their knowledge and confidence can be built up without making their deficiencies too public. Attitudes towards safety may be more appropriately dealt with by small group discussions based on documented critical incidents involving accidents in schools. The skills associated with performing a demonstration may be developed by means of students demonstrating appropriate experiments to a group of fellow students. Such an activity not only-provides individual students with an opportunity to develop skills but provides the group of students with a simulation of teaching via demonstration which can then become the subject of further analysis and discussion.

STEP [26] has devised a wide range of activities concerning areas such as: aims and objectives; the nature of science and scientific inquiry; the pupils' thinking; language in science lessons; teacher–pupil interaction; methods and techniques; resources for learning; adapting to the pupil; feedback to teacher and pupil; curriculum design; safety; laboratory design and management; and the social context of science teaching.

These activities demand student participation and are supported by resources in a variety of formats—audio/videotapes, film loops, slides, pupils' writings, case studies, simulation exercises and discussion papers.

Thus the emphasis has been on devising activities which not only cover the identified skill areas but also take into account what is likely to motivate the student teacher. There is a need for a variety of activity and resource format. Many of the activities involve dividing the students into small groups with the tutor moving around the groups, clarifying issues, relating them to experience and to theoretical perspectives, and, where appropriate, generalizing.

6.4 In-service education and training

In this selection we will distinguish two types of in-service training. One set includes examples which are organized over a long period and for a relatively small number of teachers per course. The second set is organized over a shorter period, on a large scale (country-wide), usually for a systematic introduction of a new curriculum.

The provision of in-service training must be seen from two points of view, that of teachers and that of the state, the employees and the employers. No one will need to be reminded that in-service training is expensive in terms of resources and that most effective use must be made of them. However, there is a growing recognition and acceptance of the idea that in-service education is a career-long necessity, although the means of carrying it out are not readily available.

In-service training of the first type is now seen as one part of a teacher's training which must then continue throughout the professional career. Teachers are facing changes in their work brought upon by rapid social and technological changes, on a scale never experienced before, and an adaptable and innovatory spirit must be encouraged.

From the point of view of many teachers, in-service training must be seen to be helpful towards their career. Of course, they will appreciate it as self-education, which enables them to obtain greater satisfaction from their work. However, many would also like to have the encouragement of advancement in their career, if they take the trouble to undertake in-service training.

Thus, there is a clear responsibility on employers to provide a framework in which those teachers who *are* adaptable and innovatory are rewarded, and, second, to ensure that as many teachers as possible are encouraged to take part in in-service work.

For those paying for the in-service training from a limited budget, many difficult questions need answering. Which teachers should be given priority if the system is voluntary—those who need it because they have poor ability or those who are active but need the encouragement that such courses give them? If there is a shortage of teachers, should we use resources trying to convert non-science or non-chemistry teachers into chemistry teachers or give priority to keeping a small cadre of chemistry teachers up to date? Who should organize the courses—should they be professional (say university) chemists or teacher-educators or experienced teachers? The answer is partly given once it is decided what should be taught on the course. Is it to be devoted to chemistry, bringing in new material and helping to clear up misunderstandings? Or should the course be devoted to exploring how this chemistry is taught at school level? Should it look at some of the recent findings in chemical education? Or should it be simply devoted to specific aspects of the curriculum—the use, for example, of practical work or audio-visual aids or computers, etc? Answers to these questions lead on to methods which should be adopted for the course.

The first general rule is of paramount importance. Teachers must be *involved* in the *planning* of the course. This will help to ensure that the course is what teachers need and that the level is correct. Moreover, teachers, whenever possible, should be involved in the teaching and in the administration of the course; this gives greater confidence to the other teachers and breaks down the barrier, so easily erected without realizing, between the course organizers and course participants. The second rule is that all schoolteacher participants must take an *active* part in the course. The lectures, as such, must be kept to the minimum, and discussion must be encouraged. If practical work is a theme, then the course should be centred around the laboratory. How often one has seen the advantages of practical work lectured on and discussed and how futile this is! If audio-visual aids is a theme, then the participants must make them and try them out. If assessment techniques are being discussed, then the participants must write questions and try them on the other members of the course, as so graphically described by Mathews in Chapter 5. All these are such obvious points, yet course upon course is given in a didactic fashion, in just the opposite way that the organizers would wish the teachers to behave when they go back to school. Memories come back of serried rows of teachers in a lecture room being lectured to, hour after hour.

In many countries, in-service training is a semi-voluntary activity, often taking place during school holidays. On occasions, there is an element of compulsion. For example, in Malaysia, there was a systematic and large scale in-service project when the new Integrated Science curriculum was introduced. Teachers were required to attend a two-week course annually for three successive years [30] and were expected to work through the pupil experiments, improvise apparatus and construct test items. This was also true when IPST chemistry was introduced in Thailand (Section 2).

In the USSR, it is compulsory for all state teachers to attend refresher courses every five years. The course—of 144 hours—has lectures and laboratory work organized and given by leading university teachers. Teachers are credited with their regular salaries and are provided with living quarters [31, 32]. Similarly, in-service training, at least three days a year, has been compulsory in Yugoslavia since 1972 [16].

In the United States, chemistry teachers are expected to earn a master's degree or its equivalent in postgraduate work within their first five years of teaching. Special courses, tailored to new curricula, are available, with funding from the National Science Foundation, with the academic organization and direction being the responsibility of universities. The courses could be used in advanced degree programmes and are also recognized for salary increases and professional advancement. The courses take the form of intensive six-week summer institutes, evening and weekend courses, or the regular graduate courses given at a time convenient for schoolteachers.

Nevertheless, as stated earlier, in most countries, in-service training is a spasmodic and voluntary activity. There is a wide variety of agencies and individuals arranging courses; for example, ministries of education, universities and colleges and science teacher associations. In the next few pages we will outline examples of courses from different countries to give the flavour of the variety offered.

The ministries of education, or official bodies related to them, obviously play an important part in these activities. Recently, NCERT in New Delhi conducted courses for over 500 teachers to help them with the new senior secondary-level chemistry curriculum [33]. In Japan, educational centres in each prefecture arrange summer courses of one to two weeks' duration, the lecturers for the courses being university professors familiar with the 'Course of Study' [34]. Other courses are organized for three to six months, with certification leading to enhanced status. One of the most ambitious, and expensive schemes, is also organized in Japan, in which teachers study, in groups, abroad, for up to one month. Since 1959, over 15,000 teachers have taken advantage of this scheme, studying the educational, cultural and social conditions of countries in Europe and South-East Asia, and the United States [35]. Regional centres for in-service training have been set up in the Philippines. As in Japan, enhanced chances of promotion are available to those who attend in-service courses.

In Austria, five-day seminars for senior secondary-level teachers have been organized by the Ministry of Education and Art. A recent one was devoted to reaction mechanisms, determination of the structure of organic compounds and the presentation of environmental problems in the teaching of chemistry [7]. This is typical of many courses in Europe, in the United States, in Australia and New Zealand, indeed worldwide.

School teacher associations are also active in this field. For example, the Science Teachers' Association in Nigeria (STAN) has organized teacher vacation courses for many years. The courses are organized at the University of Ibadan by lecturers in the Departments of Chemistry and Education with the important addition of supporting staff from secondary schools. These courses usually last for two weeks, with sixty teachers attending [10].

Important work in this field has been initiated by other ICASE members. For example, the Ghana Association of Science Teachers (GAST) has organized short in-service courses for chemistry teachers based on the new syllabuses for the West African Examinations Council (WAEC) [36]. The tutors were examiners of WAEC and the teachers were supported by the Ghanaian Ministry of Education. Likewise, in Hong Kong HKASME organizes courses with the support of the Hong Kong Education Authority to help teachers with the presentation of new syllabuses for the local GCE at 'O' level. At present HKASME is involved in the revision of the 'A' level syllabuses and these revisions will, in due course, call for in-service training [36].

HKASME in also very active in the design and local production of equipment for science education. Much of this is for physics, but some items for chemistry teaching are included in the list published by HKASME. Short familiarization courses are organized to help teachers in the use of the new items.

National chemical societies take part in this important work. For example, the Association of Austrian Chemists (*Verein Osterreichischer Chemiker*) has arranged several seminars, usually running for one day. The topics were, for example, modern spectroscopic methods and the theories of the chemical bond [7]. Many short courses for teachers are arranged by the Royal Australian Chemical Institute, the Royal Society of Chemistry (in the United Kingdom) and the American Chemical Society.

The universities also have a part to play. Courses from one week to one year have been arranged in Hungarian universities to help some teachers with the new curricula [37]. In Canada, universities have been involved in training high school chemistry teachers in many ways. For example, Queen's University has put on a one-week chemistry programme for high-school teachers consisting of a series of lectures and demonstrations on energy, and sessions devoted to curriculum problems [38]. The University of Brandon has conducted a science programme, including chemistry, for elementary-school teachers who live in isolated northern communities of Manitoba [39]. The instructors for this successful programme reside in Winnipeg and fly out to various centres for a month or two at a time.

Brock University has a graduate evening course, aimed at introducing teachers and prospective teachers to the investigative laboratory concept [40].

In Argentina, courses of one week's duration for teachers are held each year. A team of three university professors, in co-operation with inspectors of secondary schools (supervisors), has prepared a handbook with seventeen experiments and the logistics for their running. They organize the course for up to twenty teachers, who work in the laboratory of the best available school. There is a standard format: preliminary explanation and discussion, lab work, post-lab discussion and report. These experiments can be easily adapted by the participants for school use. The programme is supported by the Ministry of Education [41].

Internationally oriented agencies have been active in promoting in-service courses for school chemistry teachers. Mention has already been made of CREDO collaborating with teachers' courses in East Africa (Section 2.3). The British Council collaborated with STAN, in Nigeria, and with the Malaysian Ministry of Education. Many other courses, of a similar nature, have also been held world wide, among them being countries in West and East Africa, in Fiji and other Pacific islands, in Bangladesh, India, Pakistan and Sri Lanka. ALECSO has organized programmes in the Arab States. RECSAM, based in Penang, Malaysia, has been involved for ten years or so in the provision of in-service courses for selected teachers from the seven member countries of the South-East Asia Ministries of Education Organization (SEAMEO). These courses are usually of ten to twelve weeks' duration, fully residential on the campus of RECSAM, and with tutors from the staff of RECSAM together with visiting tutors provided by various aid organizations, Australia, Canada, the Federal Republic of Germany, the United Kingdom, the United States, among them. The number of participants is limited to forty or fifty and the idea is that these selected participants will return to their own countries better prepared to offer in-service training for a larger number of local teachers. There is usually an annual chemistry course for secondary-school teachers.

The second type of in-service work, as already mentioned, is where a large-scale national effort is required in a short space of time, either to introduce a new curriculum with teachers who are significantly under-prepared or simply to enhance the teacher's ability to cope with the teaching programme.

To the administrator, one of the most attractive possibilities is the multiplier effect. If one can train a cadre of 'master teachers' at a national centre, then they can organize courses in their own part of the country and stay there, as teachers in their own school, advising and helping less experienced teachers.

For example, in Finland, at the same time as the new comprehensive school system was introduced in the 1970s, a network of teacher trainers in the provinces and municipalities was created. These trainers have their own teaching positions but they also provide supplementary training for other teachers who are required to participate in this training every year [8].

Another example in Malaysia comes from a science training scheme in primary schools [42, 43]. So that a large number of teachers could be given training in the shortest possible time, experienced teachers were trained and then appointed to thirty-two centres where they, in turn, trained other teachers. Since 1975, over 40,000 teachers have been trained in this way, with international (Unesco and the Asia Foundation) and bilateral (British Council and United States Peace Corps) assistance being harnessed. A similar system was introduced successfully in Sri Lanka. From time to time, the 'master teachers' were briefed at the Curriculum Development Centre by the personnel at the centre who were responsible for designing the syllabus' [18].

In Thailand, the chemistry teachers for the chemistry programme came to Bangkok for intensive in-service training. For the integrated science (and primary science) programme, the numbers of teachers involved were too large for centralized training. Instead, science instructors in thirty-two teacher-training colleges across the country were trained in Bangkok and given responsibility for providing in-service training for all the teachers in the area in these curriculum programmes.

Let us now look more closely at two large in-service programmes in very different parts of the world, one in Peru and the other in Indonesia.

In both, the programmes were developed from some work in national curriculum development, which had had the assistance of Unesco and other United Nations agencies. It was evident from this work that there was the need, among others, to improve the education of practising teachers, to provide them with teaching aids and laboratory equipment and to provide a system whereby the curriculum could be revised and adapted with the changing demands of their societies.

The United Nation's Children's Fund (UNICEF) agreed to fund the Peruvian project in the form of laboratory equipment and material and for scholarships for the training of project personnel. The Peruvian Government agreed to provide the infrastructure, services and other personnel. The National Programme for the Improvement of Science Education (PRONAMEC) was born in June 1971, with the National Institute for Educational Research and Development (INIDE), in charge of the project.

We will concentrate here on the element of in-service work, which was concerned principally with the retraining of practising teachers through its National Centre and science units scattered through the country. Each science unit is now housed in a specially designed or adapted building, with laboratory, lecture and workshop facilities. A unit has four specialists, for the basic sciences and mathematics and a technician, all trained within the programme.

The in-service courses take a variety of forms. The National Centre trains teachers who then run the science units, and are helped in this work by lecturers from universities. These are known as Type-II courses [44] and last for twenty-two weeks, full-time. In the induction stage (two weeks), an integrated science approach is used, based on laboratory and group work. The second, specialist, stage is concerned with the single science subject and concentrates on areas which teachers find difficult. Practical work and seminars play a predominant part. The next element is the professional studies stage, in which teachers are given some up-dating in current theories of education, curriculum development and planning. The last stage is devoted to teaching materials design and trials, ending with a display of materials developed on the course. All staff of the centre and INIDE come and try to perform the experiments and exercises. During the last fifteen weeks, each participant undertakes a short project as a result of some research-type work. By 1977, ninety teachers were serving as specialists at the National Centre and in the science units, having been trained on Type-II courses. They were involved in the other courses.

Type-I are four- to five-week summer courses in the science units principally for up-dating secondary and primary science teachers. They are structured to devote 50 per cent of the programme to knowledge (of which 60 per cent is in the laboratory), 30 per cent to methodology and 20 per cent to the design and production of teaching aids.

Type-III courses, known as Zonal courses, are also conducted in the science units. They vary in character. Some are short briefing courses; some are arranged to implement curriculum changes; some are run after school, in the afternoons; others are in terms of day-release.

The Indonesian programme was launched several years later. It is described as an in-service/on-service programme for secondary-school teachers. Like the Peruvian programme, it is centred around the implementation of a new curriculum (Curriculum '75). It is concerned with teachers understanding the new curriculum's aims and content and developing skills in laboratory organization and activities. The long-term objectives are: (a) improvement in the quality of classroom instruction; (b) development of an open-minded attitude in teachers towards an acceptance of changes in the content and method of science teaching; (c) improvement in the utilization of laboratory facilities; (d) better class-room management; (e) better laboratory management.

This work is being accomplished by the Ministry of Education in collaboration with the United Nations Development Programme (UNDP) and Unesco. The organizers have been very careful to ensure that there is really effective implementation and this means that teachers must be brought into the process at as early a stage as possible. The project directors have, as will be seen below, a systems approach to curriculum development and this has allowed the teachers to take part in the development of the materials. Like the Peruvian scheme, the first stage was a special course to produce instructors from the best of the practising teachers. These courses take place at RECSAM [45] which is working in close collaboration with the project. So far, a cadre of seventy-five 'RECSAM-trained' instructors has been produced under support from Unesco.

This leads to the second stage, the national workshops, and on-the-job training of a further 250 'nationally trained instructors'. The aim was to produce a total of 325 instructors by the end of 1983. They in turn will provide one-semester in-service/on-service courses to teachers in eighteen regional centres and a rolling number of sub-centres (thirty-six at any one time), this being the third stage. This ambitious scheme hoped to give courses to 6,000 teachers by the end of 1983; by mid-1981 about 800 teachers had attended the third stage of in-service on-service.

As in Peru, the emphasis has been on local activities taught largely by local staff for it is felt that the manpower resource must operate from locally based institutions and therefore is best recruited from the regions. The recruitment is more successful from the participants of the in-service/on-service courses, and the larger the number of participants the better the quality of the recruits.

The programme begins with a two-week residential course in which the teaching of the first half of the semester is prepared, including the practical work and assessment procedures which will be used. When the teachers go back to schools, instructors visit to help solve classroom problems, audio-tape lessons for further analysis and help for the forthcoming weekly 'on-service' meetings which are held during out-of-school hours on Saturdays. A provincial science supervisor co-ordinates the programme, liaises with headteachers, arranges meetings, distributes materials, etc.

Later in the semester, the teachers return for two more discrete full-time training periods, of about a week each, to prepare the final parts of the semester.

At present, the programme is not only on schedule but the organizers feel that the 'process of review of procedures and products at all levels continues to produce genuine and undeniable improvements in teaching techniques' [46]. It is this systems approach to the programme that impresses the onlooker. Figures 1–3 describe this approach.

Figure 1 describes the activities of the national preparation workshop (Stage 2). The starting materials and outputs are shown in circles.

Figure 2 shows the activities of the regional induction and in-service workshops. As in Figure 1, the starting materials and outputs are in circles. Note that the starting materials are the output products of the national preparation workshop.

Figure 3 describes the regional on-service activities and the regional and national evaluation activities. Again the inputs are in circles and are those developed from the regional induction and in-service workshops. The outputs from the on-service activities are, in turn, reviewed at the regional evaluation workshop and thence at the national evaluation workshop.

The materials are then edited and printed at the national preparation workshop prior to the next cycle when the semester's work is taught again. There is thus continual 'on-site' evaluation of materials by teachers at all levels of the scheme with a carefully provided feedback from national level to classroom and from classroom to national level.

There are many impressive aspects of both the Peruvian and Indonesian projects. First and foremost is the insistence that in-service training is of paramount importance. The second is the careful planning prior to implementation. The third is the way in which the national organizers have used international and bilateral assistance. In Peru, there has been help from Unesco and UNICEF and Hungary has helped with equipment. The Indonesian project has Unesco and UNDP assistance. The regional centre, RECSAM, is collaborating not only with the initial training programmes for the instructors but also in evaluating the programme itself. Two of the instructors have had two-month school attachment in Australia and further attachments in Australia and New Zealand are being planned. The Indonesians are calling on the expertise of a whole region in a very carefully planned way—from Penang to New Zealand—yet are equally carefully ensuring that it is an Indonesian project.

6.5 Courses for chemistry teachers leading to further qualifications

There are in increasing number of longer courses for experienced chemistry teachers at universities and colleges which can lead to a higher degree or diploma. For example, in the United Kingdom, ten universities and one polytechnic offer courses for science teachers. Three of those lead to the M.Ed. or M.A.(Ed.) degrees and have an option in science education. The other seven are run specially for teachers of

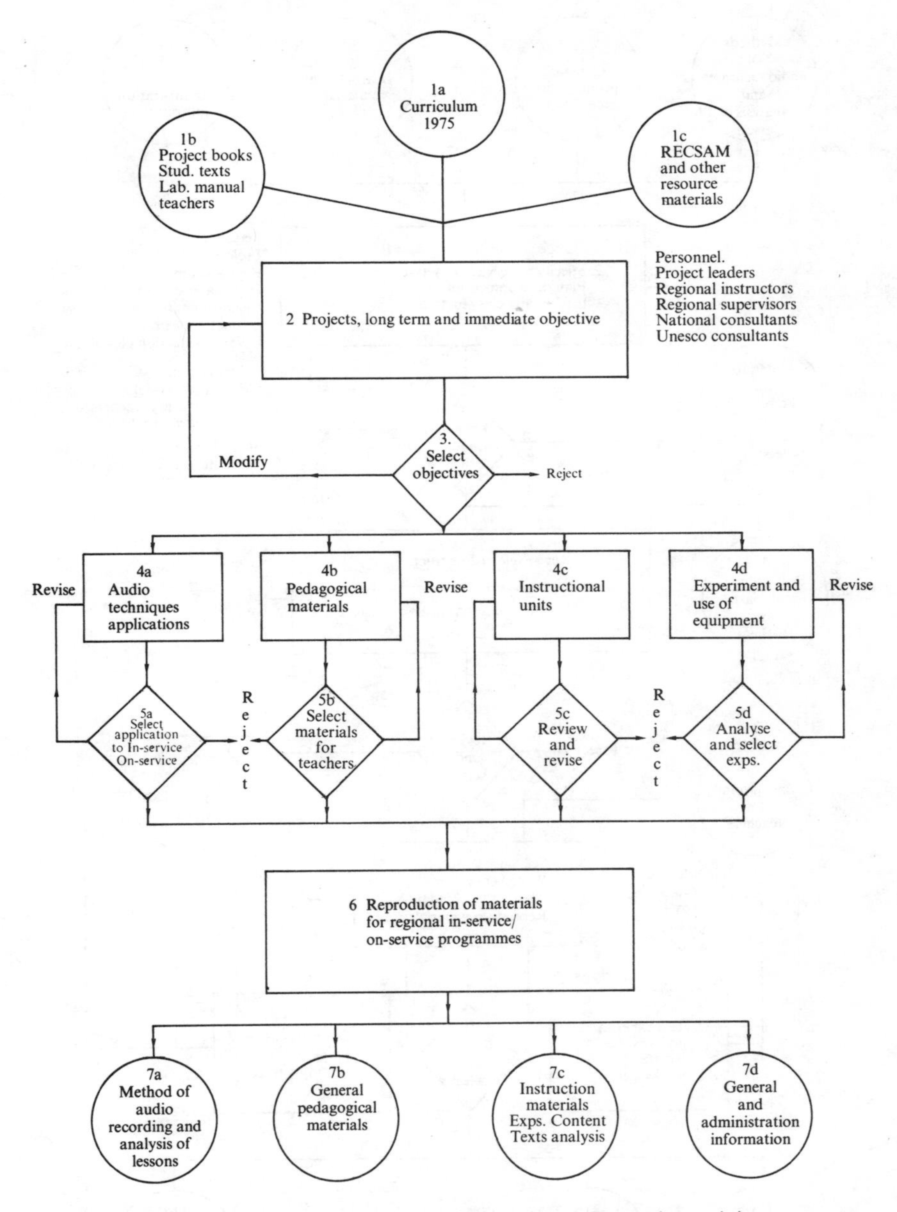

FIG. 1. Indonesia: national preparation workshop for in-service teacher training.

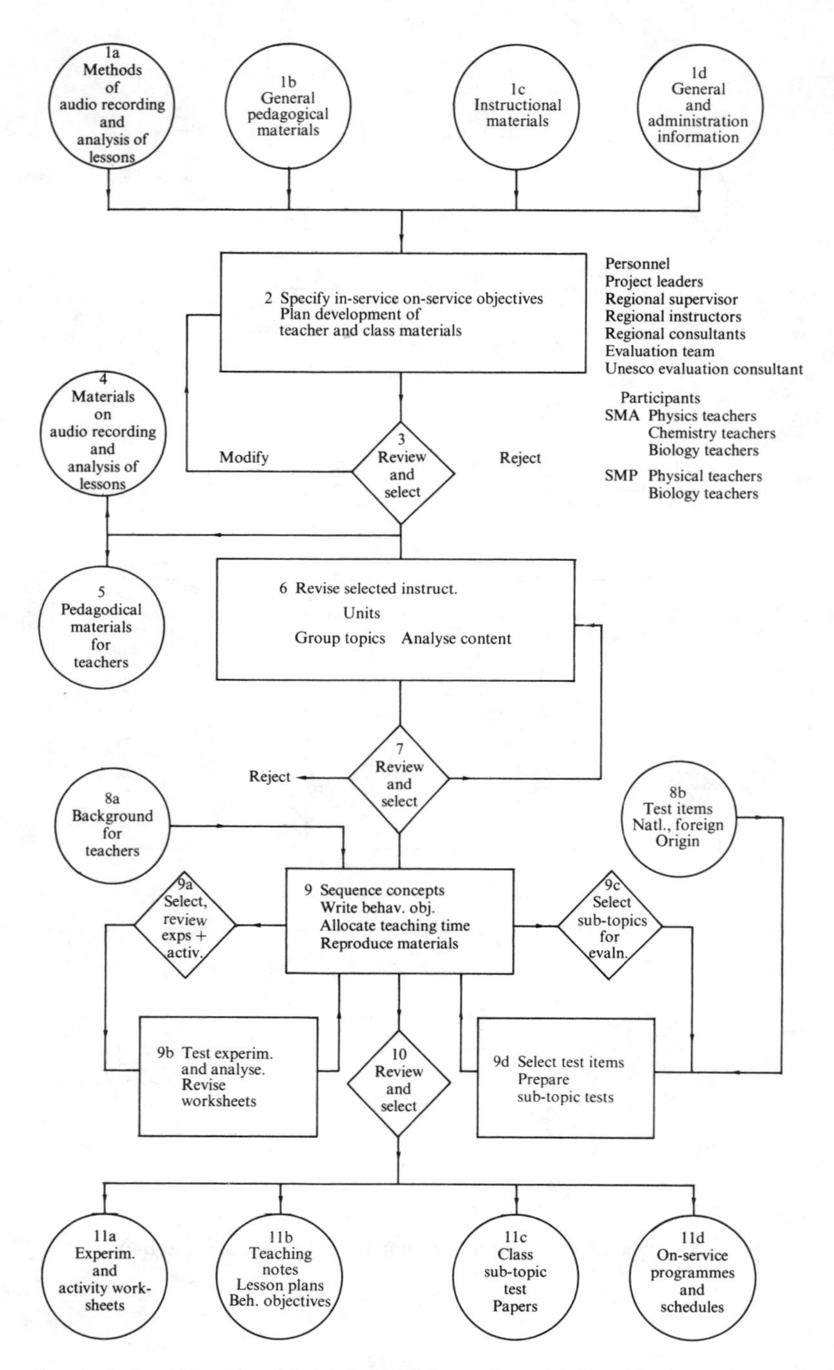

FIG. 2. Indonesia: regional induction and in-service workshops for in-service teacher training.

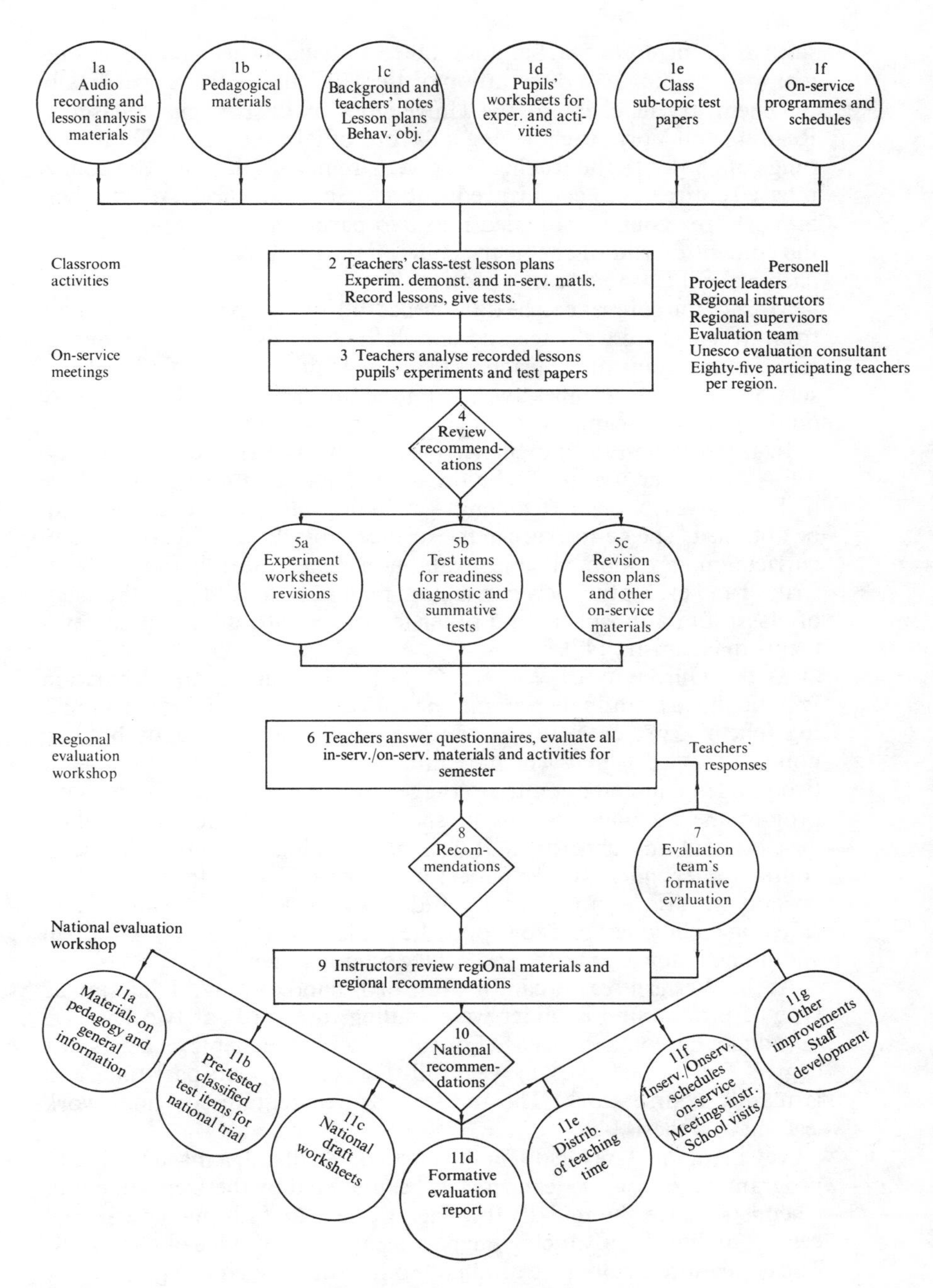

FIG. 3. Indonesia: on-service teacher training and evaluation.

science. In most, the teacher undertakes both classwork and some original work leading to a dissertation or thesis. Four are those specifically for chemistry teachers, at the Universities of East Anglia, Glasgow, Reading and York, each having a flavour of its own. They offer a wide range of choice to the teacher. At one extreme, at Glasgow, the course is based entirely on a chemical education research project. At the other, at York, the course is divided into two parts. In one, emphasis is on the content of school chemistry courses; the other part is devoted to a chemical education research project.

Most of the courses can be done either full-time or part-time; teachers from overseas are welcome and enrich the courses with their differing experiences. Many of the teachers from the United Kingdom come from schools hundreds of miles away from the university and therefore stay on the university campus in student accommodation for the course.

In Israel there are four higher institutions, the Hebrew University, Tel-Aviv University, the Technion and the Weizmann Institute, where higher degrees in science teaching—including chemistry teaching—can be obtained. The courses are in the science subjects, in education and in curriculum development and evaluation. The research ranges from writing and evaluating new curriculum materials, to studies of the effect of classroom atmosphere and other factors on attitudes and achievements in chemistry [47].

At the University of Western Ontario, the science and education faculties have a graduate programme entitled 'Master of Arts in Teaching (chemistry)', entrance to which is available to students holding honours science degrees who have taught at high school level for at least two years and intend to return to the secondary school system [48]. This programme includes courses in spectroscopy, statistics, curriculum development and experimental design, as well as a chemical projects course (independent study paper) and thesis. The main aims of the programme are: 'to up-grade the student's knowledge of chemistry and teaching techniques; and to supply the student with information of real and immediate value to the secondary school system'.

At the Western Australian Institute of Technology (WAIT) a similar type of programme is under way, leading to awards at two levels, a 'Graduate Diploma in Science Education' and at higher level a master's degree. The former takes one year full-time, or two years part-time, and is mainly course work. The master's degree requires original work leading to a thesis [49].

At Chiangmai University, in the north of Thailand, a master's degree programme for science teachers has been offered by the Department of Chemistry since the mid-1970s. This is one-year full-time course and centres mainly on advancing the participants' knowledge of chemistry. The education faculty give seminars to broaden the course [50].

In the United States, both master's and Ph.D. programmes are available through the Science Education Centres or Curriculum and Instruction departments of most universities. They usually require a major component of the course work to be done in science departments with the thesis or dissertation related to science (chemistry) teaching. Advanced training in statistics, computer science and science education are included in the course. Financial support comes from a variety of sources: government or local school system grants, teaching assistantships at the university, or personal resources of the candidates.

These courses are expensive, because the teacher usually needs a salary while away from school and the school needs a temporary replacement. This is possible, for example, on a relatively modest scale, in the United Kingdom where perhaps fifty teachers a year have this opportunity. These teachers have to apply to their local education authorities for secondment. In the Philippines, teachers with seven years' experience are allowed one year off at 60 per cent salary, an opportunity given to few teachers anywhere else in the world.

Nevertheless, further education opportunities have the following invaluable characteristics which are impossible to quantify. They are for experienced teachers and enable them to meet and discuss mutual problems; they give the teachers, over an extended period, new information about chemistry and chemical education and the chance to discuss the implication of these in terms of their teaching; they also allow teachers to become involved in chemical education research. Such research, conducted by practising teachers, has immediate use for the teachers themselves and potential value for a wider audience.

References

1. F. A. Diesterweg, *Sämtliche Werke*, Vols. 1–12, Berlin, 1956–72.
2. D. Hamcheck, *Phi Delta Kappa*, Vol. 50, 1969, p. 341.
3. A. V. Bogatski, 'Universitet i zittja [The University and Life]', *Z. autitorii—v zittja*, pp. 26–46, Odessa, 1972.
4. L. Bownan, *Journal of College Science Teaching*, Vol. 1, 1971, p. 47.
5. See for example, articles in A. Kornhauser, C. N. R. Rao and D. J. Waddington (eds.), *Chemical Education in the Seventies*, 2nd ed., Oxford, Pergamon Press, 1982.
6. C. L. Habraken, N. H. Velthorst and J. Hoekman, in ibid., p. 199.
7. H. Schindlbauer and T. Schönfeld, in ibid., p. 44.
8. G. Brunow, I. Kalkku, P. Malkönen, E. Salimen and A. Vähäkuopus, in ibid., p. 98.
9. A. Lourdusamy, *Journal of Science and Mathematics Education in Southeast Asia* (Penang, RECSAM), Vol. II, 1979, p. 35.
10. S. Tunde Bajah, in Kornhauser et al., op. cit., p. 215.
11. S. P. Koh and S. E. Loke, in ibid., p. 191.
12. C. K. Boey, *Malaysian Journal of Education*, Vol. 13, 1976, p. 93.

13. J. Zyka, in Kornhauser et al., op. cit., p. 76.
14. H. Bremer, G. Kempe, D. Krysig and D. Wagner, in ibid., p. 118.
15. D. Cros, *Etude mondiale portant sur les différents systèmes de formation initiale et permanente des maîtres scientifiques de l'école primaire et de l'école secondaire*, Paris, Unesco, 1980.
16. Kornhauser, op. cit., p. 292.
17. University of York, Heslington, York YO1 5DD (United Kingdom).
18. N. A. C. Gunatillake, in Kornhauser et al., op. cit., p. 236.
19. University of East Anglia, Norwich NR4 7TJ (United Kingdom).
20. D. Cros and M. Maurin, *International Newsletter on Chemical Education*, Vol. 8, 1978, p. 14.
21. *Guidelines and Recommendations for the Preparation and Continuing Education of Secondary School Teachers of Chemistry*, Washington, D.C., American Chemical Society, 1977.
22. B. E. Woolnough, *Education in Science*, Vol. 90, 1980, p. 27.
23. *Guidelines and Standards for the Education of Secondary School Teachers of Science and Mathematics*, Washington, D.C., American Association for the Advancement of Science (AAAS) and National Association of State Directors of Teacher Education and Certification, 1971.
24. *ASTEP—A Project in Teacher Education*, Clayton, Victoria (Australia), University of Monash, 1976.
25. The Thai Science Teacher Education Project. Private communication from Dr Nida Sapianchai.
26. 'Science Teacher Education Project' series, Maidenhead (United Kingdom), McGraw-Hill, 1974. Among the books in the series are: J. Haysom and C. Sutton (eds.), *Theory into Practice—Activities in School for Student Teachers*; C. Sutton and J. Haysom (eds.), *The Art of the Science Teacher*.
27. Teacher Education Project (Universities of Nottingham, Leicester and Exeter). The following are among the materials on classroom skills being published by Macmillan, London (1981 onwards): *Management and Control* (E. C. Wragg); *Mixed Ability Teaching* (T. Kerry and M. K. Sands); *Teaching Bright Pupils* (T. Kerry); *Teaching Slow Learners* (P. Bell and T. Kerry); *Effective Questioning* (T. Kerry); *Explanations and Explaining* (G. Brown and N. Hatton); *Handling Classroom Groups* (T. Kerry and M. K. Sands); *The New Teacher* (T. Kerry). The following is also being published: *Teaching Science* (M. K. Sands). Further inquiries to: School of Education, University of Nottingham, University Park, Nottingham NG7 2RD (United Kingdom).
28. A list is given in a pamphlet *Teacher Education Project* which can be obtained from the address given in ref. 27.
29. For further details, write to the Institute for the Promotion of Teaching Science and Technology, 924 Sukhumvit Road, Bangkok 11 (Thailand).
30. Y. Oon-Chye et al., *The Curriculum Development Centre of Malaysia*, p. 42, Bangkok, Unesco Regional Office for Education in Asia, 1977.
31. A. V. Bogatski, 'Sovremennyj universitet [The Modern University]', *Pravda*, No. 61 (19204), 2 March 1971.
32. H. J. Becher, *Naturwissenschaften im Unterricht. Physik/Chemie*, Vol. 26, 1978, p. 150.
33. K. V. Sane, ref. 5, p. 132.

34. J. T. Shimozawa, ref. 5, p. 166.
35. *Innovation in In-service Education and Training of Teachers: Japan*, Paris, OECD, Centre for Educational Research and Innovation, 1976.
36. D. G. Chisman, private communication.
37. L. Pataki and M. Palfalvi, ref. 5, p. 128.
38. *Chem Thirteen News*, Vol. 51, 1973, p. 15.
39. H. D. Gesser, see B. T. Newbold, ref. 5, p. 64.
40. *Chem Thirteen News*, Vol. 63, 1974, p. 13; Vol. 73, 1975, p. 14.
41. A. Guerrero, ref. 1, p. 30.
42. Y. Oon-Chye, et al., ref. 30, p. 23.
43. *Eighth Report for the International Clearinghouse on Science and Mathematics Curricular Developments*, p. 90, Science Teaching Center, University of Maryland, 1972.
44. C. A. Quiroz Peralta, *Programa nacional para el mejoramiento de la enseñanza de las ciencias. Conference on Integrated Sciences Worldwide—Achievement and Prospects*, London, 1978.
45. RECSAM: Regional Centre for Education in Science and Mathematics, Penang (Malaysia).
46. Improvement of Science and Mathematics Teaching in the Secondary General Schools. A UNDP/Unesco Project. Department of Secondary Education, Jakarta (Indonesia). Project director: Dr B. Suprapto.
47. D. Samuel and A. Hofstein, ref. 5, p. 149.
48. R. R. Martin, *Chem Thirteen News*, Vol. 88, 1977, p. 14.
49. D. J. Bond and J. R. de Laeter, *The Australian Science Teachers Journal*, Vol. 24, 1978, p. 69.
50. M. H. Gardner, private communication.

7 Current research in chemical education

P. J. Fensham

Many teachers, teacher trainers, curriculum planners and others interested in education find it difficult to obtain up-to-date information on current research in chemical education. The work is scattered over many journals, books and proceedings of conferences. Moreover, it is often difficult to evaluate this research work in terms of correct procedures for gathering the evidence and conclusions from the evidence.

In this chapter, Professor Fensham has outlined some of the more important areas in chemical education research and has given full references to the work.

As he says in his introduction, he has written the chapter from the viewpoint of the teachers rather than that of a research worker. Indeed, this is what he was asked to do. He was also invited to give a personal view of the field. Thus the chapter gives an opportunity to explore many of the most fascinating problems that face us both when we teach the present curricula as well as those involved in developing new courses and new techniques for teaching.

7.1 Introduction

In choosing a framework for this chapter I have been guided by the emphasis of the book as a whole. Its title, *Teaching School Chemistry*, suggests that an account of current research in chemical education from the viewpoint of a teacher rather than that of a researcher should be presented here. Such a perspective seems more than justified in a book that has as its target audiences teachers, teacher educators and educational administrators concerned with the support for chemical teaching. It will be argued shortly that this choice is more than an alternative way of presenting the same material.

However, it is first necessary to say something about the sources that have been used for this article. In other words, what has been regarded as research in chemical education. Even a cursory perusal of the published papers in chemical education (or attendance at any conference under this title) brings home that there is more assertion of personal opinions, practices and perspectives than there are *reports that gather external evidence to answer questions about chemical education or that set out to portray situations in ways that enable others to evaluate them and to draw their own conclusions.*

Because these reports have involved the thoughts or experiences of their authors and because they often involve a great deal of developmental work, they have sometimes been included as research. Kornhauser [1] did so in a broad commentary on the field.

I have chosen to be more restrictive and confine myself to reports that fit the definition of research that is embodied in the italics above. This is not to denigrate the many reports or opinions about chemical education, or the accounts of personal approaches to teaching or laboratory usage. Well-presented 'slogans' and claims, and the experiences of 'successful' teachers have always been influential on classroom practices. They are simply not the subject-matter of this chapter.

An example of the effect of this decision is the almost complete exclusion of the topic of educational technology in chemistry teaching. This has been reported in the proceedings of the major IUPAC conference in Madrid [2] and by Kornhauser, in her recent paper [1]. My point is that the reports of these innovations rarely meet the criteria for research, as explained above. They often involve chemical education beyond the school level. Extrapolation of their findings to school classrooms are so uncertain (because of the enormous problems, at that level, of existing traditions for teaching and the availability of materials and equipment) that they have been omitted.

7.2 A teacher's perspective of research

If the chapter is concerned with research on questions that have a direct interest to the teacher of chemistry, it is necessary to consider what does interest ordinary classroom teachers. Lundgren, a Swedish researcher, has introduced the idea of a classroom teacher operating within a 'frame' [3]. By this he means that there are many aspects of the teachers' task that are beyond their capacity to alter. For example, teachers are often unable to choose the textbook; this is done for them. Again the availability of a laboratory or certain apparatus or chemicals are system or school matters, as is the time available for chemistry per week, or how that time is distributed. Even acknowledging the dependence of teachers' 'frames' on their particular social context, it is surprising how much published research is concerned with issues that lie beyond the frame of the average classroom teacher in the education system under consideration.

A prime example of such 'beyond-the-teacher' research is the widely discussed and reported issue of a core curriculum for school chemistry. A number of approaches have been used in research on this issue that involve philosophical, sociological and psychological arguments. However, in the long run the decisions about what is to be the chemistry of the curriculum are taken for educo-political reasons and by powerful groups that certainly do not include the mass of ordinary classroom teachers. A small survey of chemical educators in a number of different countries in the course of preparing this paper confirmed the general irrelevance to teachers of the question, What chemistry shall I teach at what level? The chemistry to be taught is given for most teachers in the world. It is not part of their *problématique* and so this research is excluded here.

On the other hand the same survey did confirm that teachers are interested in the following:

How do pupils learn the sort of facts, concepts, principles and applications that make up chemistry?

What teaching approach is good for specific topics?

How can practical work and theory be best related?

How can chemical learning be made more relevant?

How do I motivate the uninterested pupil?

How can I relate chemistry to its application and its effect in society?

Not all of these have answers yet in research but they all do lie within a classroom teacher's frame, and this chapter concentrates upon these questions and problems.

Two other popular areas of research beyond most teachers' frames are those concerned with curriculum development and curriculum diffusion. These topics have been particularly popular for research in English-speaking countries such as the United Kingdom, the United States, Canada and Australia. These countries have rather unusual educational contexts when considered on the world scene. Each now has situations where there is some degree of freedom for schools (and possibly for individual teachers) to exercise choice over the content of their chemistry curriculum or on the materials (texts, etc.) they will use in its presentation. The diffusion and implementation of competing curricula are to these limited extents real issues in these cases. Since these countries happen also to be ones where there are more career opportunities for academic science educators, literature reporting these studies is thus widely available (e.g. Tamir et al. [4]).

On the world scene such curricular freedom at the level of the school or the teacher is so small that these research topics have no general meaning, and are not reported here. Even in the few countries that generate this type of research, the ordinary teacher is conscious of little of the freedom that underlies the issue.

Another consequence of the teacher perspective is that it requires a decision about the major dilemma that confronts so much educational research, namely, the gap between the findings of a great deal of research and the ability of practising teachers to respond to them. Clearly, given my chosen perspective, the research to be reported should concentrate on studies where this gap is likely to be least. This is not as easy to apply as one would like. Much well-conducted research has appeared to concern itself with practical classroom questions but, when finalized, the results have often not been useful to teachers who have received them.

A number of features that contribute to the gap can be identified. These are (a) the degree of teacher control; (b) the way researchers look at the problems; (c) the generality of the educational situation; and (d) the nature and communication of the findings.

Stated briefly, the resolution of this dilemma used in this chapter lies in my belief that there is a greater conjunction between teacher usefulness and research that has a case study or clinical character rather than that which is based on broad surveys and is primarily quantitative in character. These distinctions are now elaborated to show how the decision is related to the four features just listed.

The last few years have seen a lively debate on the fundamental purpose of much research into educational practice. Stake [5], an initiator of this debate, posed it in terms of Description versus Analysis, and Fensham [6] has spelt this out for science education. The dominant approach to classroom research in many countries has been analytic and psychometric. An educational situation has been perceived as dependent on certain features that could be isolated from other aspects. Furthermore, they could be abstracted from their particular context and conceptualized as variables that had influence in a population of which the original situation was now representative. By operationalizing these variables, they could be measured, and their values checked for consistent relationships with each other. Such observed relationships became the basis for explanations of the situation under study. This approach basically viewed education as if it were a social situation that could or should function by rational decisions using these analytic and atomistic relationships. The particularities of teachers, or even of schools, were largely lost in the processes of abstraction and generalization. The research into science education by the International Educational Achievement (IEA) are very obvious examples and are well known in many countries.

Stake and an increasing number of other scholars have juxtaposed another approach to research into the same situations. In it, the researchers seek to describe and round out a situation by including much of its context. They set out to portray the experience (the behaviour plus the feelings) of the participating teachers and students, etc. They seek to identify and illuminate issues or aspects of the situation where there are different perspectives operating (i.e. not agreed relationships). Their aim is not to explain, but to understand. Education for them is a process of informed judgements by particular persons in unique situations. They are portrayers of an educational situation for its participants or for people in like situations, rather than external interpreters of it for other researchers.

Both approaches have their strengths and weaknesses, but the second (or case study as it is often called) has more natural affinity with teachers than does the more traditional approach.

For example, there has been (and still is) a great deal of research into the interactions between teachers and their students that appears in researcher language (such as categories of classroom climate or of teacher questioning) and in statistically stated findings. Both of these are barriers to teachers who find it hard to identify the day-to-day variations in their classes with these high-level generalizations.

On the other hand, the reports of case studies often make use of a lot of direct quotations from teachers and their students. The fullness of these accounts and the greater detail of their findings also make them easier for teachers to read. They cannot be claimed to extend beyond the situations studied, but they may still be useful for teachers who find the findings relate in some way to their own situations.

Accordingly, considerable attention to these more clinical sorts of studies has been given in preparing this chapter because they do appear to be more directly of use to teachers.

The first teacher question to be addressed is the central one confronting all chemistry teachers: How should the knowledge of chemistry be taught?

After this major section there are smaller sections devoted to: What knowledge does a teacher need? How can chemistry teaching be more relevant/interesting? How can one make effective use of the laboratory and relate practical work to theoretical learning? How can teachers be 'researchers' in their schools?

7.3 How should the knowledge of chemistry be taught?

The central problem of chemical education is how to teach a highly developed body of knowledge so that it is learned in a meaningful way (i.e. not simply, by rote). It is heartening to be able to report that there is now much more concerted research activity into this central problem than there was ten or even five years ago.

Furthermore, in the last five years there has been a coming together of a number of hitherto separate approaches to the study of learning in the sciences (like chemistry). Some features of this contemporary research are, I believe, most exciting and significant for teachers.

Only five years ago it was necessary to report separately the contributions of science education researchers who were pursuing the theoretical ideas of say Gagné, Piaget and Ausubel. Now we find some of these scholars asking questions and using methods of data collection which overlap, and which are proving extremely fruitful to each other and to teachers. Evidence of this convergence is the designation by the National Science Foundation in the United States of cognitive structure as a priority field for research. An international seminar on Cognitive Development Research in Science and Mathematics was held at the University of Leeds in 1979 [7].

These newer research studies are exciting and significant for teachers on several grounds. First, so many of their findings are directly relevant to both teachers and pupils. Second, they raise a particular learner aspect in science teaching that has been ignored in much of the curriculum development and teaching approaches of the last twenty years. Finally, a number of their methods of data collection (and even analysis) could be used by teachers in classes, if they are convinced of their usefulness to the tasks.

Chemical learning has been the context of a number of these studies, but at this stage more of the work has involved physics and biology topics. There is thus much to be done in chemistry and there are great opportunities for both researchers and for teachers.

This broad group of studies will be discussed later under the following headings: changing questions; changing methods; changing implications. However, before this major discussion, a briefer account will be given of some of those still useful ideas from the research of Gagné, Piaget and Ausubel that were influential on a number of new science curricula between 1960 and 1980, and which can still contribute to the teaching of chemistry, whatever the curriculum confronting teacher and pupil.

7.3.1 *Learning hierarchies*

Gagné, in his view of cognitive learning, had stressed the importance of the fact that sequentially structured content like chemistry can only be meaningfully learnt if each preceding concept or intellectual skill is properly acquired by the learner [8]. His famous question for teachers who wish to plan their lessons is: What will the learner need to know or do in order to deal with the new learning I am now introducing?

Repeated application of this question leads to a pine-tree-like set of verbal knowledge and skills with the desired new learning at the top. Gagné and his followers refer to these as learning hierarchies. White, in particular, has developed the procedures for establishing a hierarchy and checking its validity in the learning situation [9]. An example of a validated hierarchy for gas law calculations is given in Figure 1 from the research of Sharma [10].

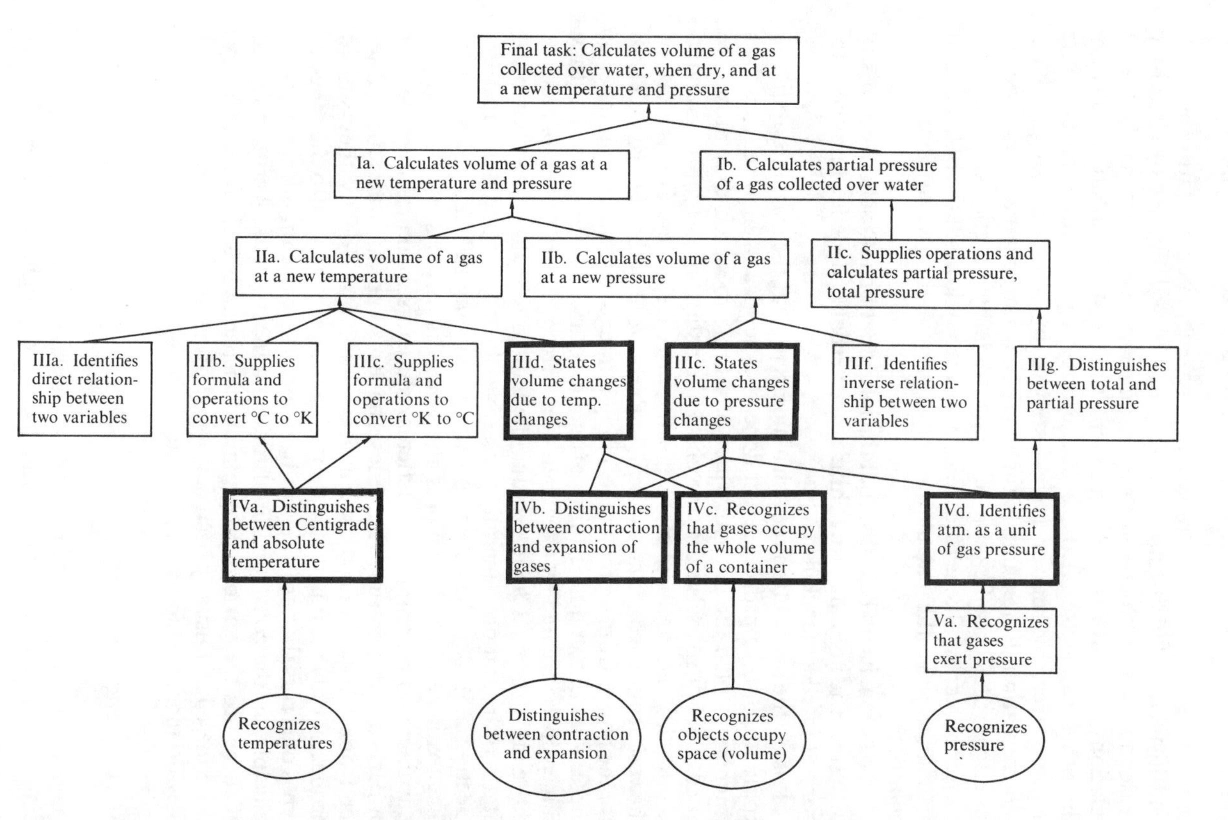

FIG. 1. Learning hierarchy for the task of calculating the volume of a gas collected over water, when dry, and at a new temperature and a new pressure (verbal knowledge: heavier boxes).

The verbal knowledge includes all the definitions that are very common in chemistry, e.g. oxidation is the loss of electrons, or aldehydes contain a $—C\begin{smallmatrix}H \\ O\end{smallmatrix}$ structure. It also includes all the particular names of elements (sodium, Na), compounds (benzene) and apparatus (burette), and historical names that are now part of the language of chemistry (Erlenmeyer flask, the Avogadro number) as well as stated laws and relationships that may be in mathematical form. Intellectual skills, on the other hand, are all those operations on pre-existing parts of the learning tree, by rule or logical step or by numerical or graphical representation. For example, they may be arranging metals in order of reactivity, the classification of compounds by functional group, the conversion of an alkene (olefin) into a carboxylic acid, the balancing of an equation or the calculation of a gas volume when temperature changes.

In passing, it can be noted that the introduction into school chemistry of systematic nomenclature shifts some verbal knowledge into the category of intellectual skill.

It is on the acquisition of such a hierarchy that the learner's ability to apply or to solve problems will depend. The idea of hierarchy enables teachers to plan the particularities of a topic to which they must give attention. It also assists them to check their starting assumptions and to identify learning failures more effectively. A Readiness Test is a test of the verbal knowledge and the skills at the bottom of the hierarchy. These are the points from which the teaching is intended to start. A number of these types of test do exist for particular courses in chemistry, but they can easily be created by a teacher, either from chemical item banks like those in Australia, Canada, India, the United Kingdom and the United States or from more generally accessible old test or examination papers (Section 5.5). Another check on where learning failure occurs is by a diagnostic test which explores the component parts of the terminal learning, and pinpoints the particular student's weakness to some part or parts of the hierarchy. These tests have a somewhat different character from most chemistry examination questions. Ideally they should probe not only the particular steps in the hierarchy but also the range of types of errors that are likely to be made. The work on misconceptions is helpful in the construction of these diagnostic tests. Another source of these are the various programmed instruction texts for chemistry, since these have usually been developed using the hierarchical idea and they do include step-by-step testing.

7.3.2 *Stages of cognitive development*

Piaget, from his monumental clinical studies of children's thinking, had long introduced the ideas of intellectual or cognitive stages (for example, concrete reasoning and/or abstract reasoning) through which children develop with age. Their significance for science curricula and science teaching has only been recently recognized. This occurred as the people working in curriculum development became aware of the fact that many of the tasks Piaget had used with children were ones which involved ideas (like conservation of matter, time, velocity, mixing of chemicals, etc.) that are part of science subjects at school.

The neatness of his idea that these stages were generally restricted to certain ages offered an enormous attraction to some of those wishing to reform science teaching. Their dictum became: Match the curriculum demand to the stage of ability of the learner and then provide a rich set of experiences that would assist and promote the development that was required for the learning in later years.

Whole science curricula were developed in the early 1970s (e.g. Science 5–13 in the United Kingdom and ASEP in Australia) on these research ideas. The curriculum materials contain a number of chemical topics and there are suggested approaches to their teaching that are consistent with the intended level of conceptual learning. For example, in ASEP a unit on Pigments and Acidity was associated with the so-called concrete stage of development, a unit on Metals with the formal or abstract stage, and one on Atoms and Molecules with a transition between these stages (Section 5.3). If the three units are compared, the suggested approaches and the learning objectives do show the increase of expectation of abstract or formal thinking.

Established chemistry courses with their own content sequences (dictated by some decision-makers' views of the logic and nature of chemistry as a discipline) are much more difficult situations in which to apply these ideas. In some courses, massive restructuring of the year-by-year topics seems to be required, and this can only occur slowly and with very strong evidence or other pressure. Shayer [11] in the United Kingdom has been a major figure in the extension of Piaget's ideas to this task of curriculum reform, both beyond and within the classroom. He has developed a series of instruments for large group or class application that enable an assessment of the pupils' *stage of cognitive development* to be measured. He also has a research procedure for the content analysis of syllabuses and curricula that enable their topics to be associated with a particular Piagetian stage. These procedures are trenchantly criticized both by other Piagetian researchers and by non-Piagetians. Nevertheless, Shayer [12] has been able to gather impressive arrays of correlated findings on these stage data and the achievement of students on test items for various topics in the science courses at middle secondary (age 11–16) schools. The general basis of this approach to curricula and to the varied abilities of their pupils does make sense to teachers when presented directly without undue jargon.

He is now engaged on the more difficult tasks of making these matching procedures available to the ordinary classroom teachers. Whether this is possible or even useful, given the rigidities of most curricula, is an arguable point. The Science 5–13 and ASEP courses are not encouraging despite their occurrence in situations where teachers had unusual curricular freedom. Espousal of the Piagetian strategies was not a feature of the uptake of these curricula by teachers.

There is, moreover, a likelihood that the response of teachers to the apparent rigidity of the Piagetian stages will be the negative ones of not trying to teach certain pupils important topics in chemistry or of even excluding them altogether. Both Jenkins [13] and Dawson [14] have recently joined the recurring voice of Novak [15] concerning this danger. These critics do not reject many other aspects of Piaget's work and indeed all now embrace the clinical methods of research he had to champion so long alone.

7.3.3 *Advanced organizers for learning*

If Gagné's emphasis was on what was to be learned, Ausubel's concern was with the process of learning itself [16]. He used the term 'subsumption' to suggest how the existing knowledge of the learner interacts with new knowledge. Ring and Novak [17] showed that learners in chemistry with differing amounts of what they called factual knowledge (non-subsuming) and concepts (subsuming) acquired new learning in the way the theory predicted. Fensham also obtained supportive evidence for the general ideas from chemistry pupils in England [18].

Ausubel [16, 19] suggested that teaching strategies that alert learners to the prior knowledge they will need to use in the new learning will be likely to optimize meaningful learning. This introductory action by the teacher he calls an *advanced organizer*. In practice, the organizer might be a piece of prose, a film, a bench demonstration, a carefully prepared short teacher talk, or a highly controlled discussion. It will explicitly remind the learners of general ideas from their earlier learning that will be useful to relate to the new knowledge which itself should also be briefly summarized. This introduction by the teacher sets the stage for the detailed presentation of the new knowledge. According to Ausubel's theory, the teacher should try following a *psychological* order (i.e. based on the learners' thinking) in this presentation, rather than a *logical* order (based on the knowledge itself, or on the teacher's understanding). The contrast with Gagné's *logical* hierarchies is striking.

A great deal of research on the use of advanced organizer has been done and West and Fensham [20], Lawton and Wanska [21] and Koslow and White [22] have provided useful summaries. Many teachers on occasion naturally use introductions to a new topic, so the general idea of an advanced organizer is not a strange one. It is, however, still not the usual approach. This is influenced by the pattern in textbooks with their emphasis on logical sequence and their summaries at the end of the chapters. The more subtle features of the advanced organizer idea could enhance the effectiveness of much teaching if they were more widely understood by teachers and textbook writers.

7.3.4 *The convergence of cognitive research*

I now turn to the ways these and other research approaches have been converging.

Changing questions

In the mid-1970s Gagné and White [23] began to turn their attention to the role of memory in the learning of new cognitive material in science. This brought them into interaction with other cognitive scholars such as Wittrock [24], Shavelson [25] and Champagne et al [26]. It also brought them into the same arena of research as the Ausubelians since it was now no longer enough to say that the required prior knowledge exists. How it is stored? Why it was so stored? and How can it be recalled for use?

These were all important questions.

Piagetians like Cawthron and Rowell [27] began to try to relate the cognitive stages of science learners and their evolution to the historical development of scientific ideas themselves. Likewise, Dierks [28] and Minssen [29] take seriously the historical aspects of particular chemical concepts. In this quest, they are in the same field as Lybeck [30, 31] and others like Champagne [32] and Jungwirth [33] who were finding problem-solving reasoning in pupils that corresponded more with earlier (Aristotelian, Newtonian, alchemical, etc.) presentations of science than they did with the textbooks they were reading or with the statements of their teachers. They were also, by associating the specificities of learning with the philosophy and history of science, following an influence that had already profoundly affected Novak [34]. Easley and Driver [35], also from the Piagetian orientation, were seeking ways of breaking out of the rigidities that had developed in the clinical interviews for Piagetian tasks. They had been struck by the likelihood that pupils' responses may be wrongly classified as lacking formal reasoning unless a more thorough exploration of the bases for their answers was undertaken. Delacôte [36] in France and Schaefer [37] in the Federal Republic of Germany, were asking similar questions in seeking to find a sounder base for science programmes for the middle years of schooling.

Among the Ausubelians, Nussbaum and Novak's [38] study exemplifies the move of this group to more profound questions about the nature of the learner's existing cognitive structure. Only vague support could be expected for Ausubel's dictum to teachers—'Ascertain what the learner already knows and teach accordingly'—unless there were better methods to ascertain and more clearly describe the existing cognitive knowledge of learners. In Australia, West and Fensham were now likewise similarly engaged [39, 40]. This greater focus on learners and on the mapping of their knowledge brought interaction with all the sorts of research described in the two previous paragraphs.

Another group of researchers could now also be recognized as being in the same set of questioning orbits. They had come more directly from science backgrounds than from learning theory. Driscoll [41], Johnstone and Handy [42] and Gilbert and Osborne [43] in their respective countries had been exploring what were then called 'misconceptions'. For example, Driscoll had used a form of diagnostic testing to explore the sorts of misconceptions significant groups of learners held about chemical equilibrium—a study that was replicated and extended by Johnstone, MacDonald and Webb [44]. Some of these scholars had found the multiple choice, paper and pencil type of testing to be restrictive and they began to explore pupils' responses more freely and to a much broader range of scientific situations.

Changing methods of research

The basic change in the methods of all these researchers is the inclusion of clinical interviewing as part of their data collection. Prior to the mid-1970s almost all of these scholars, except the Piagetians, relied on paper and pencil procedures to obtain data and large samples of respondents were always a target in their studies. They are now much more concerned with the quality rather than the extent of their data, and clinical interviewing (in a great variety of forms) has become, along with other tests and observation, a really basic part of their research methodology (Sutton [45]). The work of Pines [46] is probably the extreme form of this new enthusiasm among science educators for clinical interviewing and his methods and those of others like Brumby [47] and Tiberghien [48] have been recently reviewed by White [49].

The one-to-one questioning that is characteristic of an interview is a commonplace part of the daily transactions between teachers and their pupils. Accordingly there is now a real possibility that some of these techniques for data collection could, if useful, find their way into the regular repertoire of teachers. Two of them will thus be described in some detail.

Gilbert and Osborne [43] have developed the interview about instances approach. (This is quite similar to the Description of Events (DOE) method of Champagne and Klopfer [32].) For a particular word or concept, e.g. 'acid', 'change of state', 'chemical reaction', up to twenty familiar situations depicted by line drawings on cards are presented to the pupil. Some of the situations present an instance of the scientific concept and some do not. Children are asked, for each situation in turn, whether they consider it an instance or not and the reasons for their choice is then elicited. The interview is conducted in a way that encourages pupils to ask questions, and to clarify perceived or actual ambiguities before answering. The tape-recorded interviews are then transcribed for coding and analysis.

Concept mapping is another now common method. Shavelson [25] and Preece [50] used a paper and pencil approach based on word association responses from pupils. Novak [51] and others have worked in interviews to record the ways pupils see the concepts of a lesson or block of lessons as interrelated. The main concepts are given in a list or printed on a set of cards. The pupil is then encouraged to arrange them along a dimension such as 'general' to 'specific', or to arrange them in some way that makes sense on a large sheet of paper.

When an arrangement has been established, the pupil is encouraged to try to state the relationship between the individual concepts that are near each other or far apart. While teachers often find it challenging to think of the knowledge they are presenting in this way, learners fairly readily undertake the mapping stages and even become quite adept at identifying the new concepts in a segment of new learning.

A more elaborate technique that combines the psychology of learning with the philosophy of science is the use of Gowin's V. Gowin [52] developed his V (see Figure 2) as a heuristic device in order to expose the way the theories and concepts we hold in our minds interact with the events or objects we observe, and with the record-making (about the events or objects) we undertake. Like Karplus [53] and his Science Curriculum Improvement Study (SCIS) team in the 1960s who had tried to combat the fashion for using the idea of discovery learning in science and science education, Gowin and Novak have recently re-emphasized that *concepts (regularities in events or objects) are human inventions*.

The point of the V represents where any rational process in chemistry should begin. In chemical education, the substances, their properties, their reactions, the phenomena in which they occur, etc. would be at the point of the V. On its left-hand side are the thinking elements that are generally (by common consensus) or specifically (by particular persons) related to these events or objects. These concepts, principles and theories are all inventions; nothing on this side is discovered like gold or oil. On the right-hand side, there are methodological procedures to use on the events and objects and the products of their use. These can include experimental methods, methods of transforming data, and the value assumptions that encourage this approach and discourage others. These also are human inventions, but they are syntactical as far as the final meaning of the knowledge is concerned, rather than substantive.

Between the sides of the V are questions or problems, the answering of which requires active interplay between the sides and the bottom of the V. In their use of the V in clinical interviewing, these researchers have found that both teachers and pupils are often confused about its lower

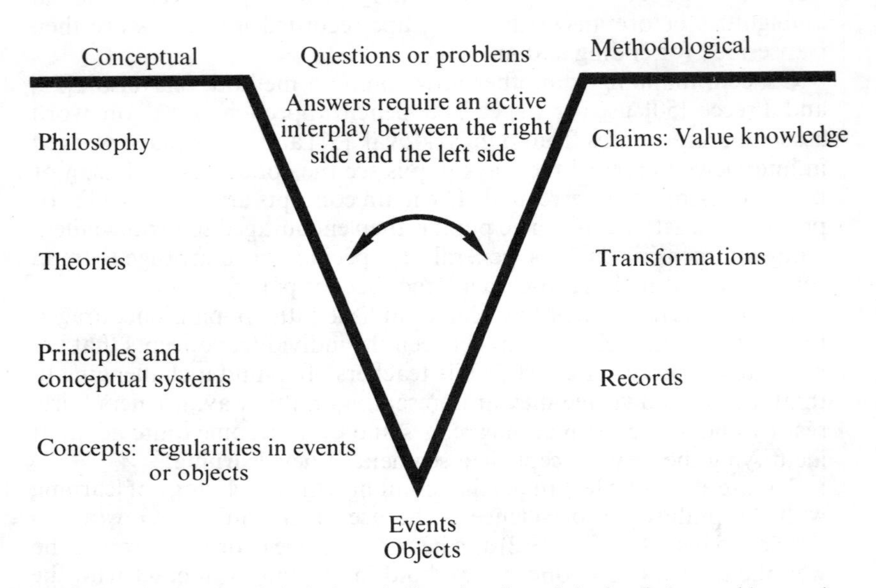

FIG. 2. Gowin's V: a heuristic device to indicate how knowledge in relation to objects and events is created.

elements—the concepts, the events and objects and the records. The V has the advantage of providing a locating structure that enables these confusions to be identified and sorted out.

The V has been used with pupils after a brief introduction to the idea of concepts as invented pieces of knowledge. An experiment, demonstration or set of observations is presented and the V is then constructed. First the pupils have to identify the events or objects under study. Then they have to state the concepts that are needed to make sense of the events, or were used in the records that were made. Principles are introduced as relationships between concepts and their role in guiding observations, record-making and record transforming is explored. Figure 3 shows the outcome of the use of the V to separate the components to be learned in a chemistry lesson in which students had been studying acidity and had observed and measured a number of solutions for concentration and pH.

In the use of each of these interviewing procedures, it is common for researchers to hear learners say, 'Ah, now I see what it is all about', or 'I hadn't really known what I was supposed to be learning till we went through it like this'. The research instruments seem to induce learning of a meaningful sort. Since many teachers in normal practice allocate quite a lot of time to the revision of topics and encourage their pupils to try to

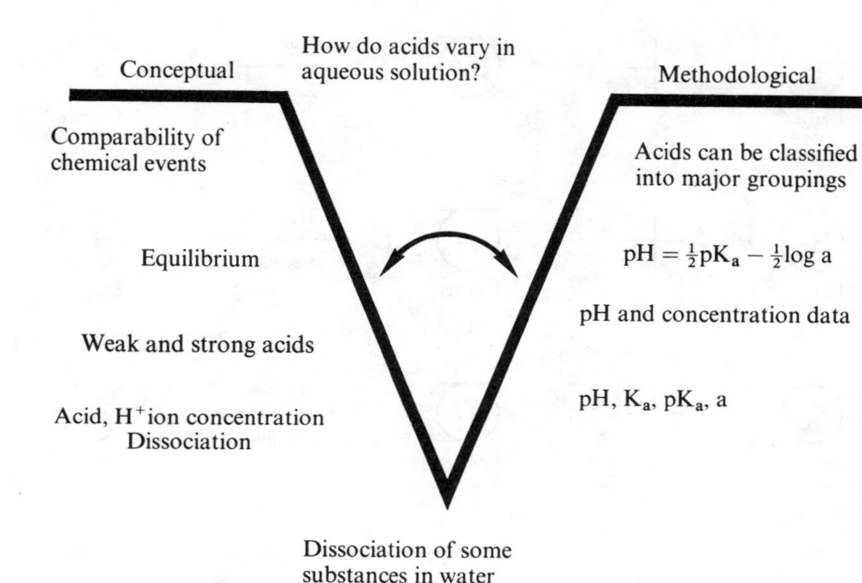

FIG 3. Gowin's V for a segment of chemical knowledge relating to acids.

use time and home for these purposes, it may well be that these techniques could make these well-intentioned hours more effectively spent.

Changing implications

Easley and Driver [35] have pointed out that whenever a teacher introduces a topic to pupils, they come to it with a framework of relating ideas which they have derived from past experience. When they are confronted with statements from their teachers, or from textbooks, or even from laboratory exercises that are not in line with their existing conceptual framework, they have to modify their own view, keep two separate views, or reject the view taught. All the new lines of research indicate that the latter two options are far more common than we have assumed throughout the great surge of science education since the late 1950s.

Diagrammatically this assumption has been either the first or second in Figure 4 in which the teacher's view of science become the learner's view.

A number of research findings now indicate how prevalent are the third or fourth situations portrayed. One of the advantages (and impressive aspects) of these newer research studies is the fact that much of it has been done with learners who are successful in terms of the usual demands the examinations in chemistry put upon them (i.e. Figure 4(c)).

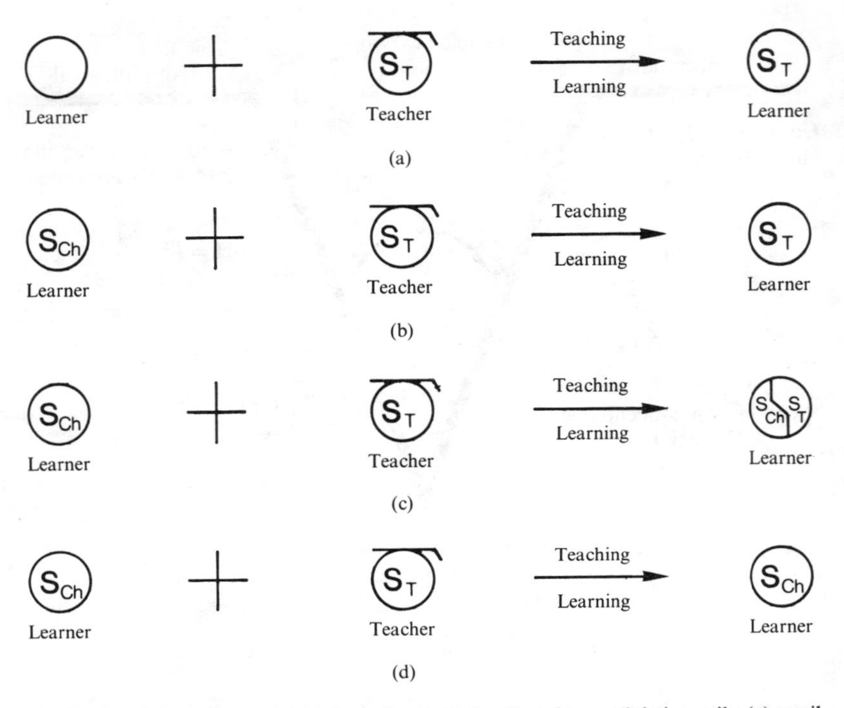

FIG. 4. Interactions between the science frameworks of teachers and their pupils: (a) pupil *tabula rasa*; (b) teacher dominance; (c) pupil two worlds; (d) pupil dominance.

The findings cannot simply be dismissed as the outcome of poor or unsuccessful teaching.

For example,

> Just after describing for me how liquid acetone evaporated if placed on the skin, a student who obtained high grades in the final school examination was unable to give me any examples of a liquefied gas. When pressed he muttered 'Solids, liquids, gases' as if this were some immutable sequence except perhaps in the eccentricities of laboratories where he thought the carbon dioxide in a cylinder might be a liquid.

As researchers increase our understanding of the conceptual frameworks pupils bring to the topics of chemistry, the term 'misconception' is being replaced by 'preconception' (Easley and Driver [35]) or 'children's or learner science' (Gilbert, Osborne and Fensham [54]). The latter is useful because it implies that often these ideas are held as a result of an interplay of a set of prior and personal experiences in the culture beyond the school and in it, that are akin to those employed by scientists in creating the knowledge of their subject. Furthermore, it has enabled

TABLE 1. Some new objectives for chemical education

Short description	Verbal description of objective	Research sources
A. Definition	To introduce pupils to examples of how chemists have defined concepts in ways that are useful to them but which conflict with common-sense experience and usage	Osborne [57] Gardner [58] Johnstone and Cassels [59]
B. Accommodation	To make explicit the world view of chemical phenomena that pupils hold and to relate these to world views held now and in the past by chemists	Cawthron and Rowell [27] Osborne [57] Champagne and Klopfer [32] Lybeck [30] Jungwirth [33]
C. Oversimplification	To enable pupils to recognize that chemists *invent* general concepts whch idealize and oversimplify real substances and phenomena	West and Fensham [40] Fensham [6]

us to see the science that is often presented by teachers and textbooks has a dogmatic character (a confusion of the elements of Gowin's V) that is quite unlike the present consensus account of the science itself. Thus C_2H_5OH on the blackboard, or a covalent bond, to many chemistry teachers, are as real as the liquid, or the properties of sugar solutions compared with those of sodium chloride. Chemistry teachers, along with their pupils, need to indulge regularly in the process of exposing their conceptual frameworks if their chemistry teaching is to have lasting effect. Karplus et al. [55] have developed an excellent set of in-service experiences for chemistry teachers which aims at this very aspect of teachers' knowledge.

The rejection of the *tabula rasa* (Figure 4(a)) view of teaching and learning opens up a number of interesting new tasks for science teachers. Fensham [56] outlined six of these, the first three of which stem from the sort of research being discussed. They are listed in Table 1.

Case [60], Rowell and Dawson [61] and Delacôte [36] have all been exploring how to exploit these *cognitive conflicts*. They set out to reveal the learner's spontaneous strategy and then, by appropriate choice of examples, expose its limitations. Teachers are more prone to stifle expression of unwanted theories and these other uses of their learners' science will require new teacher patterns.

7.4 What knowledge does a teacher need?

In delineating the scope of the section, I argued that the answer to the large-scale question, What should be the content for school chemistry? is not relevant to most teachers.

There is, nevertheless, a counterpart but much smaller scale question that teachers ought to face as they set out to teach the given curriculum: What are the consensus statements in current chemistry about the topics I am to teach?

Teachers often appear to repress too critical an examination of this question by themselves and others. However, they do recognize that some topics are difficult for their pupils and it may be that one reason for this is their own lack of complete understanding of quite familiar and apparently simple topics.

Dierks and colleagues at the Institute for Science Education in Kiel in the Federal Republic of Germany have been an active group in curriculum development and in this sort of systematic analysis of the content of chemical knowledge. Such painstaking review of the logic (and sometimes changing logic) of important chemical concepts has been undertaken with a clear hope that it will facilitate teaching and learning. As yet it has only been directly related to the difficulties of teachers and learners in a few instances. This is partly because we have almost no systematic data on the nature of teacher knowledge of chemistry. There is much anecdotal information about the deviations between teachers' science and scientists' science. Any regular classroom observer will have felt uneasy about the validity of some of the conceptual information being transmitted to the learners. In-service education of teachers seeks to remedy outdated or invalid understandings, but this is rarely systematic enough, or sufficiently specific to eradicate well-established but erroneous teacher knowledge.

Dierks has published a number of papers on the mole concept and was able to link the debate on its use in chemistry to the learning difficulties pupils experience [28]. This basic, peculiarly chemical unit in the SI system has probably been investigated more thoroughly than any other single chemical concept. Johnstone's group in Glasgow quickly established that calculations with it were a major area of learning difficulty in Scottish schools [62, 63] and a number of workers elsewhere have confirmed these difficulties. Novick and Menis [64] and Allsop [65] provided more useful evidence about the basis for this difficulty. Dierks concludes that the primary approach to teaching the mole in schools should be in terms of 'number' and not 'amount of substance'.

One more example of this type of research is that of Minssen, Buender and Walgenbach [29] who have used the historical evolution of the concept, macromolecule, as a means of providing a curriculum for senior school chemistry students and teachers. the central category of this curriculum is science as a sphere of activities—both rationalistic and aesthetic at the same time. This approach is rooted in the actualities of a field of chemistry that has become (in fifty years) very dominant in the lives of people everywhere. It may well be of very great interest to all those grappling now with the question of how to include the social consequences of science into science teaching. A few curriculum projects have done this but none seem to have had a research base for its actions.

7.5 How can chemistry teaching be more relevant/interesting?

Recurring questions for teachers in many countries are: How can I make learning chemistry more attractive/relevant/less boring/enjoyable to many students? On the whole in most countries, chemistry, as a school subject, seems to have substantial structural support for its worth as a field of study. Sometimes it is a compulsory subject in the curriculum of secondary schooling; often it is a prerequisite for the further education required by a number of very diverse and socially attractive occupations. Success in chemistry keeps many options open to the secondary-school learner.

Accordingly, it might be expected that with such extrinsic support, motivation for chemistry learning would not be a problem. Nevertheless, many teachers do still find that their pupils see chemistry as hard, dull and boring. During intuitive attempts to change this, there has been a tendency for teachers and curriculum developers to concentrate on the material to be learnt. Hence the widespread debates in recent times about the relevance of the content of chemical curricula. There are undoubtedly features of many curricula that 'ought' to be altered. Chemistry curricula are, as Fensham [66] and Layton [67] have pointed out, the products of a balance of societal demands. Whether social or historical aspects of chemistry are seen as suitable for school chemistry, and how seriously, depends on forces that are far beyond the ordinary teacher. This is even true of scientific detail such as whether alloys or other important materials like wood, ceramics, etc. is a proper topic to include in chemistry. Changes in curricular content do occur slowly, but ordinary teachers need to answer the question of interest in other ways if they are not to be frustrated by the slowness of these larger social issues.

Because of the extent of the debate about external relevance, there has been a tendency to neglect the intrinsic motivation that comes with effective learning.

If learners can be assisted, far more often than usual, to learn the meaning of the chemical content rather than merely acquire (or fail to acquire) it by rote, then we may well find the fundamental question less pressing.

Novak [51] has recently reported on some studies using Kagan's method of Inter-Personal Recall (IPR). This is a technique for making explicit the thoughts and feelings that pupils and teachers experience during an act of teaching. It is unusual and needs encouragement, but feelings certainly can be recalled and their expression has been

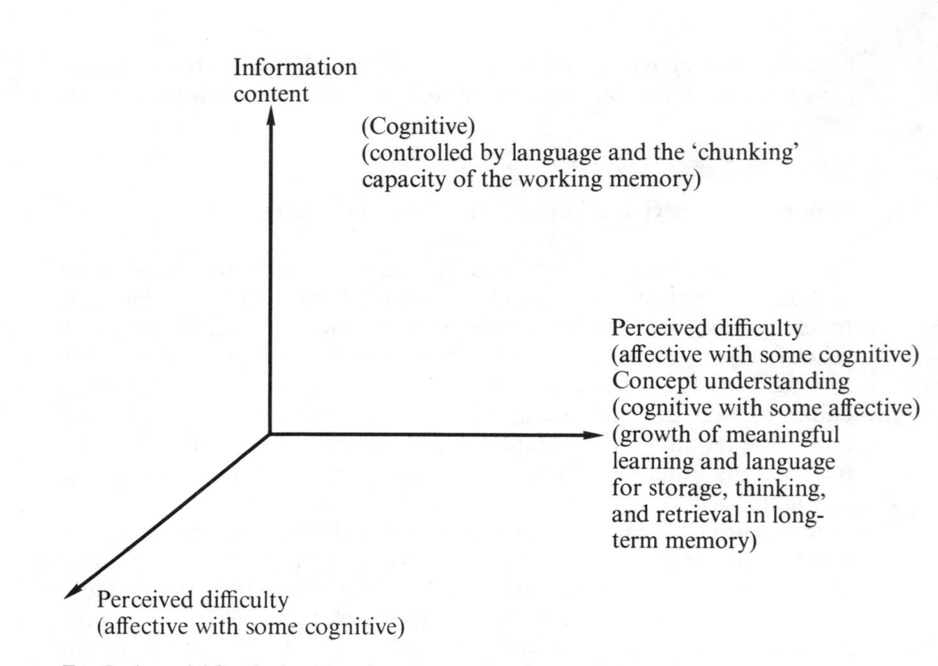

FIG. 5. A model for the learning situation confronting pupils in chemistry.

accompanied by positive learning. The ways that affective and cognitive aspects of learning are to be explored will no doubt depend very much on the particular culture concerned. A clearer understanding of their linkage probably lies at the heart of making chemistry learning more interesting.

It is thus interesting to consider the model (see Figure 5) proposed by Johnstone [63] for the situation of a learner confronted with the heavily conceptual content of a subject like chemistry. Some obvious predictions about how learners will perceive and feel about their tasks can be made from the interactions of the models' dimensions.

For example, if the information content is too high for the existing cognitive understanding, high perceived difficulty will result together with negative feeling (affect). Conversely, if the information content does not overload the concept understanding, perceived difficulty will be low and the feeling will be positive.

To avoid the less effective learning situations of this model, teachers would do well to consider the recommendations given above in the discussion of this group's specific study of organic formulae. More generally, this research suggests that 'chunking' strategies need to be very explicitly explained by the teacher, and equally well dismantled and changed when further information needs to be taken in. Consistent language use by the teacher will be a great help and there is a need to resist the temptation to provide unessential information lest the learner fail to discriminate and become overloaded again.

Gardner's review [68] of the complexities of research into attitudes and science education shows that there are no easy solutions. Chemistry teachers do have one positive asset in their search for interest and relevance. This is the laboratory and the practical side of their subject, the next section takes up some aspects of its role.

7.6 How can one make effective use of the laboratory and relate practical work to theoretical learning?

One of the distinctive features of chemistry is the diversity of its range of laboratory activities. There is a widespread belief that these activities should be reflected in school courses of chemistry. Unesco and IUPAC have on a number of recent occasions sponsored conferences that have focused on the laboratory within chemical learning. Nevertheless, the opportunities to incorporate pupil experiences of practical work in laboratories still vary enormously across and within school systems. In some countries the problems of facilities, lack of chemicals and apparatus, and unfamiliarity make practical work almost non-existent and chemistry becomes a theoretical subject. In others where practical work is possible it is often much less common than the curriculum intends. Elsewhere many hours may be spent in the laboratory, even by quite junior secondary students.

The research findings on laboratory learning, even where it is commonplace, are surprisingly disappointing. In general, the pupils' time in the laboratory does contribute positively to their enjoyment of the subject, so that any increase in the laboratory component of a course should make it more interesting. As a media for learning cognitive knowledge, however, or even for mastering the psychomotor skills of chemistry, the evidence for the laboratory is not very good. The laboratory should be the place where teachers assist students to put 'flesh on the bones' of theoretical course work, and where a sense of mastery of new and useful skills is achieved. On the evidence available schools in general have been remiss in achieving these goals and traditions, rather than honest evaluation of its usefulness, sustains a lot of school practical work. This, however, is not a very sound base from which to encourage the many school systems and teachers who would like to introduce expensive laboratory experiences.

Some glimmers of light have, however, appeared in a few recent studies. Kempa and Ward [69] and Ben-Zvi et al. [70] have developed schemes for various phases of practical work so that they can be assessed more effectively. Their four phases are: (a) planning and design; (b) performance of experiment; (c) observation of an actual phenomenon; and (d) analysis, applications and explanation. Not all phases are involved in every piece of practical work but specification of the phase involved and its sub-components makes its purpose clear to the pupil and makes its assessment possible with reliability and validity. In their research studies one exercise that emphasized phases (a) and (d) required pupils to design apparatus and methods to investigate the reaction

$$CdCO_3(s) \rightarrow CdO(s) + CO_2(g)$$

Another that involved several variations of the reaction

$$Pb^{2+}(aq) + CrO_4^{2-}(aq) \rightleftharpoons PbCrO_4(s)$$

involved the pupils in the performance of experimental work according to well-defined instructions. Their performance (largely phases (b) and (c) was then assessed using a sub-analysis of manipulative skills that covered experimental techniques, sequence of procedure, manual dexterity and orderliness.

Karplus et al. [55] developed a series of laboratory exercises for teacher in-service education that are based on Piagetian research and theory. Some of them provide excellent examples of Kempa's phases. For example, one that involves phases (c) and (d) is a phenomenon in which iron wool is moistened with vinegar and placed in a test-tube that has a stopper with a glass tube, which is then inverted over a beaker of coloured water so that the water level in the tube can be observed.

Gagné and White in an elegant series of research studies have been developing a model of ways in which memory can aid or inhibit learning [23]. They have distinguished several sorts of memory and two of these, White [71] postulates, are particularly relevant to the problem of making laboratories more effective contexts for learning. The first are called images. They are figural representations in memory of diagrams, pictures or scenes. The demonstration by the chemistry teacher in the classroom or laboratory has great potential for building up this type of memory. The second are called episodes. These are representations in memory of past events in which the individual was personally involved.

Both images and episodes are powerful aids to the recall of any knowledge associated with them. White gives the example of seeing a demonstration of carbon monoxide being ignited in a 2-metre-tall gas jar. The gas burnt down slowly with a paraboloid flame emitting blue light and a musical note which rose in pitch as the flame neared the bottom. The memorable event was associated with verbal knowledge about energy transportation and the properties of carbon monoxide—a link between two topics not normally related in chemical education.

Such a specific episode or image carries with it an emotive character which is an important component of making chemistry enjoyable or meaningful to the learner. More common, however, are the generalized episodes that have less emotive associations, but which provide a stock of concrete experiences from which meaning can be attached to new information. When pupils are being told something new about acids they will comprehend and assimilate it better if they have such a stock of generalized episodes about acids.

Laboratories on the whole for large classes of school students need to be orderly and safe. This makes them more likely to contribute generalized episodes to the pupil's memory than the emotive specific ones. However, they will not be useful for the processing of the verbal knowledge of chemistry unless the link between the practical experience and this knowledge is very explicitly made. Teachers need to explore ways of making this linkage at the *right* moment in the laboratory experience. Practical notes often include such statements, but it is clear that usually they simply operate as a set of instructions for the practical tasks that dominate the students' thinking, at least in the initial stages of a practical session. Atkinson [72] has demonstrated how impervious senior secondary-school pupils had become to such associations even when theory and practice were carefully juxtaposed in a study of the gas laws.

Since specific episodes, or the dramatic image, are going to be fairly rare events in a practical course, or in even an accomplished teacher's repertoire, this sort of research suggests that they should be chosen carefully to associate with key topics in the course of study.

However, even where the laboratory is regularly used, the learner tends to acquire a set of rather isolated episodes that are limited in their usefulness for recall of knowledge. If they can be linked, through choice of materials, or through use of laboratory apparatus outside the school, to the many more episodes that crowd into the pupil from her/his life in the world, then their power for associating and giving meaning to the abstractions of chemical knowledge is likely to be considerably higher.

7.7 How can teachers be researchers in their schools?

A suggestion that has often been made to overcome the gap between research and teachers is that teachers should conduct research themselves. However, the suggestion has usually not been taken seriously by teachers who, in most situations, are far too busy with everyday tasks to add such a different and specialized one. Furthermore, if a lot of research, as has been argued, is not directly of use to teachers, why should a teacher bother?

A number of the recent approaches to research that are reviewed in this chapter do raise the suggestion again more sensibly. For example in the section on changing methods of research (pp. 320–3), research methods that are outlined may well be powerful teaching and learning strategies. They do not involve the complexities of questionnaires, scales, sampling, control groups, etc. that makes traditional research so foreign to teachers. In addition to the obvious possibilities outlined above, there are several others that are worthy of comment because they do also seem to meet the criteria of being both possible and useful.

Keen teachers have, in many countries, been prepared to develop new experiments and new teaching approaches. With an extra effort, some of these are written down for publication in journals for science teachers. The 'how I do it' type of article has long been deservedly popular. The development of the idea of criterion reference testing is a new tool for teachers that could move these articles into the research category. Their value for other teachers could be much enhanced if these more definite measures of the effectiveness of the innovation were added to them. In a simple form, a criterion reference test is one that interprets performance on a task against an absolute standard without reference to the performance of others. In practice, the standard is often getting at least 80 per cent of the items testing this task correct. Brown [73] has recently published a comprehensive review of this much more useful approach to testing pupils. The report discussed later in this section by George and Fensham [74] is an example of the type being suggested here.

Two studies in Thailand of in-service education among science teachers involved a number of teachers in simple skills of analysing the verbal talk that went on in their lessons. Chewprecha et al. [75] and Wongthonglour [76] had chemistry teachers record the verbal exchanges of a lesson and then provided them with a simple key for analysing the exchanges in terms of encouragement of student inquiry, questioning and types of questions, and planning and checking the plan of a lesson. The Thai teachers did find they were able to follow these procedures and use this self-knowledge to modify their teaching behaviour. These procedures are now being used in Indonesia as part of a large-scale plan for the in-service training of science teachers.

An extension of this self-monitoring of their classrooms by teachers has been reported by Elliot and Adelman [77] in the United Kingdom. In this approach, the recorded version of the lesson is amplified by the contributions of another teacher who acts as an observer and who may also ask the pupils and the teacher about their perception of what is going on at various stages of the lesson. The teacher then tries to reconcile and evaluate these varying versions. From this more informed position, they then make their judgements about how they might change their lesson strategies.

Rowe [78] in the United States has been a researcher who has actively worked with teachers in classrooms in a way that has enabled them to take control of research data and modify it. The most spectacular of her findings has related to the time teachers wait for a pupil to answer a question. Not only has she been able to demonstrate to teachers that they wait much less patiently than they believe, but also that their waiting is biased against the lower achieving members of a class. A number of teachers who became aware of these tendencies have been able to extend their waiting times and thus to improve both the climate of their classes and the effectiveness of learning. Once again this type of self-investigation is open to teachers with access to a tape-recorder.

In addition to these more general ways, teachers can study their styles of classroom discourse. Several recent studies have taken serious account of the science context that the language of teachers must communicate to their pupils.

Lundgren [79], whose research has already been mentioned, has obtained evidence for two features of classrooms that likewise produce very positive response among teachers. The first it what he has called 'piloting'. Any teacher will be aware of how to ask a student a series of questions in such a way that the ultimate answer is eventually forthcoming. For example, the problem of balancing an equation like

$$ZnO + xHCl \rightarrow yZnCl_2 + zH_2O$$

can easily become

'How many Zn's on the left-hand side?'
'1.'
'So, how many do we need on the right-hand side?'
'1.'
'Right, so what is y?'
'1.'
'How many Cl's in one $ZnCl_2$ on the right-hand side?'
'2.'
'So, what is x?'
'2.'
'That's right. See if $z = 1$ works.'
'Yes.'

Such an exchange can avoid the frustration for the teacher of inadequate teaching of the basic principles and for the pupil of acknowledging ignorance. As far as learning is concerned, however, it will often be a non-event. Lundgren has found that such piloting is far more prevalent than teachers realize. He has also emphasized, along with other researchers, the way in which a small group of pupils act as a 'steering group' for the teachers' decisions when to move on to the next segment of a lesson or the next topic. Very often teachers are unaware of how much they use a positive answer from one of this group as if it means that everyone understands.

Both these ideas are again powerful and exciting criteria for the teachers to use as they listen to or reflect on their own lessons or when they observe the teaching of their colleagues.

Still more subtle aspects of classroom exchanges relate to the actual words and phrases used by chemistry teachers. Gardner [58], first in

Papua New Guinea and then in Australia, began to compile lists of commonly used words in science classrooms that were not understood by many of the pupils. The striking thing about this research was that it deliberately excluded technical words such as the names of elements, apparatus, etc. These will usually be specifically taught by the teacher if they are part of the chemistry curriculum. Gardner's list of words were ordinary words precisely used in science lessons but not usually specifically explained. For example, the italicized words in 'Gas molecules display *random* motion, we may *predict* their behaviour from *theoretical* considerations: the actual volume of the molecules may be *neglected*.' Australian pupils beginning secondary science studies at 12 years of age were checked for the meaning of these four words and 67, 12, 71 and 78 per cent were correct respectively. A cohort aged 15 in the final year of compulsory science education improved this to 78, 36, 94 and 95 per cent. Not surprisingly, pupils at the same stages in Papua New Guinea schools were less often correct since English, the language of school instruction, is a second language for most of them. Gardner's method of establishing this type of list has proved attractive elsewhere, being replicated in the Philippines, in Israel and more recently through the Chemical Society in the United Kingdom.

Both Gardner and Johnstone [62], who has been active in the British language studies, have demonstrated the non-equivalence to pupils of the same word in different contexts or of the words themselves, when they are synonymous to the teacher. For example, Gardner found that 'Carbon dioxide *consists* of oxygen and carbon atoms' was understood by 54 per cent of pupils to mean 'is a mixture', whereas 'The cloth *consisted* of wool and cotton', meant for 98 per cent 'was made from'. Johnstone and Cassels [59] have found some alarming differences in the correct responses when apparently synonymous but different wording is used in chemical test questions (see Table 2).

If language underpins all of chemistry teaching, one important aspect of it, namely the concept of structure and all stereochemical descriptions, depends also on the spatial facility of the learners. Comparatively little research seems to have been addressed to it directly.

Hill [80] showed that some specific revision and remedial work by teachers on spatial skills just prior to the introduction of various stereochemical topics may well be rewarding. George and Fensham [74] reported a successful teaching strategy in relation to structural features of primary, secondary and tertiary alcohols. Pupils were required to match representations of the same alcohol that included a variety of these-dimensional models, two-dimensional drawings of these, other written symbolisms and spectral charts.

Less positively, Nicholson [81, 82], in a series of studies with Seddon and others, found that the use of cues in two-dimensional representations was not well understood by students in Nigeria. Furthermore,

TABLE 2. Examples of the effect of mirror wording changes on the facility of pupils

Question form	Percentage correct
Which one of the following require a non-aqueous solvent to dissolve it? A. Salt, B. Sugar, C. Sodium Nitrate, D. Sulphur	34
or	
Which one of the following require a liquid other than water to dissolve it?	49
Which statement is true about the ions $^{8}_{3}Li^{+}$ and $^{8}_{4}Be^{2+}$? A. They contain the same number of neutrons B. Their atoms contain the same number of protons C. They will combine with the same number of F^{-} ions D. They contain the same number of electrons?	31
or	
$^{8}_{3}Li^{+}$ and $^{8}_{4}Be^{2+}$ have the same number of: A. Neutrons, B. Protons, C. Charges, D. Electrons.	43

Source: [59]

deliberate increases in these cues only marginally increased the pupils' facility with these representations. This finding is consistent with the growing body of evidence (alas, not yet checked in chemistry classes) that it is the convention of the representation itself that needs to be very explicitly taught. This seems to be particularly important for groups who may be learning in other than their native language, or for whom such pictorial or schematic representations have little cultural value or use. For example, Bishop [83] has reported on the high levels of recall of the structurel of actual objects that exist among pupils in Papua New Guinea. Clements and Lean [84], following up this work, have recently reported that students there, after constructing three-dimensional objects like tetrahedra from sticks and plasticine and having their two-dimensional representations or photographs associated, broke sticks (for dotted line) or added sticks (for shadow) when asked to make the object so represented. Even when so juxtaposed, each object was treated as itself.

More recently, Gardner [85] has turned his attention to another set of words that are of particular importance in the logic of scientific teaching. These logical connectives are words like (a) 'so', 'therefore'; (b) 'hence', 'besides', 'in general'; (c) 'similarly', 'that is'; and (d) 'moreover', 'if' (logical use). In a series of tests, the familiarity of 15-year-olds with these words was greater than 70 per cent, less than 70 per cent, less than 50 per cent, and less than 30 per cent for the four groups respectively. Chemical textbooks are full of examples of the use of these sorts of logical connective words. Teachers may well find that some attention to and investigation of these non-chemical components of the discourse of chemical education would be both rewarding and bear considerable fruit.

There is also some evidence of significant social class differences in these sorts of facility, and these could have widespread effect on the attempts now in progress to widen the social pool to whom chemical education is available. One potential consequence is alluded to, again by Lundgren's group, in terms of the meta-learning they believe occurs in classrooms. Unlike the intended learning which too often occurs for only a few pupils, meta-learning is acquired by all. Consider the example of piloting above. It could be argued that, regardless of the acquisition of the equation-balancing skill, pupils who experience this interchange and who were familiar with the consistent way in which the teacher uses 'So' to indicate that the required answer has already been stated, would emerge believing that the task was masterable. Others, unfamiliar with this particularity of teacher language, would meta-learn that the subject itself is unmanageable for them.

In all these varied ways, the language of instruction is proving a very rewarding field for research. So far, it has only been mined rather sparsely in chemical education. Since this is their words and their pupils' response to them, teachers really are in a strong position to study them if they wish to improve the effectiveness of their communication.

The need for chemistry teachers to teach explicitly the conventions of structural representations is a spatial counterpart to the many conventions of language use discussed above. The syntax that teachers use for presenting the substance of chemical knowledge is increasingly emerging as a barrier many students stumble at before the difficulties are even encountered. If teachers would encourage their pupils to talk about their understanding of chemistry more freely, they may well learn how to overcome many of the difficulties.

The fascinating work of Kellett and Johnstone [86, 87] on pupils' ability to read and recognize organic formulae provides another example of the ways that subsidiary instruction about learning strategies can overcome many of the difficulties facing teachers and pupils. They concerned themselves with the ability of students to recognize the important features of organic formulae, such as the symmetry of a molecule like

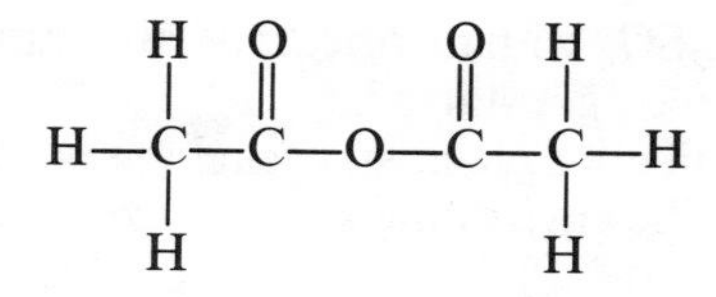

or common functional groups such as

COOH, —OH, $>C=O$.

Both these features require the whole of the structural formulae to be registered in some way, since the key functional group is usually written at the end of a formula and symmetry involves the whole formula. These authors approached this common chemical problem from information theory which has shown that many people have a capacity to handle six or seven pieces of information (such as the letters or digits in an unfamiliar word or telephone number), but beyond this number, ability diminishes rapidly. Very similar abilities were found to be operating in studying chemistry. The same formula.

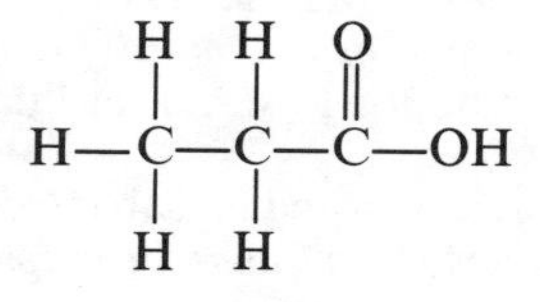

can be seen as

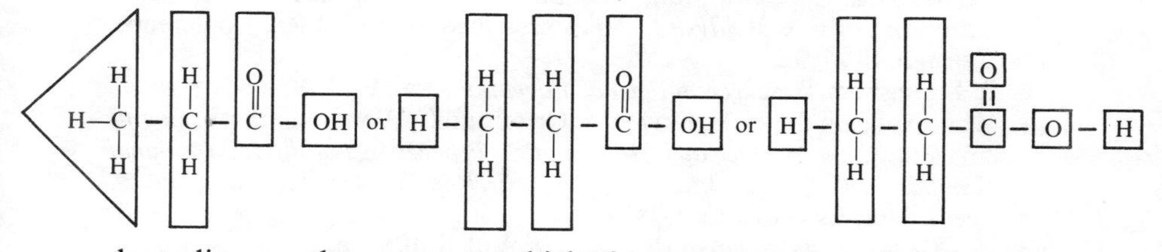

depending on the extent to which these groups of symbols have a conceptual meaning.

Kellett [88], on the basis of her studies, has argued for teaching strategies that enable learners (a) to 'chunk' more efficiently; (b) to reduce the amount of information to be considered by the learner; and (c) to assist the learner in handling information sequentially when its totality, even with chunking, exceeds his/her capacity. These three principles have much to say to chemistry teachers whose examples may often present far more information than the point intended requires, e.g. the use of $K_4Fe(CN)_6$ in order to indicate that anions can have valences as negative as 4−. Again, Johnstone's group has investigated the difficulties that arise because we write Na_2SO_4 and not $Na_2(SO_4)$, although we do write $Al_2(SO_4)_3$. The conceptual problems about cations and anions and charge balance are formidable, but a vocabulary that writes the two numbers in Na_2SO_4 with identical visual significance can hardly assist new learners to even get started.

The need for more effective chemistry teaching is steadily growing in all parts of the world. The environmental *problématique* will see to that for all of us.

Research in chemical education has undoubtedly been moving closer to the day-to-day problems of teachers. Despite the large gaps that are as yet unconsidered by research, there is now much to encourage both teachers and other research groups.

References

1. A. Kornhauser, *European Journal of Science Education*, Vol. 1, 1979, p. 21.
2. C. N. R. Rao (ed.), *Educational Technology in the Teaching of Chemistry. Proceedings of the International Seminar, International Union of Pure and Applied Chemistry (IUPAC)*. Madrid, 1975.
3. U. P. Lundgren, *Curriculum as a Context for Work*. Paper presented at the Australian Association for Research in Education (AARE) meeting, Sydney (Australia), 1980.
4. P. Tamir, A. Blum, A. Hofstein and N. Sabar (eds.), *Curriculum Implementation and Its Relationship to Curriculum Development in Science*, Jerusalem, Israel Science Teaching Center, 1979.
5. R. Stake, *An Approach to the Evaluation of Instructional Programs (Program Portrayal as Analysis)*, Paper presented at the AERA Conference, Chicago, Ill., 1972.
6. P. J. Fensham, *Research in Science Education*, Vol. 9, 1979, p. 43.
7. W. F. Archenhold, R. H. Driver, A. Orton and C. Wood-Robinson, *Cognitive Research in Science and Mathematics. Proceedings of the International Seminar on Cognitive Development Research in Science and Mathematics*, Leeds, University of Leeds, 1979.
8. R. M. Gagné, *Educational Psychologist*, Vol. 6, 1968, p. 1.
9. R. T. White, *American Educational Research Journal*, Vol. 11, 1974, p. 121.
10. S. K. Sharma, *The Effects of Programmed Learning on the Gas Laws*, Clayton (Australia), Monash University, 1974. (M. Ed. thesis.)
11. M. Shayer and R. B. Ingle, *Education in Chemistry*, Vol. 8, 1971, p. 182.
12. M. Shayer, 'The Matching of Science Curriculum to the Learner in the Middle and Secondary School', ref. 4, pp. 89–94.
13. E. W. Jenkins, *Education in Chemistry*, Vol. 15, 1978, p. 85.
14. C. J. Dawson, *Education in Chemistry*, Vol. 15, 1978, p. 120.
15. J. D. Novak, *Science Education*, Vol. 61, 1977, p. 453.
16. D. P. Ausubel, *Educational Psychology: A Cognitive View*, New York, Holt, Rinehart & Winston, 1968.
17. D. G. Ring and J. D. Novak, *Journal of Research in Science Teaching*, Vol. 8, 1971, p. 325.
18. P. J. Fensham, *Research in Science Education*, Vol. 2, 1972, p. 50.
19. D. P. Ausubel, *Review of Education Research*, Vol. 48, 1978, p. 251.

20. L. H. T. West and P. J. Fensham, *Studies in Science Education*, Vol. 1, 1974, p. 61.
21. J. T. Lawton and S. K. Wanska, *Review of Educational Research*, Vol. 47, 1977, p. 233.
22. M. J. Koslow and A. L. White, *A Meta-Analysis of Selected Advanced Organiser Research Reports 1960–1977*. Paper presented at the Boston Conference of NARST, 1979.
23. R. M. Gagné and R. T. White, *Review of Educational Research*, Vol. 48, 1978, p. 187.
24. M. C. Wittrock, 'Learning As a Generative Process', in M. C. Wittrock (ed.), *Learning and Instruction*, Glenrock, N.J., McCutcheon, 1977.
25. R. J. Shavelson, *Journal of Research in Science Teaching*, Vol. 11, 1974, p. 231.
26. A. B. Champagne, L. E. Klopfer and J. H. Anderson, *Factors Influencing Learning of Classical Mechanics*. Paper presented at the AERA Meeting, San Francisco, Calif., 1979.
27. E. R. Cawthron and J. A. Rowell, *Studies in Science Education*, Vol. 5, 1978, p. 31.
28. W. Dierks, *European Journal of Science Education*, Vol. 2, 1980, p. 145.
29. M. Minssen, W. Buender, and W. Walgenbach, 'Analysis of the Evolution of a Science Concept—Macromolecules', ref. 4, pp. 375–8.
30. L. Lybeck, 'Studies in Mathematics in the Science Lesson in Goteborg', ref. 31, pp. 331–68.
31. L. Lybeck, in H. G. Steiner (ed.), *Co-operation between Science Teachers and Mathematics Teachers*, Bielefeld, Institut für Didaktik der Mathematik der Universidad, 1979.
32. A. B. Champagne and L. E. Klopfer, *Interpretations of Observations of Motion and Semantic Knowledge Structures of Students at Four Grade Levels*. Paper at NARST, Atlanta, Ga., 1979.
33. E. Jungwirth, *Journal of Biological Education*, Vol. 11, 1979, p. 191.
34. J. D. Novak, *A Theory of Education*, Ithaca, N.Y., Cornell University Press, 1977.
35. J. Easley and R. Driver, *Studies in Science Education*, Vol. 5, 1978, p. 61.
36. G. Delacôte, 'Classroom Based Research in Science and Mathematics', ref. 7, p. 275.
37. G. Schaefer, *European Journal of Science Education*, Vol. 1, 1979, p. 87.
38. J. Nussbaum and J. D. Novak, *Science Education*, Vol. 60, 1976, p. 535.
39. L. H. T. West, 'Towards Description of the Cognitive Structures of Science Students', ref. 7, pp. 324–48.
40. L. H. T. West and P. J. Fensham, 'What is Learning Chemistry', in C. L. Fogliani and J. R. McKellar (eds.), *Chemical Education—A Review Across the Secondary, Tertiary Interface*, pp. 162–9, Adelaide (Australia), RACI Chemical Education Division, 1978.
41. D. R. Driscoll, 'Student Misconceptions in Chemical Equilibrium', Clayton (Australia), Monash University, 1966. (Unpublished B. Ed. thesis.)
42. A. H. Johnstone and J. Handy, *Education in Chemistry*, Vol. 10, 1973, p. 99.
43. J. K. Gilbert and R. J. Osborne, *An Approach to Student Understanding of Basic Concepts in Science*, Guildford, Surrey, Institute for Educational Technology, University of Surrey, 1979; *European Journal of Science Education*, Vol. 2, 1979, p. 311.

44. A. H. Johnstone, J. J. MacDonald and G. Webb, *Education in Chemistry*, Vol. 14, 1977, p. 169.
45. C. R. Sutton, *European Journal of Science Education*, Vol. 2, 1980, p. 107.
46. A. L. Pines, *Science Concept Learning in Children: The Effect of Prior Knowledge on Resulting Cognitive Structure Subsequent to A-T Instruction*, Ithaca, N.Y., Cornell University. (Ph.D. thesis.)
47. M. Brumby, *Students' Perceptions and Learning Styles Associated with the Concept of Evolution by Natural Selection*. Guildford, Surrey, University of Surrey, 1979. (Ph.D. thesis.)
48. A. Tiberghien, 'Modes and Conditions of Learning—An Example: The Learning of Some Aspects of the Concept of Heat', ref. 7, p. 288–309.
49. R. T. White, *Describing Cognitive Structure, Proceedings for the AARE Conference*, p. 198, Australian Association for Research in Education (AARE), 1979.
50. P. F. W. Preece, *Science Education*, Vol. 4. 1978, p. 1.
51. J. D. Novak, *Australian Science Teachers Journal*, Vol. 27, 1981, p. 5.
52. D. R. Gowin, 'The Domain of Education', Ithaca, N.Y., Cornell University. (Unpublished manuscript.)
53. R. Karplus and C. A. Lawson, *SCIS Teachers Handbook*, Berkeley, Calif., University of California Press, 1974.
54. J. K. Gilbert, R. J. Osborne, and P. J. Fensham, 'Children's Science and its Consequences for Teaching', *Science Education*, Vol. 66, 1982, p. 623.
55. R. Karplus et al., *Science Teaching and Development of Reasoning—Chemistry*, Berkeley, Calif., Lawrence Hall of Science, 1980.
56. P. J. Fensham, 'A Research Base of New Objectives of Science Teaching', *Science Education*, Vol. 67, 1983, p. 3.
57. R. J. Osborne, *Research in Science Education*, Vol. 10, 1980, p. 11.
58. P. L. Gardner, *Words in Science*, Australian Science Education Project, 1972.
59. A. H. Johnstone and J. Cassels, *New Scientist*, No. 432, 18 May 1978.
60. R. Case, *Review of Educational Research*, Vol. 48, 1978, p. 439.
61. J. A. Rowell and C. J. Dawson, *Research in Science Education*, Vol. 9, 1979, p. 169.
62. A. H. Johnstone, *European Journal of Science Education*, Vol. 1, 1979, p. 239.
63. A. H. Johnstone, in C. L. Fogliani and J. R. McKellar (eds.), *Chemical Education—A Review Across the Secondary, Tertiary Interface*, pp. 156–61, Adelaide (Australia), RACI Chemical Education Division, 1979.
64. S. Novick and J. Menis, *Journal of Chemical Education*, Vol. 53, 1976, p. 720.
65. R. T. Allsop, *Physics Education*, Vol. 12, 1977, p. 285.
66. P. J. Fensham, *Journal of Curriculum Studies*, Vol. 12, 1980, p. 189.
67. D. Layton, *Science for the People*, London, Allen & Unwin, 1973.
68. P. L. Gardner, *Studies in Science Education*, Vol. 2, 1975, p. 1.
69. R. F. Kempa and J. F. Ward, *Journal of Research in Science Teaching*, Vol. 12, 1975, p. 69.
70. R. Ben-Zvi, A. Hofstein, D. Samuel and R. F. Kempa, *Journal of Research in Science Teaching*, Vol. 14, 1977, p. 433.
71. R. T. White, *Physics Education*, Vol. 14, 1979, p. 384.
72. E. P. Atkinson, *Instruction–Memory–Performance. The Influence of*

Practical Work in Science on Memory and Performance, Clayton (Australia), Monash University, 1980. (Ph.D. thesis.)
73. S. Brown, *What Do They Know?—A Review of Criterion-referenced Assessments*. Edinburgh, Scottish Education Department, HMSO, 1980.
74. S. C. George and P. J. Fensham, *Education in Chemistry*, Vol. 10, 1973, p. 24.
75. T. Chewprecha, M. Gardner, and N. Sapianchai, *Journal of Research in Science Teaching*, Vol. 17, 1980, p. 191.
76. S. Wongthonglour, *Diffusion of IPST Physical Science Curriculum: An Evaluation of its Inservice Education Component*, Clayton (Australia), Monash University, 1979. (Ph.D. thesis.)
77. J. Elliot and C. Adelman, *Education for Teaching*, Vol. 92, 1973, p. 8.
78. M. B. Rowe, *Journal of Research in Science Teaching*, Vol. 11, 1974, p. 263.
79. U. P. Lundgren, *Model Analysis of Pedagogical Processes*, Stockholm, Institute of Education, Department of Educational Research, 1977.
80. D. M. Hill, *A Study of the Relationship between Performance on Spatial and Allied Perceptual Tasks and on Stereochemical Tasks*, Clayton (Australia), Monash University, 1971. (M.Ed. thesis.)
81. J. R. Nicholson and G. M. Seddon, *British Journal of Psychology*, Vol. 68, 1977, p. 327.
82. J. R. Nicholson, G. M. Seddon and J. G. Worsnop, *Programmed Learning and Educational Technology (PLET)*, Vol. 15, 1978, p. 229.
83. A. J. Bishop, *Educational Studies in Mathematics*, Vol. 10, 1979, p. 135.
84. A. McK. Clements and G. A. Lean, *Influence on Mathematical Learning in Papua New Guinea: Some Cross Cultural Perspectives*, Lae (Papua New Guinea), Mathematics Education Centre, University of Technology (Technical Report 13).
85. P. L. Gardner, *Journal of Research in Science Teaching*, Vol. 17, 1980, p. 223.
86. N. C. Kellett and A. H. Johnstone, *Education in Chemistry*, Vol. 11, 1974, p. 111.
87. N. C. Kellett, *Research in Science Education*, Vol. 9, 1979, p. 95.
88. N. C. Kellett and A. H. Johnstone, *European Journal of Science Education*, Vol. 2, 1980, p. 175.

8 The future

M. H. Gardner

After highlighting some of the many aspects of chemical education from the previous chapters, Professor Gardner ties the strands together in an attempt to 'draw inferences from events and directions worldwide in the classrooms . . . and thus look to the future'. In particular, she asks whether we can set our goals to accommodate an appropriate education in chemistry for everyone, remembering how diverse we are in ability, motivation, aims in social, political and economic terms. She looks at the future in the terms of the individual teacher, the individual student and also in the terms of the society in which they live.

Professor Gardner outlines the constraints to progress—coping in the real world—before telling us about an action agenda for the 1980s based on the recommendations and resolutions of the Sixth International Conference on Chemical Education held at her university in August 1981. These recommendations are directed at institutions, industry and individuals. It is, of course, through work by individuals, for individuals, that progress is made: the institutions provide a means within which the individual can work most effectively. Also included is a report from the secondary-school teachers at the Sixth International Conference.

As Professor Gardner writes, the future is a time of challenge, a time of promise. Her chapter will help us meet the challenge and to translate promise into progress.

8.1 Introduction

What does the future hold for chemistry teaching at school level? Or even more broadly for chemical education around the world? If our crystal ball or tarot cards were in good working order, perhaps we could predict trends, accomplishments and issues in school chemistry teaching for five years, ten years, twenty years hence. Or even more in keeping with technology and the times, if we had fully developed our software for modelling the future, perhaps we could predict a little more accurately than present guesstimates. What will be common in a pre-university classroom around the world in the year 2000?

Since neither the mysticism of crystal ball gazing nor the sophistication of our software is up to the challenge of accurately predicting the future, our educated guesses based on studies of the past and present will have to do.

In Chapters 1 to 7 of this volume, we have viewed secondary school chemistry with respect to chemistry as a discipline, curriculum, methodology, technique, technology, teacher training, assessment and research. Now let us try to tie the strands together as we examine and draw inferences from events and directions worldwide in the classrooms, in the changing social, cultural and political climates and in the burgeoning knowledge of chemistry, and thus look to the future.

In Chapter 1, Campbell challenges us immediately with the question of how to transfer the growing knowledge of chemistry from the research laboratory into the classroom. He uses a wide array of achievements of eminent chemists from around the world to illustrate the internationality and the interdisciplinarity of chemistry. He ties examples to the social scene to sensitize teachers and students alike to the need for close ties between what one learns in the classroom and what occurs in our everyday lives. He uses findings of the Nobel award-winning chemist, Ilya Prigogine, to make us aware that chemical systems are replicated in human systems and, in fact, have implications for us in curriculum change and development.

In Chapter 2, Ingle and Ranaweera utilize the curriculum projects of three decades—the 1950s, the 1960s and the 1970s—to suggest directions and strategies for the 1980s. In doing so, one can see how the development of, for example, CHEM Study and Nuffield projects in the United States and United Kingdom have influenced other curricula. First, they were translated, then adapted to new cultures and then utilized only as a model as nation after nation developed its own curricula to match its own needs and now we are seeing that other criteria are being discussed. They also trace the movement from the study of pure chemistry for the future scientist to a growing concern in developing and developed nations alike for the need for some understanding of chemistry among all members of the society. Influenced by the Unesco pilot project for chemistry teaching that was located in Asia in the 1960s and the recognition in country after country of the need for a change, the Asian and Pacific Region also became active in curriculum development during the 1970s, not allowing the western nations who took the lead in the late 1950s and 1960s to corner all the initiatives. The latter part of the chapter is devoted to some very important questions related to the present and the future of school chemistry curricula. These deal with, among other important issues, the balance of theory versus fact versus practical work, the need for citizen education in chemistry for the non-science major, the degree to which chemistry should be related to other science disciplines and the attention needed for the average and less motivated student.

Next, Kornhauser addresses methods of chemistry teaching. She defines 'methods' as arising out of research and development work in chemical education and explores four areas she considers to be most promising: the investigative method, problem solving, structuring of chemical knowledge and pattern recognition. These, in her view, relate to the desired modes of classroom teaching. This chapter represents a very sophisticated approach to the teaching of school chemistry.

In Chapter 4, Gardner, Moore and Waddington describe the potential of the microcomputer for chemistry teachers, a sophisticated technique using rapidly evolving new technology. However, it is versatile and, because of the trend in downward costs, it is moving rapidly into schools in countries with advanced technologies and may well follow in many other countries throughout the world. Teachers are playing a creative role in helping to develop programs (software) for school use. Chapter 4 also contains information about low-cost equipment and safety, two topics of importance to all teachers of chemistry if they are to realize the dreams and aspirations of the curriculum developers.

In the chapter on assessment of students, Mathews begins with definitions of norm-referenced and criterion-referenced assessment instruments and their use. He then provides practical information on planning for assessment, writing specifications and test items, and techniques of assessment, and illustrates these with a variety of styles and examples. He points out that assessment is now being seen as a valuable tool of both evaluation and instruction. Mathews concludes this chapter with ideas on training teachers in assessment techniques and defines the kind of workshop that can be used to achieve this goal.

In an introduction to the chapter on the training of teachers, Chapter 6, Bogatski Cross and Lazonby define their philosophy of education in terms of what makes a good teacher, how to select individuals for science teacher training and how they should be trained in content and in skill. The chapter then shows how various pre-service training courses are being organized and describes some in-service models for maintaining teacher competence that are being tested in different parts of the world.

Fensham in Chapter 7 provides an in-depth summary of current research in chemical education as he ranges far and wide through the professional literature to examine findings from many researchers. He presents important results related to learning hierarchies, stages of cognitive development, advanced organizers for learning and the convergence of these various ideas to provide a base for teachers to use in dealing with promoting learning among the various types of students they have in their classrooms. Later, Fensham defines what teachers need to know in order to make chemistry more interesting, more relevant and more understandable for their students. He describes the opportunity teachers have to do research with their students on the teaching of chemistry. Each classroom is, in essence, a research laboratory. With knowledge of research design and techniques, very important data and results can be accumulated by the skilled and experienced secondary-school teacher working along or in collaboration with colleagues in tertiary education.

Each author dealt in varying degrees with the evident needs in their area of presentation. In so doing, each of them evaluated the present and looked, to some degree at least, to the future.

Now let us build on these ideas and explore others that became apparent through the position papers, presentations, the discussions and the concerns of secondary-school teachers and teacher educators engaged in the Sixth International Conference on Chemical Education held at the University of Maryland in the United States in August 1981. With this base, we will look to the future with particular attention to the following areas: (a) human resources: the social scene; (b) curriculum and evaluation: the educational scene; (c) methodology, materials and techniques: the instructional scene; and (d) problems and possible solutions: the practical scene.

8.2 Human resources: the social scene

A nation's most valuable resource is its people. The intelligence, creativity and talent that resides in the human mind awaits only its release and full development through education. And an education in science that promotes the development of reasoning, prediction based on objective evidence, rational decision-making and problem solving is the type of education that can and should reach every member of society.

Chemistry, in particular, is close to a nation's health and strength and to the well-being of its people. Since chemistry touches the lives of every individual (through agriculture, industry, nutrition, medicine, the home environment, etc.), an individual's every moment, awake or asleep, at work or at play, as a youth or adult is directly influenced by the understanding and therefore the utilization he or she can make of chemistry. Scientific discoveries, technological advances, the efficiency of the work force, the exercising of citizens' rights and the quality of life are directly tied to the teaching of chemistry.

Can we set our goals to accommodate an appropriate education in chemistry for everyone? Can we establish strategies to achieve this goal? Have we periodically evaluated our progress, measured our successes, noted our failures and based our planning on feedback from the classroom? Can we set even higher goals for the future so that we are ever striving for improvement, for something better? How are we, as chemistry teachers, doing in fulfilling our role in fully developing human resources?

As we view chemistry teaching for the future, we need to consider students and their goals in a different light than in the past. Our primary concern in the past has been with educating relatively few science-talented students to become the scientists and engineers of our nations. More recently we have become aware of the importance of educating everyone, imparting the degree of knowledge of science needed for fulfilment in adult lives. To do this, it is necessary to analyse the students not as a huge amorphous group but as sub-groups with different goals and then as individuals.

8.2.1 *Diversity in student interests and goals*

There are at least five sub-groups among the students. Each needs an education in science. If we succeed in producing science-knowledgeable individuals in these five essential groups, then we can count our chemistry teaching a success.

First are the future chemists, those who will form an important part of our scientific communities. It is essential that we develop a pool of people who are talented and interested in chemistry and who have sufficient education in science and mathematics in the secondary schools to be able to advance to the heights allowed by their talents and abilities. The second group includes students who are interested in other science-based professions (e.g. the biological and earth scientists, engineers, physicians, nutritionists). A third group we need to consider are those students who will become the technical personnel, individuals who will comprise the support system for science and engineering. Usually such students go to work immediately after secondary school, often in industry, in the health sciences, or in agriculture, for example. In all instances it is essential that they can understand instructions, make good decisions and operate apparatus successfully. The next group are the potential managers in our classroom. Right now, each of us has students who will become government leaders, city officials, school administrators or industrial managers, for example. Such individuals will gain tremendously from a good education in chemistry. Not only will they develop better, more rational, decision-making powers, but they will be able to base their decisions on modern knowledge of science and technology. The fifth group, fifth in number but not in importance if we are to have a peaceful and effective scientific revolution, includes the ordinary citizen, every single individual in a nation. They each need and have a right to sufficient literacy in chemistry to be able to function effectively in societies that are increasingly being influenced by new drugs, synthetic materials, green revolutions in agriculture, micro-electronics and transportation and communication systems that we did not dream were possible ten or twenty years ago.

Equal opportunity in science and mathematics for every student in the classroom until career directions become apparent is an essential but challenging goal for the future. Some students have been traditionally counselled out of science. These might include students from minority groups or girls. Others have been unable to achieve a basic secondary-school education due to constraints of money, facilities or tradition. Yet each individual in a society matters and must be given the opportunity for a secondary education in science in order to identify and fully develop the frontiersmen, the practitioners and the managers and to adequately prepare the citizenry for life in a science-base society.

In helping to develop human resources, our goal must be that of providing high *quality* instruction in chemistry. Notice the word 'quality'. Chemistry teaching is not high enough in quality unless it has an effective laboratory component and sufficient opportunity for students to solve mathematical and intellectual problems. Ideally, they should be encouraged to investigate, to explore, to use the library, to use the natural environment and to discuss chemical concepts and issues in order to provide them sufficient opportunity and experience to cope with and benefit from the products and processes of chemistry throughout their lives.

In the past, chemical educators have concerned themselves primarily with the groups of students preparing for science-based careers and the curriculum has been designed for this group. Currently, and increasingly in the future, it will be necessary for us as teachers to provide a much different chemistry curriculum which may take the shape either of a broader course suitable for everyone at the secondary level or of several streams, but we can expect to have the obligation of providing chemical education for everyone. This new responsibility will demand of us substantial changes in our philosophy, our methodology, our content and our teaching materials.

For one thing, we must move to a more student-centred programme. Since there is no global authority to make decisions or guide us, this will be necessary from country to country and classroom to classroom as each teacher ultimately implements the curricula in an individualistic way.

Evaluation and research are essential to learn what the various types of students need to know about chemistry, how they learn and what they need to be able to do both in terms of intellectual and laboratory practical skills at the conclusion of secondary school. The answers may differ from country to country and even from locality to locality but the obligation for chemistry for all will be constant across the world in this coming decade.

Ketudet defines such an education as one that

> cultivates and develops knowledge, intelligence, skills and attitudes. It enables one to learn about oneself as well as to learn and understand life, society, nature and the environment of which he is a part, so that he can apply scientific knowledge and understanding to solve problems of everyday life and can make his society a better place, all in harmony with nature [1].

Now, let us look more intently at the three major components in chemical education of our human resource base. These include the science students, the non-science students and the teachers.

The science students

Science students will need to be able to manage qualitative and quantitative aspects of chemistry. This means a rigorous course with mathematics through introductory calculus in the secondary school. In addition, the science students can accept a more theoretical, more abstract approach to understanding chemistry than will interest the non-science oriented majority. However, they also need practical experiences in the laboratory and experience in predicting, presenting data, designing experiments, measuring and evaluating results, redesigning, etc. The verification experiment is not as important for these students as open-ended exploratory practical work. They should be challenged to the extent that requires them to use references, data books and the library. They should be encouraged to scan textbooks and journals to find ideas and answers, plan extensions of experiments from the initial design given to them, ask good questions and defend their results. Building oral and written communication skills is a necessity.

The non-science students

While 15 to 20 per cent of the students of secondary-school age may have an interest in chemistry or a science-based career such as medicine, engineering, agriculture, nutrition, etc., 80 to 85 per cent of the students—the large majority—plan to spend their adult lives pursuing other interests. Yet these non-science students have as high a need for chemistry in the secondary school as those who will pursue science-based careers. For the non-science student, the need is scientific literacy. If they understand advances in chemistry, the advantages and side-effects of chemical materials and the economic roles of the chemical industries in their nation, for example, their lives will be enriched. They also need such cognitive skills learned in chemistry classrooms as decision-making, problem solving or predicting for use in their every-day lives.

Literacy in general and scientific literacy specifically is on the increase in most nations. In the USSR, 90 per cent of Soviet students complete the ten-year pre-tertiary education which ends at the age of 16 [2]. In Japan, 92 per cent of students complete secondary education [3]. In the United States, 80 per cent of students complete the pre-tertiary programme through the age of 18 [4]. Similar high percentages of students complete secondary school in Canada, the countries of Western and Eastern Europe and in Australia and New Zealand. Many of these countries have intensive instruction in the sciences and mathematics for all or most of the secondary school students. In Japan, more than 70 per cent of the students study chemistry in the secondary school and over 90 per cent in the USSR. While the statistics are less available but almost certainly lower in developing countries, percentages there also are on the rise.

Why should chemical education for all be a national concern? One reason is the close relationship of chemistry and chemical engineering to the health and strength of a nation. A country cannot have a strong scientific and technological enterprise without a base in chemical education/science education. That is absolutely fundamental. Both economic productivity and the quality of life depend directly on the quality and quantity of science instruction offered in the schools. Each nation needs creative scientists, engineers, science teachers and technologists. Each also needs individuals trained in the processes of science and possessing an understanding of science to serve as the school administrators, legislators and business and industrial leaders. Each needs an informed citizenry. At the international level, the degree of scientific literacy possessed by the world population will help to determine the outcome of global questions related to food, population control, pollution, energy and peace.

8.2.2 *The chemistry teachers and their needs*

The key to high quality chemical education is the chemistry teacher in the secondary school classroom. It is in this setting that students become interested or lose interest in chemistry. They establish a base of concepts and skills that are dependable or unreliable. They develop a friendliness towards or a dislike of chemistry, specifically, and the scientific enterprise in general. For these and many other reasons the importance of the competent chemistry teacher cannot be overestimated.

There are at least five qualities that a chemistry teacher will need as we look to the future and the challenges associated with providing a chemical education for all. These include curiosity, competence, commitment, creativity and compassion. The first four are more familiar to us than the last, yet the newly emerging sense of compassion, the humanization of science, is very important. It is this quality that brings science into harmony with the society and the culture that it impacts daily and in a very personal way. Curiosity is the spirit of inquiry, the asking of question, the desire to know. Competence is the know-how, understanding of chemistry and the ability to teach it with enthusiasm and to bring the latest research results into the classroom in an effective way. Creativity is a special capacity within each of us in varying degrees. It involves insight, intuition, the ability to make intellectual leaps, to see spatial relationships, to transfer knowledge to new situations, to think new thoughts, to develop ideas and to design and produce new instructional materials, whether scientifically or artistically based. Commitment means a sensing of the values in chemistry as a discipline and in the worth of a life spent working with young people. To feel pride, satisfaction, dedication and achievement and to exude enthusiasm. Compassion is concerned with feelings, with concern for the individual and with the interface between science and social responsibility.

Teacher status, rewards and the feeling of worth matter greatly. Chemistry teachers must both earn and demand respect, remuneration and recognition sufficient to keep good teachers in the classroom. A sensitive and responsible public must become advocates of such rewards to maintain and extend the quality of chemistry teaching it wishes for its striplings and for its society.

Supply and demand exert heavy influences on the quality of chemistry teaching. When there is an over-supply of chemistry teachers, then young people move into other professions and a shortage develops. Temporarily at least classrooms are filled with under-qualified teachers. If adequate teacher training is not available at the pre-service and in-service levels, our classrooms are then, too, populated by teachers who are not fully competent. The salaries teachers are paid, the reward system and the degree to which they are held in esteem in a society are contributing factors to quality of teaching, as are the degree to which they are allowed to focus their talents and training on the actual teaching of chemistry and not be disrupted by extraneous responsibilities such as monitoring cafeterias, supervising transportation, preparing administrative paper reports, etc.

The effective chemistry teacher takes the students frequently into the laboratory and to the field to learn chemistry through doing chemistry. This means that they must be well educated in the use of laboratory equipment and procedures, in safety practices and actions in case of an accident, and in managing students individually and in small groups in order to stimulate the greatest learning from each. The skill of creative questioning that allows the student to think through concepts and situations and respond out of his own background of experience and intellect is essential.

The education of students and of teachers will require precious and incredibly complex sets of materials, procedures and practices. Content, curricula and evaluation systems for students and teachers will be explored in the next section.

8.3 Curriculum and evaluation: the education scene

This section will be divided into three parts:
1. Chemistry content and curriculum for science and non-science students.
2. The chemistry teacher curriculum, pre-service and in-service.
3. Evaluation of students, teacher and the curriculum.

Prior to suggesting some specific ideas for the future, let me outline some general goals.

8.3.1 *The student curriculum in chemistry*

One essential is finding a method to bring new chemical knowledge from research into the classroom more quickly. The gap between research findings and classroom content is wide with from five- to thirty-year time lags evident in various parts of the world. Students are often being taught erroneous facts, concepts and theories when more accurate ones exist. Speeding up the development of teaching materials and communication systems will be an essential part of this process.

Another goal is to build on the experiences of the past and to fully utilize the findings from educational research. There is no need to re-invent the wheel. Costly lessons were learned during the curriculum development days of the 1960s and 1970s. Expertise was developed; an experienced cadre of chemists with curriculum development experience now exists. Researchers have new findings to report on human development and learning. The need is to mobilize and utilize these resources.

A third goal is broadening the chemistry curriculum. It is essential that an interdisciplinary-oriented chemistry curriculum comes into being to satisfy our larger audience of students and the needs of the world. This is already happening in some countries, as has been pointed out in Chapter 2. It is neither realistic nor academically wise for chemists to ignore the larger world. Science does have close relationships with mathematics and with other science disciplines. Chemistry has economic, social and political overtones, as well. They cannot and should not be ignored in the chemistry curriculum.

And the fourth goal for the future is to prepare to utilize the new technologies and methodologies as they become available in nations around the world.

As stated earlier, there are five somewhat overlapping audiences for chemistry courses at the secondary school level. If we consider their future and the knowledge of chemistry that should be beneficial, it will help us to determine what the content should be, how much emphasis to place on various topics, and to what degree interdisciplinarity, relevancy and social–economic interfaces should enter into the curriculum. As a reminder, these five audiences to whom we address our instruction in chemistry whether together or as separate groups are the following:

1. Those who will become the chemists or the physical scientists as adults and devote their lives to extending the frontiers of knowledge in these areas. Innovation and progress deeply depend on our success in producing a constantly renewable pool of talented scientists and engineers.
2. Those who will use chemistry as a base for their further learning in a science-related profession. These include engineers of all types, agriculturalists, nutritionists, the physician, dentist, nurse, etc.
3. Technicians and other support staff in the health and food industries, for example. It is critical to the health and safety of the entire population that these individuals possess a functional literacy in chemistry.
4. Those who will become the managers, the decision-makers (e.g. industrial managers, government officials, school administrators); their decisions frequently require a knowledge of science as well as the rational decision-making processes learned through science.
5. The citizenry who need a basic level of literacy of chemistry and in science for their everyday living, as well as to use in making responsible decisions as citizens when their vote is called for.

The needs of all of these groups must be taken into consideration when designing chemistry courses. Can one course be designed to serve all of these groups or are several needed? If one can be effectively designed to serve all of the students at the secondary-school level, prepare them for the use of chemistry in any walk of life and help them to keep all of their career options open that is the preference. In practice, a science stream and a non-science stream may be necessary.

Content of the chemistry curriculum

Secondary school chemistry in many countries is oriented to inorganic and physical chemistry. However, students need an introduction to organic and biochemistry. And another area that is essential for scientific literacy and understanding is that of environmental chemistry. With organic chemistry, it is possible to give the students an introduction to, and some understanding of, petroleum chemistry, pharmaceutical chemistry and polymer chemistry (essential areas of importance in our economic growth), for example. Through biochemistry, a number of important principles can be consolidated and extended. They include catalysis, rates of reaction, types of reactions, mechanisms of reactions and chemistry related to the student as a human being. Topics from environmental chemistry will help students understand chemical weathering, pollution, effects of fertilizers on the land and in the water, etc. Some attention to colloidal and surface chemistry, to photochemistry and to nuclear chemistry will be attractive to students and broaden their perspective and view of their own world as well as that of the entire chemical enterprise.

However, there are severe problems associated with trying to develop either interdisciplinary chemistry or to address the economic, social and political interfaces. These include our lack of knowledge and expertise in other science and social areas, the difficulty in determining what the delicate balance among disciplines should be and the problem of how to best use the brief time available to us for teaching chemistry. Perhaps the answer lies in teaching chemistry but in using the relevant (as opposed to the classical) examples and anecdotes to do so. As we teach catalysis, acid/base chemistry or the Haber process, for example, teachers can use interesting anecdotes that relate the chemistry being taught to the news of the day, the importance of the product or something of industrial processes. We can also use snippets of humour and history to enliven the presentation of facts, concepts and theories. It is particularly important to relate all these as far as possible to the experiences and needs of the student. This is a major hurdle in countries where, for example, the chemical industry and research is still embryonic. Nevertheless, there are some excellent but under-utilized sources of contemporary material available. These include newspapers and popular magazines. In the United States, for example, *Chemical and Engineering News* publishes profile data on petrochemicals and other high-volume chemical compounds. This is a source of up-to-date information on production, use, import–export data, market trends, etc. for teachers who wish to bring an economic dimension into the classroom. Articles from magazines such as *Nature* or *Science '84* on topics such as acid rain, planetary chemistry, new polymers or the chemistry of fireworks can bring a relevant interface to chemical education, while remaining within the context of chemistry. In countries where there are relatively fewer sources and it is very difficult to obtain these materials, Unesco has attempted to bridge the gap through its *New Trends in Chemistry Teaching*, but an even greater international effort, in which organizations such as IUPAC must play the leading role, is needed.

Organization of the chemistry curriculum

Any new efforts towards chemistry curriculum development should establish chemistry as a basic part of the secondary-school curriculum a required, not an elective subject) for all students. Once chemistry is a permanent part of the curriculum, a core and options approach in place of the more traditional scope and sequence approach is an option to consider. In the core and options approach, it is possible for chemical educators (a working group of chemists and high school chemistry teachers in any country) to determine a minimum set of facts, principles and laboratory experiences that should become a part of every student's study regardless of career goals. Such a core could comprise no more than 60 per cent of the time available for chemistry teaching, leaving 40 per cent for options designed to meet the differing needs, talents, and interests of this large universal group of secondary-school students. Such options can be directed to interdisciplinary relationships within the world of chemistry (e.g. biochemistry, geochemistry, environmental chemistry, nuclear chemistry) or to differences in abilities (modules that require higher mathematics including calculus versus those that are primarily qualitative) or to differing interests (modules on rice farming, food preparation, medical tests, industrial use of metals or the chemistry of oils and pigments for the artist).

If a core and options approach is not feasible, another way of organizing the curriculum for the diversity of student interests, abilities and career goals is to design a two-stream curriculum. One is mathematically rigorous, reasonably abstract, relatively pure chemistry for students who clearly have science-oriented goals in mind. The second can be a very broad interdisciplinary chemistry based on an integrating theme such as 'Investigating the Chemistry of Planet Earth' and including sections on cosmo-chemistry, atmospheric chemistry, chemical oceanography, geochemistry, biochemistry, environmental chemistry, nuclear chemistry, etc. The Chemistry Committee of the Australian Academy of Science is experimenting with such an approach under four main headings, Earth, Fire, Water and Air [5]. This is back to the Ancient Greeks, but in modern costume.

Another example of emerging work in chemistry is a new set of modules developed under the auspices of the American Chemical Society (ACS) entitled, *Combatting the Hydra* [6]. Designed for the lower secondary schools, these are relevant units that have a goal of acquainting students with the understanding that for every new development there are unanticipated side effects, some good, some bad. They use advances in fibres, pest control, food additives and energy as examples. These small modules have been field-tested and published for possible integration into general science curricula in the United States.

A more ambitious new project, Chemistry in the Community, has been funded by the National Science Foundation. This programme, under the sponsorship of the High School Office of ACS, is directed by W. T. Lippincott of the University of Arizona, formerly editor of the *Journal of Chemical Education*. It will develop a series of consumer-oriented chemistry modules to be classroom-tested and integrated into existing chemistry curricula. The fact that both these programmes are modular and that most of the new work being done around the world is modular is simply recognition that it is close to impossible to replace an existing curriculum or to add in any major degree to them. The strategy for curriculum development then, for the present and foreseeable future, has become the development of materials that can be inserted into existing curricula with minimum disruption to the system. Other examples are described in Chapter 2.

Designing curricula for the future

There are a number of questions that need to be explored in the next decade and for which answers must be found through experimentation with carefully designed projects under the direction of skilled curriculum innovators and experienced classroom teachers. What is the role of chemistry in the total pre-tertiary curriculum from the first primary grade through to the end of secondary school? How shall materials be developed and sequenced by level, and what in fact is the correct amount of chemistry to be balanced with other subjects in the primary school, the middle school, the junior secondary and the senior secondary systems? Should chemistry be presented as a separate discipline or as part of an integrated science until the upper secondary school?

How shall chemistry and mathematics best be related? One of our persistent problems is that the mathematical principles taught in mathematics classes do not transfer well to chemistry classes. Students who have learned to do algebra, ratio and proportion, and logarithms in mathematics classes, for example, do not seem to be able to use them in problems dealing with chemistry. Therefore, the transfer of mathematical knowledge to the study of chemistry is a basic area of concern that needs attention in the near future.

The question of how much laboratory or practical work can provide maximum value in a secondary curriculum is still open for investigation and answer. To what degree can demonstrations replace experimental work in times of financial stringencies? What is the appropriate mix of demonstration and experiment to meet optimum conditions of safety for students, reasonable costs and minimum hazards, which will promote the most learning for the students and best utilize the talents and experience of the teacher?

How can we best define the particular goals of specific groups and therefore design a curriculum that will benefit all? What is the appropriate balance of facts, mathematics, laboratory experiences, reasoning skills and theories or models in secondary-school chemistry? To what degree should the secondary-school curriculum be descriptive as opposed to theoretical? [7]

To what degree should the teaching of chemistry in the secondary school be based on concrete experiences as opposed to abstract ideas? Certainly the learning theorists and the educational psychologists urge us to be concerned about this last question and to base new curricula on ideas that have been recently developed and tested as discussed by Fensham in Chapter 7. We also need to be concerned about motivation, industrial ties and articulation with the tertiary level.

Is a new round of curriculum development due? I say yes. The curriculum development move in the late 1950s and 1960s in the United States and western Europe and the further intensive activity, particularly in Asia, Israel and Australia during the 1970s, have all had sufficient trial for us to learn from them and to move on. This will take all of the creativity, competence and commitments that can be mustered around the world.

8.3.2 *Curricula for the education of a chemistry teacher*

One human resource that is vitally important to the health of the scientific enterprise and therefore to the development of a nation is the science teaching personnel from primary school levels to graduate levels. Since we are focusing our attention on the secondary level, let us consider the directions we must take in the near future to prepare highly competent, curious, committed, creative, compassionate teachers of chemistry for the secondary level.

The training of a chemistry teacher has not changed substantially in any area around the world during the past thirty years and very little since the beginning of the century. In nation after nation, teachers are trained in chemistry and education either concurrently or consecutively as noted in Chapter 6. But in either case, the pre-service curriculum includes a substantial professional educational component prior to the beginning of a career as a teacher. Those outside this pattern are teachers who come into chemistry teaching, in essence through the back door or by chance, by being employed first and then earning the necessary education or chemistry credits while they teach. A curious effect noted anecdotally is that many of the best teachers are exactly those who received at least part of their training in education after some teaching experience.

Correlated with this is the fact that surveys in the United States and in other countries find that experienced teachers, in retrospect, rate their education in chemistry, the methods classes and the practice teaching experience as valuable and rate most of the other components of their pre-service education, (i.e. educational psychology, social foundations, comparative education, introduction to education, etc.) of little value. Why is it then that we have not been able to change the teacher training curriculum? Why are we so reluctant to experiment with dramatically new modes of pre-service education? Are we bankrupt in ideas, bankrupt in financial resources or unable to cope with tradition and the administrative and bureaucratic systems of our nations?

Characteristics of a successful chemistry teacher include enthusiasm for chemistry, love and understanding of the students and the challenge of developing future potential through education. An effective teacher is up to date and well grounded in chemistry facts, theories and laboratory procedures. A successful chemistry teacher knows how to manage a laboratory, guide experimental work, follow safe procedures, use emergency care in case of accident and orchestrate changes of pace in the classroom. In order to motivate students and facilitate their learning, teachers must possess in their repertoire of skills a mix of mini-lectures and interaction modes through skilled questioning, facilitation of discussions and encouragement of problem-solving, and then employ these judiciously. The use of audio-visual aids can expand a student's understanding by bringing into the classroom vicarious experiences that cannot be enjoyed otherwise. Overhead projectors, video recorders and other equipment make for more interesting lessons. In many countries, they are not available. Indeed, teachers may be lucky to have electric power and running water. But visual aids, diagrams for the wall, home-made or professionally made, are still a valuable asset to many lessons. Teachers must also be prepared to assist individuals gifted, average or slow in learning effectively, utilize the newest technologies when they become available (computers, calculators, interactive television, etc.) and be able to design or adapt instructional materials and assessment instruments so that they are appropriate to the interests, abilities and career aspirations of the students in a particular classroom and suit the requirements of society. These expectations place heavy demands on a teacher and on our training programme.

It is becoming increasingly apparent that a chemistry teacher's education has only begun with the completion of a pre-service education and teaching credentials. Through in-service education, the teacher must extend his or her learning throughout a lifetime. Let us explore some directions for the future by considering the career-span education necessary [8].

While practical realities may dictate against some of the recommendations that follow, the education of secondary school chemistry teachers for the future might ideally be considered to consist of four phases:

Phase 1. A four-year concurrent university education in science and education to include (a) a science component that includes a full major in chemistry, mathematics through calculus, and a foundation in physical, biological, environmental and earth sciences and (b) education courses to include methods of teaching chemistry, supervised practice teaching and an integrated introductory professional education course (sociology, history and philosophy of education, educational psychology) at the most.

Phase 2. At least two years of probationary teaching to gain experience and to make a definite decision about a long-term career as a teacher of chemistry; the decision should be made jointly by the individual involved and the officers of the system.

Phase 3. An in-service education that extends to at least the equivalent of one year of advanced study in a university and includes such elements as theory courses on teaching, learning and human development, courses in curriculum and instructional methods and materials, in the preparation and use of audio-visuals, evaluations, history, philosophy and sociology of education, and statistical and research measures. At least half of this in-service effort should be devoted to increasing competence in chemistry, including the laboratory component, through advanced studies in chemistry.

Phase 4. Continuing education throughout a career through self-education, professional activity and formal coursework.

If these recommendations are to be followed, they clearly place new demands on governments, on educational systems, on universities, and on the classroom teachers themselves.

Pre-service curricula

New instructional materials and courses must be designed to assist the prospective chemistry teacher in gaining professional competency, in particular, since the chemistry and science courses available are generally reasonably adequate for our needs.

Innovative approaches to field experience that precedes the practice teaching component must provide opportunities for prospective chemistry teachers to study and participate in not one but a variety of school classrooms when student learning is guided by teachers using widely differing approaches. The goal is to make it clear that there are various 'right' ways to teach successfully. At the same time, pre-service students can gain practical experience in interacting with high school students, in preparing reagents for laboratory experiments, in assessing student learning in a laboratory, and, very importantly, in discussing chemistry teaching with experienced teachers and learning that there is a clear philosophy of chemical education in the mind of each teacher who succeeds.

This type of field experience prior to the practice teaching experience is rated by students as one of the most valuable experiences of their pre-service curriculum. Some decide a teaching career is not for them, but most develop a commitment and 'teaching readiness' of great value during their practice teaching experience. Such an experience is individualized; it takes the additional time and effort of the students, the co-operating schoolteachers and the teacher educators but it pays high dividends. Can we move to a more individualized teacher preparation programme using this or some other model as part of the curricula? Since chemistry teachers are so vital to the community, such expenditures of time and effort and funding could easily be justified.

Preparation for laboratory teaching is an indispensable part of pre-service chemical education. Although it is time-consuming and relatively costly, as well as physically, mentally and emotionally taxing for teachers, this component of a student's education in chemistry is critically important if the spirit of science and the problem-solving, reasoning skills are to be conveyed to the younger generation. Also, if one of our goals is to develop in students the capacity for independent learning much more attention must be paid to preparing prospective teachers for this aspect of their responsibility.

In-service curricula
This formal component should be completed within five years following the beginning of the teaching career. It can be accomplished through evening, weekend, summer or full-time studies, as circumstances permit. The co-operation of university chemistry departments and education departments is essential in making such opportunities available at a time convenient for school chemistry teachers. Alternative modes of delivery should also be considered. These might include home study, mini-courses, self-instructional packages, use of radio and television and other modes of mass communication.

Efforts have been made in a number of countries to produce improved methodology and materials. Some of these include the STEP materials [9] developed in the United Kingdom during the 1970s which then spawned similar projects in a number of countries including ASTEP, THAI-STEP, and work in Israel, I-STEP. Some of these are discussed in Chapter 6.

In the United States, the National Science Foundation funded the Undergraduate Pre-service Science Teacher Education Project (UPSTEP) in various universities across the nation. These were intended to develop innovative approaches to science teacher education. One set of materials that can serve as an example of this effort is the work of the Iowa UP-STEP [10].

These projects have attempted to establish a new dimension in the pre-service and/or in-service training of teachers through the development of questioning skills and laboratory and field competence, for example. Central co-ordinating groups such as the Committee on the Teaching of Chemistry of IUPAC could collect these innovative materials developed in various regions of the world and select from them the most promising examples, organize them into teaching packages with new promise and disseminate them to chemical educators on the international roster for use in the various countries around the world [11]. We must take steps not only to improve our teacher education but to share our ideas more effectively and such an effort would be one of value.

An example of an innovative and very powerful mode of in-service education is the Chemistry Teaching Associate (CTA) programme developed and implemented at the University of Maryland from 1968 to 1973 [12]. The CTA progamme brought together experienced high school teachers and college professors in a team effort to improve instruction and curricula at the secondary and tertiary levels. The secondary teachers were released by their schools to spend one academic year as associated members of the chemistry department faculty. At the university, they worked in partnership with lecture professors, provided leadership for junior instructors and pursued their own coursework designed to improve their competencies and to advance them professionally. At the end of the year they returned to their secondary school classrooms to share the benefit of their new learnings with their students and to serve as important and knowledgeable liaison agents, thus providing improved articulation and ties between secondary and tertiary chemical education.

In addition to dramatic multiplier mechanisms (30 CTAs influenced more than 50 chemistry professors, 200 junior instructors and 10,000 tertiary-level students and then carried the benefits to approximately 150 secondary students per CTA per year through the remainder of their careers) there were positive side effects (development of new courses, curricula, resources for students, professional development programmes for teachers and the building of mutual respect among secondary and tertiary teachers). Models like this need to be explored further with co-operative funding from secondary- and tertiary-level institutions, government and industry.

Continuing education

A teacher's education is never complete. It must be constantly updated and upgraded. This can be accomplished in many ways and is largely the responsibility of the individual teacher. Formal coursework leading to advanced degrees is one possibility. Informal workshops, seminars or short courses offered by higher education, industry or the education system are other choices. Keeping current through the reading of professional journals, where available, participating in chemistry and education meetings and conferences, reading the newspapers, popular science magazines and viewing television presentations also contribute to keeping the teacher up to date, enthusiastic, knowledgeable and exciting to the students.

The responsibility for such continuing education exists throughout an individual's teaching career to retirement and beyond. Its rewards are in professional satisfaction, the excitement of seeing talented young people develop and in the love and respect accorded a teacher who keeps contemporary and competent.

8.3.3 *Evaluation*

The terms 'evaluation' and 'assessment' are often used interchangeably. To some they have similar meanings. To others, evaluation is a broader term incorporating assessment. It is the latter interpretation of evaluation that will be used to set the context for looking into the future.

In the past we have focused our attention primarily on assessing student learning in chemistry. For the future, that is only one-fourth of the equation that equals success. We should also be evaluating curricula, instruction and the assessment instruments and systems. Informally, at least, we have been doing this. There are always popular opinions about the effectiveness of a course, a curriculum, a particular teacher and a test. But we need to formalize these results, if possible, to move them from the realm of hearsay, myth and popular lore to documented evidence, no matter how crude. Early efforts will set the stage for improved documentation year-by-year into the future.

In addition to the need for improved assessment techniques to evaluate these four areas of chemical education, there is another aspect of the changing scene that must be high on the priority list of goals and strategies for the future teacher training in assessment and all areas of evaluation. This arises from the fact that external assessment at the secondary level is on the decline and internal assessment is increasing. While external matriculation (university entrance) examinations will surely continue, more and more teachers are being requested and required to assess student progress at every stage in the pre-university system.

In addition, higher education institutions in many developed and developing countries, too, are moving strongly to formula admissions that include more elements than test scores. These might include the scholastic average from secondary school coursework, if a grading and credit system exists, letters of recommendation from teachers and school administrators, evidence of leadership and special talents through activity in student organizations and student activity in community or religious organizations. Interviews, awards and honours, and the results of papers or student projects are also entering into the broadening evaluation system being utilized by tertiary education and potential employers as well. This trend requires at least two steps forward for the future: (a) improved education in evaluation theory and assessment techniques for the classroom teacher; and (b) improved assessment instruments for teacher use that satisfy either norm-referenced or criterion-referenced modes of assessment.

Assessment of student learning

Since students, and teachers too, have different abilities and talents and are able to demonstrate their accomplishments in various ways, true assessment requires a variety of techniques. Instruments must be developed to better assess the levels of cognitive learning (facts, principles, concepts, theories, mathematical relationships, etc.) Assessment of the learning from the laboratory is equally important. What manipulative skills (operating a burette, using a balance, setting up apparatus for an experiment, etc.) and intellectual skills (interpreting data, calculating, designing graphs and tables, analysing results, stating and revising hypotheses, designing an experiment, etc.) have been accomplished? Measuring the attitudes, opinions and interests of students in order to design better courses is important as well. The function of the first two assessment modes is to determine student learning and competence. The last should never be used to assess students, but only to provide guidance to curriculum developers and classroom teachers so that they may improve the course, the curriculum or the quality of teaching. If our goals in chemical education include teaching students to estimate, to predict, to interpret, to analyse, to solve problems and to make rational decisions, we must learn to assess these skills. If they are not included in our assessment programmes, they will be relegated to second-class priority in the classroom or ignored entirely.

Evaluation of curriculum and instruction

Computer analysis of textbooks and other instructional materials is another possibility for the future. Through the development of the necessary software (computer programming), instructional materials can be analysed for reading level, quantitative and reasoning levels, gaps and overlaps in content, or writing style, for example. Such analysis, impossible to complete manually, will, if thoughtfully used by authors, editors and publishers, substantially improve instructional materials for chemistry education. Such assessment of courses and curricula through their instructional materials has potential for long-range significant benefits if utilized effectively.

Evaluation of curriculum and of instruction will inform the public, who have a right to know about the quality of education being provided for their children. It will provide for parents the comfort (or conversely, recognition of the need for revision) of an up-to-date curriculm that provides learning in areas considered basic by chemical educators. And it will provide for classroom teachers a profile of their strengths and weaknesses so that they can build on the strengths, eliminate the weaknesses or leave the profession.

Tests and assessment systems

Advances in assessment should also permit the development of tests that fit the full range of student cultural/social backgrounds. Assessment instruments have been generally designed for urban students from families of the average or above average income segments of the society.

They measure unfairly against students of high ability from rural backgrounds or low socio-economic families who have the native intelligence to become outstanding scientists but may not be able to demonstrate this on the current tests due to cultural or educational disadvantages. This talent can be identified through modelling tests to suit particular cultures, thus providing the opportunity for disadvantaged students to demonstrate their talents and competencies.

Assessment instruments are powerful ones. In many nations they control the curriculum and instruction more forcefully than any other element. Regardless of what the instructional materials and the syllabus recommend, teachers are under pressure to teach to external examinations if they exist. If the external examinations are not in harmony with the educational objectives, great difficulties arise. Confusion, disillusionment and frustration result for student, teacher and parent alike. The curriculum is often re-ordered to suit the examinations. A better fit of testing to educational objectives must be a goal for the future.

In a report on testing, teaching and learning from the National Institute of Education in the United States, a panel of experts recommend that testing, and therefore education, can be substantially improved through (a) better fitting of testing to cultural background of students; (b) better fitting of testing to educational objectives; and (c) combining testing with teaching [13]. To accomplish these goals, three important categories of effort will be required:

1. The current movement towards criterion-referenced testing should be strengthened.
2. The potential of the new technology should be exploited in such areas as creation of computer-based test item pools, testing of problem-solving processes and testing tailored to the individual.
3. Better information must be developed and made available on tests and testing for school systems and classroom teachers.

Our assessment goals for the future should encompass increased research and development on the theory and practice of criterion-referenced tests specifically designed to determine what a student knows and can do in the discipline of chemistry. Increased development and use of the new information-handling technology so that computer-based banks of test items produced by professional test makers can be available to be shared among chemistry teachers is also necessary. Such banks allow teachers, schools or school systems to create tests that are reliable, valid and closely matched to the objectives they have set for their chemistry students.

Individualized testing is now possible through use of manually selected items from test banks or the microcomputers that are currently moving rapidly into the schools. As the cost of technology decreases and its effectiveness increases through improved software, the computer-based systems will become a part of education at the secondary level in many nations in varying degrees of development. Computer technology should also permit the design of sophisticated items that have the characteristic of testing student understanding as they apply learning to new situations. Variation in items and in scoring such items is now possible through computer-based technology.

Similarly, computer technology allows improved assessment of existing tests, of new tests, of scoring, item analysis, reliability and validity characteristics, content coverage or cognitive level of thinking. Such technology when fully developed will allow more extensive use of short-answer and essay questions that can be computer-graded to relieve the current dependency on multiple-choice items when large numbers of tests must be given and scored.

In order to carry out criterion-referenced testing, it is essential to define a set of basic skills that are necessary and the level to which they must be attained. Taking an example from the work of the National Council of Supervisors of Mathematics (NCSM), it is possible in most fields to define a set of basic skills that every student, regardless of career goals, might be expected to obtain [14]. Their set includes the following: problem-solving, applying mathematics to everyday situations, alertness to the reasonableness of results, estimation and approximation, appropriate computational skills, geometry, measurement, reading, interpreting and constructing tables, graphs and charts, using mathematics to predict and computer literacy.

If we change the word 'mathematics' to 'chemistry' in the above set, they are also basic goals of chemical education at secondary level, with the possible exception of geometry. In 1981, the National Council of Teachers of Mathematics (NCTM) took another step in defining their educational objectives [15]. They determined that school mathematics at all levels should be organized around the development of problem-solving abilities and that school mathematics programmes should take full advantage of the power of calculators and computers wherever available at all grade levels, so that computer literacy should be part of the education of every student. These same ideas are as appropriate in chemical education and should be used in planning assessment and setting education objectives for assessment.

An even broader view of assessment that gives even more authority to school and teachers at the local level is encompassed in a package entitled *School Controlled Evaluation* developed by the Department of Education for the Government of South Australia [16]. This concept involves a close look at some aspects of the school's activities, describing what is happening and what its effects are. The main purpose of such an evaluation is to improve the quality of teaching and learning in the school. Controlled evaluation is seen as a professional responsibility of the school, its administrators and its teachers. It is aimed at improving the curriculum and instruction within a school. Therefore it is clearly distinguishable from external evaluation or system-wide action. A single aspect of the school such as the school's science programme or the audio-visual centre's service to the entire school might come under scrutiny. Action groups which include faculty, administrators and community members all participate.

This Australian model is a very broad base internal assessment system that involves professionals and interested parties as well. It may be the wave of the future. If so, more research on test construction and more training of evaluators will be highly essential. More development of software and more computer literacy among teacher evaluators will be important, too.

The trend of the future is towards more internal assessment and broader interpretation of what needs to be assessed to include curricula, instructional materials, the teachers and assessment system, as well as student learning. Even the concept of student learning is being broadened to measure various levels of cognitive achievement, the skills and knowledge gained from practical work and the informal learnings in the form of opinions that can be useful in redesigning and improving courses.

In order to document improvements in assessment and gains in learning and to be accountable to our governments or our public or private supporters, intensive research and development of the theory and practice of assessment will be necessary. This means development of further human resource potential in the professional evaluators, in the curriculum developers and in classroom teachers.

8.4 Methods, techniques and materials: trends in the instructional scene

The practice of the past and the present in too many chemistry classrooms around the world has and continues to be teacher talk about chemistry while the students take notes and prepare to give the same information back on assessment instruments. The intellectual interactions that promote learning have been too scarce. These include questioning techniques, wait time awareness, laboratory experiences and other interactive modes among the teachers, the students and the discipline of chemistry.

There are certain goals we should set in our attempts to improve chemistry teaching for the future. One is closer integration of classroom and laboratory activities and concepts.

A second is a continuing search for challenging, relevant, interesting 'hands-on' experiments for students that reach beyond verification of known information and allow them to explore reactively.

Third, a continued push for the development of low-cost equipment, manufactured within a country through using indigenous materials and talents whenever possible, is essential in developing countries. Developed countries face budget stringencies, too, and will have to cut back on their instructional materials and instrumentation. Yet for those teachers who have access to the new technology available for teaching (e.g. hand calculators, microcomputers, electronic balances and pH meters, interactive television and videodisc equipment), better ways of using these adjuncts to teaching must be found. If the costs of such instrumentation continues to decline, the economy of a nation improves and the sophistication of software, methodology and techniques grows, it will be advantageous for these new support systems for instruction to have been sufficiently classroom-tested that the best, most effective use will be determined and this information shared with colleagues throughout the world.

Trends and new directions

In developing this section I will utilize a format that looks briefly at trends in methodology and the related technology that seems to be developing and outline directions that could be useful for us to move towards during the 1980s.

8.4.1 *The role of the teacher*

The teacher should develop from classroom lecturer to facilitator of learning, from teacher-centred instructional strategies to student-centred learning strategies. Motion in this direction can be accomplished through strategies such as 'mini-lectures', interactive, inquiry-oriented questioning, laboratory exercises, student-designed reports and projects, simulations and games, independent study, auto-tutorial instruction and attention to 'wait time'.

8.4.2 *The role of the student*

The student should progress from talking about science to doing science; increase his experimental work and demonstrations as well as more open-ended investigations. In fact, as student interest increases in the chemistry of the environment and other areas of chemistry closely related to the natural world, field work, too, becomes an important component of high school chemistry teaching.

8.4.3 *Laboratory costs and safety factors*

There should be a concern for cost and safety: as laboratory experimentation, demonstration and field work increase, so does the cost of chemicals; more instrumentation is necessary and there is greater risk involved as the possibility of laboratory or field accidents increase. These circumstances create a need for finding local sources of inexpensive chemicals that are pure enough to be used in secondary school chemistry, teacher training in safety procedures and development of low-cost equipment.

8.4.4 *Low-cost equipment*

Use should be made where possible of endogenous rather than imported equipment. The contribution of chemistry to national development cannot be over-emphasized. It would be shortsighted on the part of both the profession and a nation if the quality of instruction in chemistry were limited unnecessarily by the lack of equipment. In fact, the challenge of producing equipment from local materials, equipment that can be built and maintained by a classroom teacher, is gaining favour around the world. This trend stimulates excitement and fires the imagination of committed teachers.

The development of low-cost equipment using indigenous materials and talent is evident in the new chemistry curriculum developed by IPST in Thailand, the work of NCERT in India and the IUPAC/ Unesco group on low-cost equipment working in India. Efforts in Sri Lanka, Nigeria, Kenya and Ghana provide further examples of what can be done.

Let us use the Thai example to illustrate. Recognition that new laboratories could not be built and equipped for all of the high schools did not lead to the usual decision to abandon or severely curtail laboratory experimental work or to limit laboratory activity to a few selected schools. Instead, the Thai chemistry team decided to include experimental work intensively in the new curriculum by integrating it into the textbooks and developing portable laboratory kits. These kits are so designed that they contain the equipment and chemicals needed for groups of students to carry out all of the experiments required in a course.

Ingenious apparatus such as test-tube holders, burners and balances were developed. Teachers were trained in workshops to repair the apparatus, if necessary, and to build additional pieces as required. If only one item per class was needed (e.g. a periodic table to hang on the wall of the classroom, or a bunsen burner for high temperature demonstrations), it was included in a teacher's kit.

The pieces of laboratory apparatus developed by the chemistry curriculum team working in co-operation with the laboratory equipment team were tested in schools to determine durability, effectiveness, accuracy, acceptance by teachers and students, etc. When they had been modified, based on classroom feedback, and determined to be effective, the prototype equipment was turned over to industry in Thailand to be developed *en masse* for distribution across the nation.

No one claims that this approach solved all of the problems, but they can document the fact that students in Thailand are now 'doing chemistry' which was not possible until low-cost equipment of this type was developed and teachers trained to use it. Pride is taken in the fact that nearly all of the materials and supplies can now be produced in Thailand without the high costs associated with import. In fact, not only has new equipment been developed where indigenous equipment did not previously exist, but new factories have come into being to produce this, a superb accomplishment for a developing nation.

8.4.5 *Simulations and games*

When experimentation and demonstration are not feasible, a move from textbook narratives and teacher talk to vicarious experience through simulations and games is helpful. When it is not possible to carry out an experiment, learning can occur through simulations which may require the student to take data from a real experiment, plot it in graph or table form, design an experiment around it, make predictions about extensions and have them verified by research laboratories or computer use. Simulations that deal with how much chemical fertilizer should be placed on a field to reach optimum crop yield is one type of simulation. The students, given enough data, can choose different doses and determine the results.

There are many games available. Some are in card format and help students to learn the names of elements and compounds, symbols, formulae, equations, etc. Some are in forms that require game boards, dice and cards, and follow a Monopoly-like format. In this case, students must make decisions as they move towards desired goals. Such games concerned with pollution are examples that bring chemistry into play. Nearly everyone has played bingo or similar games. Getting to know the periodic table can be facilitated by such games. Chemical-reaction rummy, a quiz show format or crossword puzzles promote further incentives through the game techniques.

8.4.6 *Field-work and use of resource persons*

The teacher can move from a course limited to himself and the classroom to a wider world of instructional activities and resources. Sometimes it is not possible for the teacher to provide certain experiences in the classroom that would be of value to chemistry students. It may be possible to take the students to the school grounds or a nearby park to do some work in environmental chemistry or to a nearby industrial or government laboratory to extend and enrich the learning on a particular topic in the classroom (catalysis, for example). Or, in many cases it is easier to have one resource person come into the classroom than to take a large class elsewhere. Chemists from industry, from universities and from government laboratories are honoured by an invitation to come into a secondary school classroom and spend a day with students. They may be reluctant to initiate such a request but generally are very responsive when invited. Field experiences and resource people are as available in rural as urban areas. The key to success is the creativity of the teacher in identifying and using them. More of this should be happening in the future.

8.4.7 *Audio-visual materials and equipment*

A move from the use of motion picture films, film loops and slides to the shorter single-concept films, short television segments and interactive television should be made. An hour-long film is often a waste of time in a high school classroom. Students become bored, unless it is an unusually fine film, and lose interest. Even twenty-minute films are questionable when time is so precious. Short single-concept films, three to five minutes in length, that present an idea clearly and raise questions for discussion are far more powerful. Teaching posters are also effective audio-visual aids. They are less expensive and quite feasible to produce in nearly every country. A short television segment can be very effective since it can be contemporary and bring into the classroom events that cannot be experienced otherwise. With interactive television, it is possible to plan programmes in short segments that pose questions to be explored in the laboratory, through discussion or reading. Students find answers, check them against the programme or expert on television and continue on. Two-way television permits questioning the expert as ideas are developed. Videodisc systems allow random access and controlled sequencing of many thousands of separate frames of full colour slides or data displays. As this high technology becomes more widely available in more countries, teachers will need to prepare for its advent in their situation. In the meantime, the potential of the chalk board, the poster, the flip-board and the overhead transparencies is unfulfilled. Opportunities for creativity in the classroom abound.

8.4.8 *Micro-electronic technology*

A change in instrumentation is taking place in the secondary classroom from the abacus and slide rule to calculators and computers. Slide rules are being replaced by hand-held calculators. Portable microprocessors no larger than a typewriter fitted with a small video screen are moving into the chemistry classrooms as are electronic balances and pH meters.

In countries where this is possible, such instrumentation should be welcomed and fully utilized. In fact, the recognition that a growing number of students have these in their homes already and will want the opportunity to use them in the classroom, too, puts a whole new complexion on individualized instruction in chemistry and the training of new chemistry teachers. It was estimated that in 1981 25 per cent of secondary schools in the United States had micro-electronic capabilities in the science departments. In the United Kingdom, every secondary school has at least one microcomputer, some having as many as ten or twenty, and the trend is spreading around the world. As costs come down, the new technologies will become more available.

8.4.9 *The availability of software*

The change from little or no available software for secondary education to a drive for new programming and networks to share it is now taking place. While the instrumentation (pH meters, balances, computers, etc.) is advancing rapidly, one barrier to full utilization of such resources is a shortage of appropriate software for interfacing apparatus and microcomputers and for micro-electronics-assisted instruction for the secondary schools. This will require the creative talent of experienced classroom teachers. It opens a challenging new avenue for contribution from those who are especially interested. Software generated nationally and internationally will help; it can be shared and made available to local schools systems. However, teachers recognize that software must be modified and individualized to their own teaching styles and their students' needs. No two teachers teach exactly alike in the classroom and no two teachers will find the same software completely satisfactory without putting their own trademark on it through some kind of modifications. Some of the best and most creative software is being generated by classroom teachers and shared through networks with colleagues who have similar competencies and interests. This is a great new development because it opens opportunities for creativity that the teachers need to increase their professional satisfaction and status.

8.4.10 *Printed instructional materials*

Since the cost of producing textbooks is high and since textbooks cannot be revised and reprinted in cycles much shorter than five years, there is a growing trend towards the use of the more flexible short unit, or module. These can be developed, kept up to date and used in the curriculum to suit the different needs, interests and abilities of students in different locations in a nation (e.g. rural or urban, mountain or marine environment). Modules can be tailored to career goals with some more appropriate for vocational students, commercial students or humanities students than others. Even within the sciences, modules that will interest students with physics inclinations as opposed to those who have biology inclinations can be designed.

8.4.11 *Individualized instruction*

Development is now from class or group teaching formats to those that cater more to individual differences. There is a growing concern for differences in individuals and in finding ways to attend to individual needs. For example, the physically handicapped (blind, deaf, crippled) have frequently in the past been excluded from chemical education. With newly devised equipment, laboratory facilities and special training for teachers in working with the physically handicapped, this is no longer necessary. A growing humanitarian concern for giving these individuals an education as a human right is evident (see discussion in Chapter 2).

Another group, the mentally handicapped, the slow learning and mentally retarded, still need an education in chemistry in order to carry out their functions as adult members in a society. The use of remedial self-paced modules that take into account the learning deficiencies of such students and allow them to move at a much slower pace through either laboratory experience or classroom problem-solving exercises is increasing. These will be of value to all nations since the need exists everywhere.

The science-prone students who are often the gifted, the especially talented in perceiving quantitative relationships and in abstract thinking, have the promise of becoming the theoretical scientists who open new frontiers for the world. They need to be challenged and have the opportunity for enrichment activities during their secondary school years that keep their motivation and interest high, stretch their minds and extend their learning to the maximum degree. They should not be held back—in fact, they should be accelerated through the use of advanced placement, programmed learning, individualized instruction, the new instructional technologies or any of the many other modes and methodological strategies available. Their potentials must be developed as early as possible so that they might make the maximum contribution to society. The average student should not be viewed as part of an amorphous mass, but given the same compassionate, committed care as well.

8.4.12 *Reports and projects*

A change from delegating responsibilities for developing communications and quantitative skills to the language and mathematics teachers to accepting some in chemical education is required. Giving each student the individual responsibility to complete a small research study or project creates within them some independence as a scholar and commitment to science as they invest themselves in preparation and presentation. Experience in presenting their ideas logically and clearly in written, quantitative and/or oral from is invaluable. Such competencies can be promoted through giving students opportunities to write, develop case studies, participate in panel discussion, debates, science fairs or student-led seminars.

8.4.13 *The role of research*

In our efforts to improve methodology and techniques and in designing materials and equipment for the future, it is essential that we be knowledgeable about and willing to use the latest findings from research that deal with how students learn and how best to teach. These findings include help in setting our educational objectives, assessing them, developing instructional materials, improving classroom teaching through the use of questioning, knowledge of human development and learning, problem-solving techniques, decision-making skills and concept mapping as examples. This direction for the future also means that there must be more researchers in chemical education. One very important group of individuals who can contribute enormously to our stock of findings are the classroom teachers. Through utilizing themselves, their students and the curriculum as an experimental laboratory, they can mine new knowledge for future generations.

8.5 Problems and potential solutions: coping in the practical world

In the first part of this concluding section, twelve of the top priority problems that plague schoolteachers of chemistry will be briefly noted. Some of these arise out of the economic, political and social environment. Some emerge from the need for further advances in the teaching of chemistry.

In part two, some possible solutions, partial at least, will be identified. In particular, attention will be paid to the role of responsible organizations that have the authority to help effectively. Specific actions that can be taken will be summarized in the form of the Resolutions and Recommendations from the 6th IUPAC International Conference on Chemical Education held at the University of Maryland in the United States, 9 to 14 August 1981. A supplementary set of recommendations developed by secondary-school teachers at the conference extends the possible actions to improve the teaching of chemistry.

What are the constraints to good chemical education into the foreseeable future? What challenges do classroom teachers and chemical educators associated with them face? Some of the most important problems that must be solved during the next decade are briefly summarized below:

Economic constraints. Insufficient money available for purchase and maintenance of conventional equipment and materials; cost of chemicals rising; budget needed for new technology and software development; textbook publishing costs and purchase prices continue to climb.

Political and social constraints. Uncertainty of manpower needs; competition from industry for trained teachers; uncertain priority of science education in the society; insufficient public understanding of science/chemistry.

Bureaucratic burdens. Increasing record-keeping and clerical workload; growth in rules and regulations that inhibit creativity and experimental work.

Teacher status and rewards. Low status generally; declining further in some countries; salaries lower than government and industry; tendency for some of the better teachers to be drawn into administrative posts or into various aspects of tertiary education (for example, teacher training colleges); even the cost-free and most-prized reward, professional satisfaction, is being sapped by outside influences.

Teacher training. Necessity for in-service education (in both content and method) throughout a teaching career; re-examination and innovation in teacher education needed; special needs for training in laboratory safety, laboratory management, doing demonstrations, asking questions, constructing tests.

Curriculum renewal. Gap between research results and classroom utilization. Uncertainties with respect to (a) organization (sequential or block; textbook or module); (b) streaming or mixed ability grouping; (c) same content and approach for all students or tailored for prospective scientists, technicians, non-science students, etc.; (d) proportions of theory, descriptive and practical work; (e) development of new curricula or adaptation of imported curricula.

Student learning. Widely varying range of student abilities, interests and career goals; how to achieve universal science education/chemical education; facilitating student-centred learning; motivational problems.

Facilities and resources for practical work. Inadequate laboratory space; unsafe ventilation in teaching and storage area; insufficient working equipment; need for development of low-cost equipment and new experiments; concern for carcinogens and toxic or explosive substances.

Internal and external assessment. Need for improved test-item banks and item-writing skills; more modes of testing different abilities and skills; greater acceptance of schoolteachers' assessment of students' learning and talents; elimination of coaching schools for external examinations; lack of correlation between university admission and other external examinations and educational objectives of secondary schools.

Interfacing and articulation. Need for closer secondary/tertiary co-operation; need for interfacing with industry; elimination of gaps between expectations of industry or higher education and the curriculum and educational objectives of the schools; need for recognition of interdependencies, for mutual respect.

Communication. Need for increased sharing of information, materials and ideas between and among schools, regions and nations, with higher education, industry and with the public.

Research. Lack of theoretical base for science/chemical education; insufficient understanding of human growth and development, teaching and learning.

These areas of concern demand concerted efforts towards solution. Each individual can play some part but the collective action of organizations with power, resources and authority is also essential.

8.6 Some strategies towards solution of problems

8.6.1 *Clarifying the role of responsible organizations*

There are at least five types of organizations that have responsibility for improving chemical education at the school level. Each of these must assess their past activities and achievements, set task groups to work at planning specific solutions to science education problems and organize specific and substantial goals for the future. These organizations include professional societies such as the umbrella-type American Association for the Advancement of Science (AAAS), those specific to chemistry such as the Federation of Latin American Chemical Societies (FLAQ) and teachers' organizations such as the Association for Science Education (ASE) in the United Kingdom. Also supremely important are the international organizations such as IUPAC, Unesco, ICSU and ICASE. As mentioned earlier, government agencies, industry and tertiary education all have responsible roles to play. Each must clarify goals and roles and commit resources and personnel sufficient to the accomplishments envisioned.

8.6.2 *Recognition of interdependencies and creative use of them*

Chemistry departments in universities depend on the secondary schools for their product; hence, they have a vested interest in facilitating school chemistry to ensure a high quality student body. Industry is dependent also on a good base in secondary school chemistry for their technicians, chemists and chemical engineers. The government and the public need to be made more aware of the importance of the school-level education in chemistry for the scientific and technological advancement of their nation, that their economic and social progress depends on high quality results.

In the past each of the responsible sectors has operated relatively independently of one another. Creative co-operation would be more cost-effective, efficient and satisfying, even exciting, for all. It is time to close these gaps in communication and co-operation.

8.6.3 *Towards improved teacher training*

Internships, study leaves, teaching associate opportunities in tertiary education [17], and short-term experiences in industry or in government laboratories are all possible means of improving the work of in-service teachers through relevant practical experience. Opportunity to build on the base of knowledge obtained in the pre-service education through advanced coursework leading to master's and other advanced specialist degrees is also necessary. Short courses that convey the latest knowledge and techniques to teachers will be needed, as will materials for home study and access to professional journals and books.

8.6.4 *Towards improved curriculum and evaluation*

An endless searching for better content, better methods of instruction and better modes of assessment will be required. Despite two decades of curriculum development activity around the world, we have not yet found ways to devise and revise the curricula, facilitate learning and evaluate student achievements that are satisfactory. A search for even higher quality must continue.

8.6.5 *Towards a stronger research base*

The research base for chemical education is woefully weak. This is partially true because such research was only begun within the past decade. There are few trained researchers and little in the form of either a theoretical base or baseline data on which to build. Research designs generally follow the patterns set by educational psychologists and statisticians. Yet these may not be appropriate and surely are not adequate to conduct effective research related to the teaching and learning of chemistry. More case studies, clinical studies and classroom-oriented studies need to be done. Research on problem-solving, decision-making, learning patterns and structure of chemical knowledge is essential. Research must be directed at the curricula, instructional materials, assessment systems and pedagogical methods as well. The results of many individual studies from various countries around the world must be analysed and synthesized for general use. We must learn how to use the school classroom and laboratory for research without jeopardizing the education of the students. The real results come from the field, not from artificial or ivory-tower situations or environments. Recognition and measurement of side effects and unanticipated results is a neglected area of great promise.

8.6.6 *Towards increased national and international communication*

Research and development results must be communicated worldwide so that schoolteachers and their trainers may share the knowledge, ideas and products generated worldwide. The Newsletter of the IUPAC Committee on Chemistry Teaching (IUPAC/CTC) should be expanded or published more frequently to carry detail on research design so that studies may be replicated and results verified internationally. Curriculum developers must plan within their budget for the printing of additional copies of textbooks, modules and supplementary materials for distribution to colleagues in other countries, using the international roster of chemical educators, further developed by each International Conference on Chemical Education (ICCE) and maintained by the IUPAC/CTC. Journals such as the *Journal of Chemical Education*, *Education in Chemistry*, the *European Journal of Science Education*, or the *Journal of Science and Mathematics in Southeast Asia* and *Chem-13 News* must serve as major resources for international communication. The 'IUPAC/CTC Anthology for Chemical Education' that appears in the *Proceedings* of the 6th ICCE (available separately from the IUPAC/CTC office at the University of York in the United Kingdom) is another impressive source of ideas. The Unesco International Centre for Chemical Studies could serve as a clearing-house of information on research and development in chemical education and issue quarterly bulletins to disseminate such information to each country.

In addition, mechanisms must be established to translate research and development findings into practical articles or packages for classroom teacher use. Individuals or organizations in each country must accept this responsibility and develop strategies and resources to disseminate pertinent chemical education information directly to schoolteachers.

Opportunities for researchers, developers and practitioners to meet in seminars and conferences must be available. Since the IUPAC international conferences occur biennially, regional and national conferences could be held in the in-between years. One excellent model for successful school chemistry teacher conferences is the CHEM-ED conferences held at the University of Waterloo, Ontario, Canada. The REACTS (Regional Assembly of Chemistry Teachers) Day at the University of Maryland, an activity-packed professional day, is a shorter version that has enjoyed high impact and success. Conferences that bring chemistry teachers together with other science teachers are held in many countries.

8.7 Recommendations and resolutions: an action agenda for the 1980s

In preparing for the 6th ICCE held at the University of Maryland, 9 to 14 August 1981, it was decided that an attempt would be made to draft a set of resolutions and recommendations that went beyond the type of cliché that often arises if this activity begins in the course of a conference. In order to attempt a more in-depth look at directions for the future and at agencies who should take responsibility, the organizing committee requested Malcolm Frazer of the University of East Anglia to serve as chairman of an international task force that would evaluate all resolutions and recommendations from previous conferences and select from among them those that should be deleted, those that should be retained, those that should be modified. They could also suggest new ones to be added.

For an entire year in advance of the conference, Professor Frazer corresponded with people in approximately fifty nations around the world. The results were published in the June 1981 issue of the *Journal of Chemical Education* and served as a working paper for the full conference body meeting in August.

During the conference, under the leadership of the National Representatives to IUPAC/CTC, task groups reworked this draft. The secondary-school teachers devoted additional time to the development of supplementary sets of recommendations that were more specific to their own concerns. In the final plenary session these recommendations and resolutions were presented to the entire body and, after discussion and some modification, were unanimously approved for implementation; these now follow.

RESOLUTIONS AND RECOMMENDATIONS SIXTH IUPAC INTERNATIONAL CONFERENCE ON CHEMICAL EDUCATION UNIVERSITY OF MARYLAND, 1981 [18]

It was recognized that recommendations can only be converted into action by individuals (school and university teachers, inspectors, curriculum planners, research workers in industry, etc.). The recommendations are therefore directed for action at appropriate individuals and at the organizations where they work. Each recommendation is followed by one or more letters A, B, C, D and E to indicate the individuals or organizations to which it is principally directed.

(A) Unesco or agencies of local, regional and national governments.

(B) IUPAC, regional federations of chemical societies, national chemical societies or national and international associations of science teachers.

(C) Universities and other institutions of higher education.

(D) Schools and secondary level chemistry teachers.
(E) Industry.

The following resolutions were presented to the final plenary session of the Conference.

1. *Exchange of information.* There is a need for exchanging information about chemical education (e.g. current problems, new approaches, new research results). It is therefore recommended that:
   - 1.1 The publishing, selective republishing and appropriate distribution of source books on chemical education in different languages should be encouraged and supported. (*Directed to A and B*)
   - 1.2 National, regional and international conferences, workshops, etc. on carefully selected topics of current concern in chemical education should continue to be organized. (*Directed to A and B*)
   - 1.3 The International Newsletter of IUPAC-CTC should continue to be supported and developed further. The distribution should be increased and where possible articles should be republished in national journals and newsletters. (*Directed to A and B*)
   - 1.4 Further attention should be given to ways of assisting countries experiencing currency exchange problems related to items 1.1, 1.2, and 1.3. (*Directed to A*)
2. *New organizations.* In every country there should be an association for all those individuals active and interested in working for the improvement of science education including chemical education. There is also a need for regional co-operational between such organizations. It is therefore recommended that:
   - 2.1 Encouragement should be given to create associations for science teachers (or similar organizations appropriate to local conditions) in those countries which do not have them. (*Directed to B, C and D*)
   - 2.2 Support should be given to international agencies such as the International Organization for Chemical Sciences in Development and to regional federations of chemical societies; such federations should be encouraged to collaborate with the IUPAC Committee on Teaching of Chemistry. (*Directed to A and B*)
3. *Teaching at School level.* Education of students at school level in science is fundamental to the development of the individual as a scientist and as a constructive member or society. It is therefore recommended that:
   - 3.1 Professional training appropriate for the task of primary and secondary school teaching be provided. (*Directed to C*)
   - 3.2 Opportunities for chemistry teachers to receive in-service education should be increased. (*Directed to A, B, and C*)
4. *Teaching at tertiary level.* There is a need for greater effort in evolving new approaches to curricula, and in improving teaching, at tertiary level. It is therefore recommended that:
   - 4.1 Continued support should be given for courses and other activities aimed at improving teaching of chemistry at tertiary level. (*Directed to A, B, and C*)
   - 4.2 The career structure of chemistry teachers at tertiary level should be more related to contributions to teaching and to chemical education research and development. To this end further efforts should be made to develop objective methods of evaluating contributions to teaching. (*Directed to A and C*)
5. *New experiments and low-cost equipment.* For laboratory work it is desirable to use local materials and simple equipment wherever possible. Furthermore it is cheaper. It is therefore recommended that:
   - 5.1 Continued efforts should be made to design low-cost equipment and to produce new experiments using locally produced equipment and local or less expensive chemicals. It is also important that curriculum development should take into

account the problems of the cost of apparatus and chemicals. (*Directed to A, B, C, D and E*)

5.2 Initial and in-service teacher training couses should: (i) include the design, construction, adaptation and use of locally produced, low-cost equipment; and (ii) encourage replacing expensive chemicals by alternative locally produced materials (*Directed to A, B and C*)

5.3 Particular attention needs to be given to publishing and exchanging information about new experiments employing simple apparatus or local materials. (*Directed to A, B, C and D*)

6. *Education–industry co-operation.* The need for better understanding and exchange of information and personnel between industry and educational institutions at all levels is widely accepted. It is therefore recommended that:

   Continued encouragement from industry and Government agencies should be given to realistic schemes for promoting education–industry co-operation locally, nationally and internationally. (*Directed to A, B, C and E*)

7. *Attitudes to chemistry.* Attitudes to science are largely formed before the age of 14 and the influence of home, local community and primary school are important. It is therefore recommended that:

   There should be continued support for programmes aimed at (i) improving science teaching in primary school (particularly those concerned with science courses in primary school teacher training); and (ii) bringing a balanced appreciation of science to parents and to the general public especially through the media. (*Directed to A, B, C, D and E*)

8. *Chemical education and society.* A current trend, which will continue in the 1980s, and which should certainly be encouraged, is the close relationship of chemical education to society and to future needs. It is therefore recommended that:

8.1 Ways should be found for including more material on the social, economic, technological, legal (including patents, licences, technology transfer, etc.), cultural and ethical aspects of chemistry in curricula at both school and university levels. (*Directed to A, B, C, D and E*)

8.2 Chemistry courses at all levels should make students aware of the role of the chemist in society and of the present and future role of chemistry in society. (*Directed to A, C, and D*)

8.3 In curriculum planning attention should be paid to the possible future directions of chemistry so that students learn what the challenges will be tomorrow as well as what they were in the past. (*Directed to A, B, C, D and E*)

8.4 Chemistry courses should include instruction on information storage and retrieval. (*Directed to B and C*)

8.5 Encouragement should be given for the development of chemistry courses and media presentations for adults (particularly those with no formal scientific education who through their work are involved with chemicals). (*Directed to A, B, C and E*)

9. Scientists are becoming increasingly aware of the need for improved safety practices and their consistent use. These are essential to the health and to the education of the student. It is therefore recommended that:

   Consistent guidelines for the safe use and disposal of materials, appropriate for chemistry teaching, be developed and widely distributed. (*Directed to A, B, C and E*)

REPORT FROM THE SECONDARY SCHOOL INTEREST GROUP [19]

The following items were identified as major concerns of secondary teachers from countries represented at the Sixth International Conference on Chemical Education.

A. *Safety*

Scientists and teachers are becoming increasingly aware of the need for improved safety practices and for their consistent use. Establishment of such guidelines are essential to the protection of health and to the education of the students.

1. Safety courses are not generally a part of the established curriculum for prospective teachers. Teachers, therefore, find themselves setting up a storeroom and laboratory without the necessary safety background.
   Recommendation: Safety courses should be an integral part of teacher training.
2. There is a lack of continuing education programmes in safety for all science teachers.
   Recommendation: In-service training in safety be provided for *all* science teachers.
3. Scientists are becoming increasingly aware of the need for improved safety practices and for their consistent use. Establishment of such guidelines are essential to the protection of health and to the education of the student.
   Recommendation: Consistent guidelines for the safe use and disposal of materials appropriate for chemistry instruction be developed with the aid of the chemical educator. These guidelines should then be widely distributed to various science organizations, schools, university administrators and teachers.
4. There seems to be a lack of safety programmes in the schools to increase *student* awareness of safety precautions.
   Recommendation: Safety programmes for students should be developed and shared within the science education community.
5. Many schools are experiencing an overcrowding in the chemistry laboratory. The safety of the students in the lab seems to be related to class size.
   Recommendation: The number of students assigned to the laboratory should neither exceed the number that the laboratory was designed to accommodate nor should it exceed the number that can be safely supervised by the teacher.
6. The question of liability was also addressed, but insufficient time and information was available to thoroughly investigate this problem.

B. *Curriculum development*

There is a concern that many established curricula succeed only in motivating a few students to pursue a career in chemistry or related fields. Often the interest of the student is actually decreased or destroyed altogether, and the negative feelings are all too readily transferred to science.

1. All students should be made aware of the processes of chemistry and the many links between chemistry, society and the environment.

   Recommendation: Instead of offering one highly theoretical course for a select group and a 'weaker' course for the rest, it is recommended that at the introductory level only one course be offered which would meet the needs of all. Such a course might emphasize the role of chemistry in life, business and relaxation.
2. The needs of students change as society and chemistry evolve. The chemistry curricula must thus be periodically evaluated and revised to reflect these changes.

   Recommendation: Constant communication between the pre-secondary, secondary, and tertiary level teachers should be encouraged, so that pre-tertiary level teachers can make important decisions in curriculum development at all levels.

C. *Professional concerns*

There were several concerns relating to the profession of chemistry teaching.

1. There is a need for a greater awareness among teachers of the conditions under which members of the profession operate in other countries. These include the constraints of finance and accommodation under which we work and the status of the teaching profession within the community.

   Recommendation: It is proposed that a series of articles should be written by teachers in different systems of secondary education throughout the world. These articles would describe the constraints and resources within the teachers' own system.
2. There is a need for a continued increase in the contributions made by secondary-school teachers to the next and all future International Conferences on Chemical Education.

   Recommendation: Teachers in secondary schools should continue to be encouraged to be actively involved in the organization of conferences. Organizations concerned with the teaching profession and with chemistry should encourage teachers to attend international conferences.
3. There is a need to prevent the increasing cost of chemicals and apparatus from decreasing the quality of teaching.

   Recommendation: Chemical suppliers should be encouraged to make chemicals available in suitable quantities for school use. This applies particularly to chemicals which are expensive, difficult to store, or difficult to dispose of. Local groups should investigate the possibility of sharing apparatus and chemicals. Local industry should be contacted and encouraged to donate used equipment or sell it at a substantial reduction to schools.

D. *Multi-national concerns*

1. There is a need to narrow a wide gap between teachers' salaries and salaries for chemists working in industry.

   Recommendation: Salaries should be increased to more closely parallel their industrial counterpart. Well-paid summer work which would provide appropriate industrial experience for the teacher should be made available.

2. There is a need for teachers to have greater prestige within their communities.

   Recommendation: The quality of teaching is important and it is desirable for the teacher to interact with the community by way of lecture demonstrations, parent–teacher groups, etc. Increases in the financial status of teachers is also warranted.

3. There is a need for an increased participation by the secondary-school teacher in the chemical community. The communication between secondary and tertiary education should be improved. Teachers should be encouraged to join organizations for teachers and chemists. The possibility of lower membership fees for educators should be investigated. Chemical organizations should be encouraged to provide more activities appropriate to the needs of the secondary-school teacher.

   Recommendation: Opportunities for continuing education should be improved. Funded summer institute programmes and local university courses at appropriate times should be provided.

E. *Characteristics of students*

Although specific recommendations for the solution of the following concerns were not elicited, these items are worth noting with the hope that individuals working in these areas will share their recommendations with the international community.

1. *Communication skills*. Ways must be found and implemented to improve the mathematics and other language skills.
2. *Attrition of students*. It was noted that there seems to be an increasing growth of attrition amongst our students. This is reflected in drop-out rates from courses.
3. *Women in science*. There is much concern in some parts of the world that insufficient numbers of women are entering scientific professions (e.g., engineering, medicine, and chemistry). There is often insufficient opportunity to study science, particularly chemistry, at school.

8.8 Concluding comments

At the beginning of the decade of the 1980s, we seem to have reached a third critical moment or period of time in chemical education. The first was the beginning of chemistry teaching as an academic subject in the school curriculum. This began at different times in various countries ranging from the late 1800s to approximately 1950. (In a few countries of the world, chemistry is still not available as a school subject.) Generally, the initiation of a school chemistry curriculum was motivated by pressure from the tertiary system to provide introductory work at the school level. In the United States, for example, some chemistry was taught in a didactic lecture course earlier, but the real beginning of chemistry as an academic subject with practical work included began about 1880 when Harvard College set pre-college study in science as a prerequisite for admission to Harvard and defined the experiments they expected students to have done in the secondary school laboratory. That action had momentous impact as other tertiary institutions followed Harvard's lead and the secondary schools adjusted their curriculum accordingly.

The second period of great importance encompassed nearly three decades from the mid- or late-1950s to the end of the 1970s. During this period, as Ingle and Ranaweera note in Chapter 2, chemical educators around the world became aware that the school chemistry curriculum was out of date with respect to content, too fact- and lecture-oriented, and that students were not gaining enough experience in thinking, in problem-solving and in learning from doing experiments. This lead to the curriculum reform movement which began in the United Kingdom and United States and spread to all regions of the world. Great strides were made during this time, not only in bringing new content, experiments and methods into the classroom, but also in the training of teachers, improvements in equipment, instructional materials and tests and in developing experienced leaders.

Now, it is more than a quarter of a century since the curriculum reform movement began. It is now the third critical moment. Strengths and weaknesses in the school chemistry curriculum have been identified; gaps between school level and tertiary approaches and expectations exist. Much more is known now about the science of chemistry (e.g. catalysis, reaction mechanisms, atomic and molecular structure). We have further ideas and experience with facilitating student learning through relevancy, attention to levels of maturity, use of educational technology, etc. Another major initiative, building on past achievements and employing the co-operative efforts of the most qualified chemists and educators, is in order. Between now and the end of this century, another round of curriculum development must take place. A giant step forward is urgently needed, one that involves scientists and educators, researchers and practitioners (the classroom teachers), not only within a country but between and among countries. IUPAC and Unesco can play lead roles in organizing such activity.

What direction will the new curriculum reform movement take? What will distinguish it from the past? The move to a school chemistry curriculum designed for general educatión for all, scientific literacy *per se*, is my prediction. Specialization can occur at the tertiary level or in post-secondary vocational training. The need of the public at large to achieve a functional level of understanding of the concepts and processes of chemistry and the role chemistry plays in their daily lives will be their goal. 'Better Living Through Chemistry' may well become the slogan for a new round of reform.

In a recent report from the National Academy of Sciences in the

United States [20], the purposes and importance of providing an education in science (chemistry) for all the students was outlined as follows: (a) the need for scientifically enlightened citizens to help implement the pluralistic functions of citizen decision-making about science and technology; (b) the need for knowledgeable graduates of at least our secondary school systems to serve as opinion leaders in the political structure to help the public at large understand science and technology; and (c) the belief that well-prepared non-specialists can provide leadership in professions other than science and technology if they have a command of essential scientific concepts and processes.

Some of the associated reasons for providing an education in science for the non-specialist include overcoming fears, exploding myths, developing critical thinking competency, learning how to seek reliable sources of scientific and technical information and how to use them throughout a lifetime and possessing enough scientific and technological know-how to perform more competently in their chosen professions and to fulfil their civic responsibilities in an increasingly scientific and technological society.

To accomplish these educational goals in our various nations, possible strategies include: (a) increasing science requirements at the secondary level for school-leaving certificates and university admission; (b) increasing the rewards for good teaching and the status of science teachers; (c) improving the use of non-traditional teaching aids such as computers, interactive television and videodiscs; (d) designing courses and teaching methods to address more effectively the needs of the non-science student; (e) ameliorating the restrictive influence of external examinations; (f) appropriating the necessary financial support for high quality research and development work in science education.

Our governments must recognize their obligation and responsibility for effective teacher development and improved curricula and assessment systems that evaluate more effectively attainment of educational goals.

The future is a time of challenge, a time of promise. The same characteristics of competence, curiosity, creativity, commitment and compassion will be required in larger measure as we accept the much larger and more demanding goal of providing a high quality education in chemistry for everyone.

References

1. Ketudat, Sippunondha, in W. T. Lippincott (ed.), *Proceedings of the Sixth International Conference on Chemical Education*, p. 83, College Park, Md., University of Maryland, 1982.

2. I. Wirszup, *Educational Leadership*, Vol. 38, 1981, p. 358.
3. *American Association for Higher Education Bulletin*, July 1981, p. 7.
4. M. H. Gardner, *Today's Education*, Vol. 70, September–October 1981, p. 54 (Mathematics/Science edition).
5. D. W. Watts and R. Bucat, *Chemistry Curriculum Committee Report*, Canberra City, Australian Academy of Sciences, 1981.
6. *Combatting the Hydra*, Washington, D.C., High School Office, American Chemical Society, 1981; *Chemistry for the Community*, Washington, D.C., High School Office, American Chemical Society, 1983.
7. R. J. Gillespie and D. A. Humphrey, *Journal of Chemical Education*, Vol. 57, 1980, p. 348.
8. M. H. Gardner, in A. Kornhauser and A. N. Kholodilin (eds.), *Natural Sciences–Social Sciences Interface*, p. 422, Ljubljana, Unesco-IOCD International Centre for Chemical Studies, 1981.
9. *Science Teacher Education Project*, Maidenhead, McGraw-Hill, 1974.
10. J. E. Penick, *Formative Evaluation of I-UPSTEP Model*, Iowa City, University of Iowa, 1978 (Technical report, 17); D. S. Sheldon, *The Development and Maintenance of a Unique and Effective In-service Teacher Education Model*, Iowa City, University of Iowa, 1979 (Technical report, 19).
11. M. H. Gardner, *Journal of Chemical Education*, Vol. 59, 1982, p. 84.
12. M. H. Gardner, *Journal of College Science Teaching*, Vol. 2, 1972, p. 20.
13. *Testing, Teaching and Learning*, Washington, D.C., National Institute of Education, Government Printing Office, 1979.
14. *The Mathematics Teacher*, Vol. 71, 1978, p. 147.
15. *Priorities in School Mathematics*, Reston, Va., National Council of Teachers of Mathematics, 1981.
16. *School Controlled Evaluation*, Canberra City, Education Department of South Australia, Commonwealth Schools Commission, 1979.
17. M. H. Gardner, *Pure and Applied Chemistry*, Vol. 50, 1978, p. 563.
18. 'Resolutions and Recommendations', *Journal of Chemical Education*, Vol. 59, 1982, p. 87.
19. 'Report to the ICCE Conference Editors from the Secondary School Interest Group' *Journal of Chemical Education*, Vol. 59, 1982, p. 116.
20. *Science for Non-Specialists*, Report to the National Science Foundation, National Research Council, National Academy of Sciences, Washington, D.C., 1981.

Bibliography

BRASTED, R. (ed.). *Japan–USA Seminar on Fundamentals of Chemistry for the Non-Major Student in Tertiary Education*. Minneapolis, Minn., University of Minnesota, March 1982.

Guidelines and Recommendations for the Preparation of Secondary School Teachers of Chemistry. Washington, D.C., American Chemical Society, 1977.

KORNHAUSER, A.; KHOLODILIN, A. N. (eds.). *Natural Sciences–Social Sciences*

Interface. Ljubljana, Unesco-IOCD International Centre for Chemical Studies, 1981.
KORNHAUSER, A.; RAO, C. N. R.; WADDINGTON, D. J. (eds.). *Chemical Education in the Seventies*, 2nd ed. Oxford, Pergamon Press, 1982.
Science Education Databook. Washington, D.C., National Science Foundation, 1980.

Notes on contributors

David WADDINGTON is currently Chairman of the Committee on Teaching of Chemistry of the International Union of Pure and Applied Chemistry (IUPAC/CTC) and President of the Education Division of the Royal Society of Chemistry.

He is an author of several well-known chemistry textbooks used in schools and universities worldwide, editor of the Modern Chemistry series for Bell and Hyman, editor of the *International Newsletter on Chemical Education* and editor of two Unesco publications, *Sourcebook of Chemistry Experiments*, Vol. I and *New Trends in Chemistry Teaching*, Vol. V.

He has travelled widely on behalf of Unesco, the British Council and the various British Government offices. His interests include technician training, where he heads the ICSU/CTS sub-committee as well as being a member of the executive committee of ICSU/CTS.

His chemical research interests include work on gas-phase kinetics with special reference to atmospheric reactions.

A. V. BOGATSKI graduated from the Odessa State University, was appointed to a lectureship at the university and became Chairman of the Organic Chemistry Department in 1959, a post he still holds. He has also served as Dean of the Department of Chemistry and pro-Rector and Rector of the University. Since 1975 he has been Chairman of the Ukrainian Academy of Sciences' Southern Scientific Centre and, since 1977, Director of the Academy's Institute of Physical Chemistry, and a Member of its Praesidium. In 1976, he became a Member of the Academy and in 1979 an Honoured Scientist of the Ukrainian SSR. He is the author of over 400 books and papers and among his fields of interest are stereochemistry and conformational analysis and the chemistry of biologically active compounds.

J. Arthur CAMPBELL has been Professor of Chemistry and Chairman of the Department since the start, in 1957, of the Harvey Mudd College of Engineering and Science. Once the College was under way he also undertook, from 1960 to 1963, the direction the Chemical Education Material Study. Supported by the National Science Foundation of the United States, CHEM Study produced a set of materials (books, films, charts) designed to explore experimentally based methods of teaching a first course in chemistry. CHEM Study has had its materials used in many ways in many countries, with translation of the books and films into more than twenty languages.

Professor Campbell has published four textbooks plus about 100 articles both on problems of chemical education and on his research interests in the structural chemistry of liquids and crystals.

Danièle CROS is Maître-Assistante at the University of Montpellier, Montpellier, France. Her chemical research interests include charge transfer complexes and, more recently, the transfer of ions through lipid membranes.

Her work in chemical education is concerned with pre-service and in-service teacher training, and she is responsible for the national working group in ReCoDic on teacher training. She has organized national courses and an international course, on behalf of Unesco, on the teaching of university laboratory courses.

Dr Cros was awarded a prize by the teaching commission of the French Chemical Society in 1976.

Peter J. FENSHAM has been Professor of Science Education at Monash University in Melbourne, Australia, since 1967. Previously, he was Reader in Physical Chemistry at the University of Melbourne with special interests in catalysis and the solid state. He has contributed to a number of areas of research in chemical education, in particular student learning and its interaction with the social context. His science education group has been innovative in the initial and continuing training of chemistry teachers and their introduction to research findings in their field. He has served on several national curriculum committees for new approaches to the teaching of chemistry in secondary schooling. Well known internationally, he has been awarded the Nyholm Medal by the Royal Society of Chemistry and was one of the plenary speakers at the Sixth International Conference on Chemical Education.

Marjorie H. GARDNER, Professor of Chemistry at the University of Maryland, received her B.S. degree from Utah State University and her M.S. and Ph.D. degrees from The Ohio State University.

Professor Gardner has been a teacher of chemistry in high schools in Utah, Nevada and Ohio as well as the Ohio State University and the University of Maryland. Within the American Chemical Society, she has served as Chairman of the Curriculum Committee of the Division of Chemical Education and on the Council's High School/College Interface Committee. Currently a member of the Society's Committee on Chemical Education and the International Activities Committee, she was Conference Organizer for the Sixth International Conference on Chemical Education, August 1981.

Other activities include the membership of the Councils of the American Chemical Society and American Association for the Advancement of Science. She is President of the Fulbright Association, Director of the Interdisciplinary Approaches to Chemistry (IAC) project and of NSF-funded curriculum development, teacher education, leadership and international activities.

Richard B. Ingle is currently lecturer in science education at the University of London Institute of Education. He graduated in the physical sciences at Durham University and then taught science in secondary schools for fourteen years in Scotland, England and Uganda. He subsequently held posts in chemical education at Makerere University College, Uganda, and at the Centre for Science Education, Chelsea College, University of London. During the 1970s he undertook an evaluation of Nuffield Chemistry and subsequently became general editor of the 'Revised Nuffield Chemistry' series. He was, for a time, education adviser at the Ministry of Overseas Development. His current interests include the pre-service and in-service education of science teachers, cultural aspects of science education and studies related to learning difficulties that pupils face in using mathematics in the course of their science education.

Aleksandra Kornhauser started her professional life just after the Second World War as a non-qualified primary-school teacher, graduated in chemistry as a part-time student, taught chemistry at secondary schools and became, in the second decade of the working period, professor of organic chemistry and Dean of the Teacher Training College in Ljubljana. She was also active in research of alkaloids and in 1969 became university professor of organic chemistry at the University of Ljubljana. Today she teaches chemistry of natural products and is Director of the Unesco/IOCD International Centre for Chemical Studies at Edvard Kardelj University in Ljubljana.

She has published over 100 papers on alkaloid chemistry and methods of chemical education, as well as eight textbooks for organic chemistry for secondary and tertiary level. For her work she has received the highest national awards.

Professor Kornhauser was also chairperson of the working party on chemical education, Federation of European Chemical Societies and is now a member of the IUPAC Committee on Teaching of Chemistry and of the Committee on the Teaching of Science, International Council of Scientific Unions.

John N. Lazonby graduated in chemistry from the University of Durham and, after taking a diploma in education, taught in secondary schools for thirteen years. He has since become a senior lecturer in the Department of Education at the University of York where he teaches on undergraduate, postgraduate certificate and higher degree courses in chemical education.

His publications include research papers on chemical education, an International Bibliography of Chemical Education Journals and chemistry books for secondary schools. His latest book is *Chemistry in Today's World*, published by Bell & Hyman.

John C. Mathews took a degree in chemistry at the University of Liverpool in 1943. After war service, he took the Diploma in Education and later was elected Fellow of the Royal Society of Chemistry. He taught chemistry in schools until 1969 when he joined the Department of Educational Research at the University of Lancaster where is now Senior Lecturer.

His work in curriculum development began in 1963 with the Nuffield Science Project for which he was responsible for designing and operating the O and A level examinations. Since 1967, he has worked for short periods in India, South-East Asia and Africa, mainly as a consultant. He is Chairman of the Research Advisory Committee of an examining board in England and is closely connected with the development of a common system of examining at 16+. His publications include two Schools Council bulletins, various books and papers on examinations, and the Modern Chemistry series published by Hutchinsons.

John W. MOORE received his A.B. from Franklin and Marshall College and his Ph.D. from Northwestern University, concentrating in physical inorganic chemistry. During 1964/65, he held a National Science Foundation Postdoctoral Fellowship at the H. C. Oersted Institute of the University of Copenhagen, working with Professor Carl Ballhausen. He has taught at Indiana University, Bloomington (1965–71), and at present at Eastern Michigan University; during 1981/82 he was visiting professor at the University of Wisconsin, Madison. He is the author of numerous publications in inorganic chemistry, chemical education, computer applications in chemistry and environmental chemistry. These include *Environmental Chemistry* (with Elizabeth A. Moore) and *Chemistry* (with W. G. Davies and R. W. Collins). He recently completed, with Professor Ralph Pearson, the revision of *Kinetics and Mechanism*. He has also produced numerous audio-visual teaching aids, including films, overhead-projection transparencies and computer graphics.

Dr Moore is currently editor of the computer series feature of the *Journal of Chemical Education* and has just received a National Science Foundation grant to produce and disseminate microcomputer-based instructional materials. From 1975 to 1980 he was co-editor (with Elizabeth A. Moore) of the column 'Science/Society Case Study' which appeared in *Science Teacher*. In 1979, he was appointed to the Michigan Environmental Review Board and during 1979/80 served as a consultant to the Ecology Center of Ann Arbor working on problems related to the disposal of toxic and hazardous wastes. He has been active in the American Chemical Society Division of Chemical Education for several years and at present serves as its programme chairman.

A. Mahinda RANAWEERA is the Deputy Director-General of Education, Ministry of Education, Sri Lanka, in charge of the national curriculum development and teacher education programmes. As the Director of the Curriculum Development Centre, he was responsible for directing the chemistry curriculum development programme at the GCE Ordinary and Advanced levels. He was directly involved in designing the new curricula, preparation of teachers' guides, pupils' textbooks, conducting in-service education courses for teachers and redesigning the GCE O and A level examinations in chemistry in Sri Lanka since 1962.

He is the author of several publications and has contributed articles to international journals on science education. He is well known in Sri Lanka as a pioneer in teaching science in Sinhala (one of the national languages), a radio broadcaster, lecturer and writer in popularization of science in the Sinhala language.

[II 60] ED. 82/XIV.II.n/A